PROLOG

Reproductive Endocrinology and Infertility

SEVENTH EDITION

Critique Book

The American College of
Obstetricians and Gynecologists
WOMEN'S HEALTH CARE PHYSICIANS

ISBN 978-1-934984-44-4

Copyright 2015 by the American College of Obstetricians and Gynecologists. All rights reserved. No part of this publication may be reproduced, stored in a retrieval system, posted on the Internet, or transmitted, in any form or by any means, electronic, mechanical, photocopying, recording, or otherwise, without prior written permission from the publisher.

12345/98765

The American College of Obstetricians and Gynecologists
409 12th Street, SW
PO Box 96920
Washington, DC 20090-6920

Contributors

PROLOG Editorial and Advisory Committee

CHAIR

Ronald T. Burkman Jr, MD
　Professor of Obstetrics and
　　Gynecology
　Department of Obstetrics and
　　Gynecology
　Tufts University School of Medicine
　Baystate Medical Center
　Springfield, Massachusetts

MEMBERS

Louis Weinstein, MD
　Past Paul A. and Eloise B. Bowers
　　Professor and Chair
　Department of Obstetrics and
　　Gynecology
　Thomas Jefferson University
　Philadelphia, Pennsylvania

Linda Van Le, MD
　Leonard Palumbo Distinguished
　　Professor
　Division of Gynecologic Oncology
　University of North Carolina School of
　　Medicine
　Chapel Hill, North Carolina

PROLOG Task Force for *Reproductive Endocrinology and Infertility*, Seventh Edition

COCHAIRS

Bruce R. Carr, MD
　Department of Obstetrics and
　　Gynecology
　University of Texas Southwest
　　Medical Center
　Dallas, Texas

Daniel R. Grow, MD
　Professor and Chair
　Department of Obstetrics and
　　Gynecology
　Tufts University School of Medicine
　Baystate Medical Center
　Springfield, Massachusetts

MEMBERS

Victor E. Beshay, MD
　Department of Obstetrics and
　　Gynecology
　University of Texas Southwestern School
　　of Medicine
　Dallas, Texas

Krystene B. DiPaola, MD
　Assistant Professor
　University of Cincinnati
　UC Center for Reproductive Health
　West Chester, Ohio

Erika B. Johnston-MacAnanny, MD
　Medical Director
　Wake Forest Center for Reproductive
　　Medicine
　Assistant Professor, Reproductive
　　Medicine
　Office of Women in Medicine and
　　Science
　Wake Forest Baptist Medical Center
　Winston-Salem, North Carolina

J. Ricardo Loret de Mola, MD
　Professor and Chairman
　Department of Obstetrics and
　　Gynecology
　Southern Illinois University
　Springfield, Illinois

Continued on next page

PROLOG Task Force for *Reproductive Endocrinology and Infertility*, Seventh Edition *(continued)*

Jennifer E. Mersereau, MD
   Assistant Professor
   Division of Reproductive
     Endocrinology and Infertility
   Department of Obstetrics and
     Gynecology
   University of North Carolina
   Chapel Hill, North Carolina

Steven T. Nakajima, MD
   Clinical Professor and Director of the
     IVF Outreach Program
   Co-Director of Clinical Operations,
     Reproductive Endocrinology and
     Infertility
   Department of Obstetrics and
     Gynecology
   Stanford University
   Palo Alto, California

Beth J. Plante, MD
   Fertility Centers of New England
   Reading, Massachusetts

Thomas M. Price, MD
   Associate Professor
   Department of Obstetrics &
     Gynecology
   Division of Reproductive
     Endocrinology & Fertility
   Duke University School of Medicine
   Durham, North Carolina

John T. Queenan Jr, MD
   Professor, Obstetrics and Gynecology
   University of Rochester Medical Center
   School of Medicine and Dentistry
   Rochester, New York

Spencer S. Richlin, MD
   Surgical Director
   Reproductive Medicine Associates of
     Connecticut
   Division Director
   Reproductive Endocrinology & Infertility
   Norwalk Hospital
   Norwalk, Connecticut

COLLEGE STAFF

Sandra A. Carson, MD
   Vice President for Education

Erica Bukevicz, MBA, MS
   Senior Director, Educational
     Development and Testing
   Division of Education

Christopher T. George, MLA
   Editor, PROLOG

CONFLICT OF INTEREST DISCLOSURE

This PROLOG unit was developed under the direction of the PROLOG Advisory Committee and the Task Force for *Reproductive Endocrinology and Infertility*, Seventh Edition. PROLOG is planned and produced in accordance with the Standards for Enduring Materials of the Accreditation Council for Continuing Medical Education. Any discussion of unapproved use of products is clearly cited in the appropriate critique.

Current guidelines state that continuing medical education (CME) providers must ensure that CME activities are free from the control of any commercial interest. The task force and advisory committee members declare that neither they nor any business associate nor any member of their immediate families has material interest, financial interest, or other relationships with any company manufacturing commercial products relative to the topics included in this publication or with any provider of commercial services discussed in the unit except for **Steven T. Nakajima, MD**, who is a consultant for Actavis PLC and Previvo Genetics LLC. All potential conflicts have been resolved through the American College of Obstetricians and Gynecologists' mechanism for resolving potential and real conflicts of interest.

Preface

Purpose

PROLOG (Personal Review of Learning in Obstetrics and Gynecology) is a voluntary, strictly confidential self-evaluation program. PROLOG was developed specifically as a personal study resource for the practicing obstetrician–gynecologist. It is presented as a self-assessment mechanism that, with its accompanying performance information, should assist the physician in designing a personal, self-directed lifelong learning program. It may be used as a valuable study tool, a reference guide, and a means of attaining up-to-date information in the specialty. The content is carefully selected and presented in multiple-choice questions that are clinically oriented. The questions are designed to stimulate and challenge physicians in areas of medical care that they confront in their practices or when they work as consultant obstetrician–gynecologists.

PROLOG also provides the American College of Obstetricians and Gynecologists (the College) with one mechanism to identify the educational needs of the Fellows. Individual scores are reported only to the participant; however, cumulative performance data and evaluation comments obtained for each PROLOG unit help determine the direction for future educational programs offered by the College.

Process

The PROLOG series offers the most current information available in five areas of the specialty: obstetrics, gynecology and surgery, reproductive endocrinology and infertility, gynecologic oncology and critical care, and patient management in the office. A new PROLOG unit is produced annually, addressing one of those subject areas. *Reproductive Endocrinology and Infertility*, Seventh Edition, is the third unit in the seventh 5-year PROLOG series.

Each unit of PROLOG represents the efforts of a special task force of subject experts under the supervision of an advisory committee. PROLOG sets forth current information as viewed by recognized authorities in the field of women's health. This educational resource does not define a standard of care, nor is it intended to dictate an exclusive course of management. It presents recognized methods and techniques of clinical practice for consideration by obstetrician–gynecologists to incorporate in their practices. Variations of practice that take into account the needs of the individual patient, resources, and the limitations that are special to the institution or type of practice may be appropriate.

Each unit of PROLOG is presented as a two-part set, with performance information and cognate credit available to those who choose to submit their answer sheets for confidential scoring. The first part of the PROLOG set is the Assessment Book, which contains educational objectives for the unit and multiple-choice questions, and an answer sheet with a return mailing envelope. Participants can work through the book at their own pace, choosing to use PROLOG as a closed- or open-book assessment. Return of the answer sheet for scoring is encouraged but voluntary.

The second part of PROLOG is the Critique Book, which reviews the educational objectives and items set forth in the Assessment Book and contains a discussion, or critique, of each item. The critique provides the rationale for correct and incorrect options. Current, accessible references are listed for each item.

Continuing Medical Education Credit

ACCME Accreditation
The American College of Obstetricians and Gynecologists is accredited by the Accreditation Council for Continuing Medical Education (ACCME) to provide continuing medical education for physicians.

AMA PRA Category 1 Credit(s)™
The American College of Obstetricians and Gynecologists designates this enduring material for a maximum of **25 *AMA PRA Category 1 Credits*™**. Physicians should claim only the credit commensurate with the extent of their participation in the activity.

College Cognate Credit(s)

The American College of Obstetricians and Gynecologists designates this enduring material for a maximum of 25 Category 1 College Cognate Credits. The College has a reciprocity agreement with the American Medical Association that allows *AMA PRA Category 1 Credits*™ to be equivalent to College Cognate Credits.

Fellows who submit their answer sheets for scoring will be credited with 25 hours. Participants who return their answer sheets for CME credit will receive a Performance Report that provides a comparison of their scores with the scores of a sample group of physicians who have taken the unit as an examination. An individual may request credit only once for each unit. *Please allow 4–6 weeks to process answer sheets.*

Credit for PROLOG *Reproductive Endocrinology and Infertility*, Seventh Edition, is initially available through December 2017. During that year, the unit will be reevaluated. If the content remains current, credit is extended for an additional 3 years, with credit for the unit automatically withdrawn after December 2020.

Conclusion

PROLOG was developed specifically as a personal study resource for the practicing obstetrician–gynecologist. It is presented as a self-assessment mechanism that, with its accompanying performance information, should assist the physician in designing a personal, self-directed learning program. The many quality resources developed by the College, as detailed each year in the College's *Publications and Educational Materials Catalog*, are available to help fulfill the educational interests and needs that have been identified. PROLOG is not intended as a substitute for the certification or recertification programs of the American Board of Obstetrics and Gynecology.

PROLOG CME SCHEDULE

Reproductive Endocrinology and Infertility, Sixth Edition	Credit through 2015
Gynecologic Oncology and Critical Care, Sixth Edition	Credit through 2016
Patient Management in the Office, Sixth Edition	Credit through 2017
Obstetrics, Seventh Edition	Reevaluated in 2015– Credit through 2018
Gynecology and Surgery, Seventh Edition	Reevaluated in 2016– Credit through 2019
Reproductive Endocrinology and Infertility, Seventh Edition	Reevaluated in 2017– Credit through 2020

PROLOG Objectives

PROLOG is a voluntary, strictly confidential, personal continuing education resource that is designed to be stimulating and enjoyable. By participating in PROLOG, obstetrician–gynecologists will be able to do the following:

- Review and update clinical knowledge.
- Recognize areas of knowledge and practice in which they excel, be stimulated to explore other areas of the specialty, and identify areas requiring further study.
- Plan continuing education activities in light of identified strengths and deficiencies.
- Compare and relate present knowledge and skills with those of other participants.
- Obtain continuing medical education credit, if desired.
- Have complete personal control of the setting and of the pace of the experience.

The obstetrician–gynecologist who completes *Reproductive Endocrinology and Infertility*, Seventh Edition, will be able to

- establish a differential diagnosis and screen with appropriate diagnostic tests for specific gynecologic conditions.
- determine the appropriate medical management for specific gynecologic conditions in adolescents and adult women.
- identify appropriate surgical interventions for various gynecologic conditions and strategies to prevent and treat surgical complications.
- apply concepts of anatomy, genetics, pathophysiology, and epidemiology to the understanding of diseases that affect women.
- counsel women regarding treatment options and adjustment to crises that may alter their lifestyles.
- apply professional medical ethics and the understanding of medical–legal issues relative to the practice of gynecology.

Reproductive Endocrinology and Infertility, Seventh Edition, includes the following topics (item numbers appear in parentheses):

SCREENING AND DIAGNOSIS
Abnormal uterine bleeding (109)
Abnormal uterine bleeding in an adolescent (92)
Androgen disorders (61, 103)
Androgen insensitivity syndrome (48)
Anorexia and nutrition (54)
Antiphospholipid syndrome (3)
Assisted reproductive technology and ovarian hyperstimulation syndrome (8)
Congenital adrenal hyperplasia (60)
Cushing syndrome (110)
Depot medroxyprogesterone acetate and unscheduled vaginal bleeding (49)
Differential diagnosis for hirsutism conditions (149–151)
Elevated dehydroepiandrosterone sulfate (93)
Evaluation of gonads in androgen insensitivity syndrome (128–130)
Evaluation of male factor infertility (5)
Hereditary cancer syndromes (131–133)
Hyperthyroidism (74)
Hypothalamic amenorrhea (117)
Hypothyroidism (114)
Late-onset congenital adrenal hyperplasia (81)
Macroadenoma and galactorrhea (10)
Maternal virilization in hyperreactio luteinalis (100)

Metabolic syndrome (104)
Müllerian anomaly (76)
Normal menopausal transition (120)
Polycystic ovary syndrome and estrogen breakthrough bleeding (73)
Preimplantation genetic screening (71)
Premature adrenarche (62)
Premature ovarian failure (83)
Premature thelarche (20)
Primary hypogonadotropic hypogonadism (88)
Recurrent pregnancy loss (21, 69)
Sheehan syndrome (70)
Testing for insulin resistance in polycystic ovary syndrome (91)
Testing for ovarian reserve (95)
Testing for thyroid disease (152–154)
Turner syndrome (33)
Ultrasonographic findings in a patient who takes tamoxifen citrate (2)
Virilization due to Sertoli–Leydig cell tumor (107)

MEDICAL MANAGEMENT
Acne in an adolescent patient (84)
Alternative therapies for menopause (50)
Ambiguous genitalia (115)
Amenorrhea and galactorrhea (97)
BRCA mutations (43)
Breast cancer (127)
Bulimia nervosa and binge-eating disorder (52)
Chronic pelvic pain (96)
Condom failure and morning-after contraception (65)
Congenital adrenal hyperplasia (45)
Contraception for a patient with *BRCA* gene mutation (22)
Contraception for a patient with diabetes mellitus (44, 67)
Contraception for a patient with systemic lupus erythematosus (66)
Contraception for a patient with venous thromboembolism (19)
Contraception for an older patient who smokes (55)
Donor egg in vitro fertilization (29)
Dysmenorrhea (42)
Ectopic pregnancy (64, 99)
Endometrial polyp (4)
Endometriosis (37)
Fertility options after tubal ligation (47)
Fertility preservation in a patient undergoing cancer treatment (32)
Fertility preservation techniques (75)
Functional hypothalamic amenorrhea and osteoporosis (108)
Heavy menstrual bleeding (18, 80, 106)
Hirsutism (112)
Hydrosalpinges and in vitro fertilization–embryo transfer (12)
Hysterosalpingography complications (89)
Intrauterine device complications with infection (102)
Intrauterine microinsert and pregnancy (35)
Intrauterine microinsert follow-up (87)
In vitro fertilization and intracytoplasmic sperm (11)
Labial adhesions in children (30)
Leiomyoma in infertility (36)
Leiomyomas and heavy menstrual bleeding (56)

Low bone mass and bone physiology (94)
Low-dose estrogen therapy (72)
Luteal phase deficiency (118)
Macroadenoma (46)
Male factor infertility (38)
Maternal virilization in hyperreactio luteinalis (68)
Mechanism of action of hormonal contraceptives (146–148)
Migraine and use of oral contraceptives (7)
Müllerian dysgenesis (27)
Obesity and insulin resistance (78)
Obesity and pregnancy (51, 125, 155–158)
Osteoporosis and celiac disease (101)
Ovarian androgen-secreting tumor (98)
Ovulation induction in a patient with polycystic ovary syndrome (59)
Ovulation induction treatment options (90)
Patient with gastric bypass (9)
Pelvic inflammatory disease and reproductive dysfunction (124)
Phantom human chorionic gonadotropin results (40)
Postmenopausal uterine bleeding and hormone therapy (34)
Postpartum thyroiditis (53)
Precocious puberty (82)
Pregnancy termination (26)
Primary amenorrhea due to gonadal dysgenesis (86)
Psychologic effect of infertility (85)
Retrograde ejaculation (111)
Salpingitis isthmica nodosa (79)
Unexplained infertility (16)
Use of stem cells in reproduction (122)
Uterine fibroid embolization (63)

PHYSIOLOGY
Ambiguous genitalia (115)
Hormonal changes in pregnancy (139–142)
Low bone mass and bone physiology (94)
Mechanism of action of hormonal contraceptives (146–148)
Normal menstrual cycle (41)
Perimenopausal changes (28)
Physiology of the menstrual cycle (25)
Reproductive function, nutritional status, and protein levels (15)

SURGICAL MANAGEMENT
Bilateral tubal ligation and future fertility (123)
Complications of robotic-assisted surgery (105)
Endometriosis and infertility (14)
Hysteroscopic complications (24)
Laparoscopic surgery complications (159–161)
Müllerian anomalies (23)
Oligospermia (77)
Ovarian cancer in a *BRCA1*-positive patient (13)
Patient with gastric bypass (9)

EPIDEMIOLOGY AND BIOSTATISTICS
Absolute risk versus relative risk (57)
Breast cancer (127)

Lifestyle choices and pregnancy outcome (134–138)
Risks and benefits of hormone therapy (31)
Statistical analysis (143–145)

COUNSELING
Bilateral tubal ligation and future fertility (123)
Bioidentical hormones (113)
Condom failure and morning-after contraception (65)
Contraception for a patient with *BRCA* gene mutation (22)
Contraception for a patient with diabetes mellitus (44)
Contraception for a patient with systemic lupus erythematosus (66)
Contraception for a patient with venous thromboembolism (19)
Contraception for an older patient who smokes (55)
Depot medroxyprogesterone acetate and bone loss (58)
Donor oocyte use (126)
Donor sperm use (119)
Intrauterine device complications with infection (102)
In vitro fertilization and intracytoplasmic sperm (11)
Luteal phase deficiency (118)
Multifetal pregnancy reduction (116)
Normal menopausal transition (120)
Obesity and contraceptive choices (17)
Primary ovarian insufficiency (6)
Uterine fibroid embolization (63)
Window of fertilization (121)

ETHICAL AND LEGAL ISSUES
In vitro fertilization (1)
Use of donor sperm for same-sex couples (39)

A complete subject matter index appears at the end of the Critique Book.

1

In vitro fertilization

A 44-year-old woman comes to your office to discuss in vitro fertilization (IVF). She underwent a tubal ligation 15 years ago. She now has a new partner with whom she would like to conceive. She and her partner are not interested in using donor eggs for an IVF cycle. You tell her that in her age group your clinic's live-birth rate for IVF without donor eggs is less than 1% per attempt. The most significant guiding issue to consider before deciding whether to proceed with IVF is

    (A) her financial resources
    (B) concerns about legal ramifications
    (C) nonmaleficence
*   (D) patient autonomy

Patient autonomy is defined as an individual's right to hold views, make choices, and take actions based on his or her own personal values and beliefs. A medical ethics framework can assist patients and health care providers in making difficult medical decisions. In a principle-based health care ethics system, autonomy is one of the four principles used to identify and analyze ethical issues in a medical dilemma:

- Autonomy
- Beneficence
- Nonmaleficence
- Justice

In the case described, the most significant guiding issue before deciding whether to proceed with IVF is patient autonomy.

In certain cases, the reproductive goals of a particular patient have a low or nonexistent likelihood of success. Generally, the term "futile" refers to treatments that have a 1% or less chance of success per cycle. The term "very poor prognosis" refers to low odds, ie, a 1–5% chance of live birth per cycle. Most patients and health care providers are able to reach a consensus about the realistic chances of success and to redirect efforts to options such as third-party assistance in reproduction (eg, egg or sperm donation), adoption, or acceptance of a child-free life. However, there are instances where individuals or couples are interested in pursuing whatever treatments are available to achieve their reproductive goals, regardless of the odds of success. In these situations, ethical issues may arise that force choices between patient autonomy and the health care provider's medical recommendation.

Because of her prior tubal ligation, the described patient is unable to attempt conception without medical assistance. However, given her age, her chance of pregnancy and subsequent live birth with assisted reproductive technology, such as IVF, using her own eggs is less than 1% per attempt. The chance of live birth with donor egg IVF, however, is more than 50% per attempt. The decision to proceed with third-party reproduction is a complicated one for many patients and in itself may have ethical, legal, financial, and emotional ramifications. In this case, the couple is not interested in considering donor egg IVF.

A complicating factor when considering futility is that patients and health care providers may interpret the statistics in different ways. Whereas a health care provider may feel that the risk of harm from IVF outweighs the likelihood of live birth, an individual patient may interpret a 1% chance of live birth to be hopeful odds. In addition, patients and health care providers may interpret the final goal differently. For some patients, attempted conception with IVF, even if the odds of pregnancy are exceedingly low, may provide them with information about whether or not they can produce eggs and, therefore, could help them feel that they have tried everything before they accept the reality of their situation.

A health care provider's professional interest may influence their decisions about cases of medical futility. Ethical concerns arise about the cost of fertility treatments when the odds of live birth are exceedingly low and in instances when charging patients or insurance providers could become fraudulent. However, given the complexities with interpreting "futility" and "very poor prognosis," there are no clear guidelines about when it is unacceptable to receive payment for treatments that have little chance of success. A patient's ability to pay for fertility treatments, regardless of the odds of success, should not guide the clinician in making medical recommendations.

---

* Indicates correct answer.
Note: See Appendix A for a table of normal values for laboratory tests.

Additional questions may arise about the rights and legal duties a health care provider has in regard to proceeding with or refusing treatment. The relationship between the health care provider and patient is consensual. That is, the patient and the health care provider have the right to refuse to pursue treatment with which they do not feel comfortable, as long as they do not violate laws about impermissible discrimination. In fact, health care providers are permitted to terminate the patient–physician relationship as long as they provide timely notice and information about other health care providers who may have better success rates for individual situations, if applicable.

Nonmaleficence is the principal of avoiding medical care that may cause harm to a patient. The actual medical risks of IVF treatment are minimal. Many of the medications are injectable and have a small risk of bruising and pain. A remote risk exists of complications, such as ovarian hyperstimulation syndrome, pelvic injury during an egg retrieval, infection, ovarian torsion, ectopic pregnancy, and multiple gestation. However, the risk of serious complications from an IVF cycle is low enough that the principal of nonmaleficence should not be a guiding factor when deciding about potentially futile IVF treatments.

The ethics committee at the American Society for Reproductive Medicine recommends that decisions about treating or refusing to treat patients should be patient-focused to promote patient autonomy. If a patient has been counseled adequately about the low chance of success and has reviewed the risks and benefits of treatment, it is considered reasonable for health care providers to proceed with fertility treatments. Strong consideration should be given to referral for psychologic counseling. Clinics should consider developing explicit policies to guide decisions about pursuing futile fertility treatments. This will allow consistency and prevent having to make decisions on a case-by-case basis. Ideally, these policies should be discussed with patients when the initial decision about treatment is being made.

Ethical decision making in obstetrics and gynecology. ACOG Committee Opinion No. 390. American College of Obstetricians and Gynecologists. Obstet Gynecol 2007;110:1479–87.

Fertility treatment when the prognosis is very poor or futile: a committee opinion. Ethics Committee of American Society for Reproductive Medicine. Fertil Steril 2012;98:e6–9.

2

Ultrasonographic findings in a patient who takes tamoxifen citrate

A 52-year-old single nulligravid woman is undergoing chemoprophylaxis with tamoxifen citrate after being diagnosed with a ductal carcinoma in situ of the right breast 9 months ago. She has had persistent vaginal bleeding for the past month, and you decide to order transvaginal pelvic ultrasonography. The most common endometrial abnormality that is likely to be identified is

    (A) hyperplasia
    (B) cancer
    (C) cystic atrophy
* (D) polyps
    (E) submucosal leiomyoma

Tamoxifen citrate is a selective estrogen receptor modulator with a complex mechanism of action including antiestrogenic activity in the breast and estrogenic effects in other tissues such as the endometrium. It is widely used for the treatment of breast cancer and for chemoprevention in high-risk premenopausal and postmenopausal women. Tamoxifen is approved by the U.S. Food and Drug Administration for breast cancer prevention in women aged 35 years or older. Tamoxifen reduces the risk of breast cancer by nearly 50%. This effect is observed even in women with up to three first-degree relatives with breast cancer, but the medication only reduces the risk of estrogen receptor-positive breast cancer.

The finding of a decrease in contralateral breast cancer incidence after tamoxifen administration for adjuvant therapy has led to its use in breast cancer prevention. Four large prevention trials of tamoxifen and cancer have demonstrated the benefits of using tamoxifen. In the National Surgical Adjuvant Breast and Bowel Project's Breast Cancer Prevention Trial, tamoxifen reduced the risk of invasive breast cancer by 49%. A recent study suggests extension of therapy to 10 years.

One important determinant of tamoxifen safety is the possible effects of the drug on the endometrium. The first report to connect tamoxifen and endometrial cancer was published in 1989, and since then, several reports have described the endometrial changes in postmenopausal breast cancer patients treated with tamoxifen. The results also indicated that even though the risk-reducing effect of tamoxifen appeared to persist for at least 10 years, most of the adverse effects did not continue after the 5-year treatment period. The most common endometrial changes include endometrial polyps (with the incidence between 8% and 36% versus 0–10% in untreated women) and endometrial hyperplasia (1.3–20% versus 0–10% in untreated women). The risk of endometrial carcinoma for treated women is 1.3–7.5-fold higher than the risk for untreated women; cystic atrophy, adenomyosis, and leiomyomas also have been reported. For the described patient, the most common abnormality that is likely to be identified by means of ultrasonography is endometrial polyps. The most appropriate treatment for her is hysteroscopic resection.

Detailed guidelines are available for the follow-up of patients who take tamoxifen for the treatment of breast cancer. Currently, there is no active screening for asymptomatic patients treated with tamoxifen other than routine gynecologic care.

Davies C, Pan H, Godwin J, Gray R, Arriagada R, Raina V, et al. Long-term effects of continuing adjuvant tamoxifen to 10 years versus stopping at 5 years after diagnosis of oestrogen receptor-positive breast cancer: ATLAS, a randomised trial. Adjuvant Tamoxifen: Longer Against Shorter (ATLAS) Collaborative Group [published erratum appears in Lancet 2013;381:804]. Lancet 2013;381:805–16.

Palva T, Ranta H, Koivisto AM, Pylkkanen L, Cuzick J, Holli K. A double-blind placebo-controlled study to evaluate endometrial safety and gynaecological symptoms in women treated for up to 5 years with tamoxifen or placebo—a substudy for IBIS I Breast Cancer Prevention Trial. Eur J Cancer 2013;49:45–51.

Pinkerton JV, Goldstein SR. Endometrial safety: a key hurdle for selective estrogen receptor modulators in development. Menopause 2010;17:642–53.

Polin SA, Ascher SM. The effect of tamoxifen on the genital tract. Cancer Imaging 2008;8:135–45.

Tamoxifen and uterine cancer. Committee Opinion No. 601. American College of Obstetricians and Gynecologists. Obstet Gynecol 2014;123:1394–7.

3

Antiphospholipid syndrome

A 28-year-old woman, gravida 3, para 0, comes to your office for evaluation of recurrent pregnancy loss. Her history is significant for a deep vein thrombosis at age 21 years. You are concerned that she might have antiphospholipid syndrome (APS). In addition to testing for lupus anticoagulant and anticardiolipin antibodies, the factor that you should screen for is

    (A) phosphatidylserine antibodies
* (B) anti-β_2-glycoprotein I antibodies
    (C) annexin V antibodies
    (D) phosphatidylinositol antibodies

Antiphospholipid syndrome is an autoimmune disorder characterized by venous or arterial thrombosis, recurrent fetal loss, or placental insufficiency in the setting of antiphospholipid antibodies. Antiphospholipid antibodies are present in approximately 5% of the general reproductive-aged population and in 15% of patients with recurrent miscarriages. The syndrome is characterized by at least one clinical feature plus at least one specific laboratory criterion. It can occur on its own or together with another autoimmune disease, infectious disease, malignancy, or certain medications; up to 37% of patients with systemic lupus erythematosus also will have APS.

The diagnostic criteria for APS have changed over time but are currently based on a consensus statement published in 2004 and revised in 2006 (Box 3-1 and Box 3-2). In addition to including one clinical feature, the criteria for APS require the persistent presence for more than 12 weeks of medium-to-high titers of at least one specific autoantibody of immunoglobulin G (IgG) or immunoglobulin M (IgM) isotype for anti-β_2-glycoprotein I or anticardiolipin, or the presence of lupus anticoagulant. Transient positive tests may occur and, therefore, two positive laboratory tests at least 12 weeks apart are required to diagnose APS. Many different antiphospholipid antibodies have been identified that are related to each other but still distinctly different. Only lupus anticoagulant, anticardiolipin, or anti-β_2-glycoprotein I can be used in diagnosing APS. Testing for other antibodies, such as phosphatidylserine antibodies, annexin V antibodies, or phosphatidylinositol antibodies, is not recommended.

Lupus anticoagulant is associated with thrombosis and not anticoagulation. The presence of lupus anticoagulant is assessed indirectly with a series of tests being required to confirm its presence. Initially, a combination of sensitive clotting assays are used to screen for lupus anticoagulant. Lupus anticoagulant, if present, prolongs clotting times by interfering with the assembly of the prothrombin complex. A number of other factors can cause prolonged clotting times, so additional testing is necessary if the initial screening is positive. Lupus anticoagulant cannot be quantified and, thus, is ultimately only reported as present or absent.

Anticardiolipin antibodies and anti-β_2-glycoprotein I antibodies usually are detected using enzyme-linked immunosorbent assays. The IgG and IgM isotypes should be measured. Results are reported in international standard units: G phospholipid and M phospholipid for IgG

BOX 3-1

Laboratory Criteria for the Diagnosis of Antiphospholipid Syndrome

1. Lupus anticoagulant is present in the plasma on two or more occasions at least 12 weeks apart. It is interpreted as either present or absent. Testing for lupus anticoagulant is ideally performed before the patient is treated with anticoagulants, or
2. Anticardiolipin antibody of immunoglobulin G and/or immunoglobulin M isotype is in the serum or plasma, present in medium or high titer (ie, greater than 40 G phospholipid or M phospholipid, or greater than the 99th percentile) on two or more occasions at least 12 weeks apart, or
3. Anti-β2-glycoprotein I of immunoglobulin G and/or immunoglobulin M isotype is present in the serum or plasma (in titer greater than 99th percentile for a normal population as defined by the laboratory performing the test) on two or more occasions at least 12 weeks apart.

Modified from Miyakis S, Lockshin MD, Atsumi T, Branch DW, Brey RL, Cervera R, et al. International consensus statement on an update of the classification criteria for definite antiphospholipid syndrome (APS). J Thromb Haemost 2006;4:295–306.

> **BOX 3-2**
>
> **Clinical Criteria for the Diagnosis of Antiphospholipid Syndrome**
>
> 1. Vascular thrombosis
>   One or more clinical episodes of arterial, venous, or small vessel thrombosis, in any tissue or organ, or
> 2. Pregnancy morbidity
>   a. One or more unexplained deaths of a morphologically normal fetus at or beyond the 10th week of gestation, with normal fetal morphology documented by ultrasonography or by direct examination of the fetus, or
>   b. One or more premature births of a morphologically normal neonate before the 34th week of gestation because of eclampsia or severe preeclampsia, or features consistent with placental insufficiency, or
>   c. Three or more unexplained consecutive spontaneous pregnancy losses before the 10th week of gestation, with maternal anatomic or hormonal abnormalities and paternal and maternal chromosomal causes excluded.
>
> Modified from Miyakis S, Lockshin MD, Atsumi T, Branch DW, Brey RL, Cervera R, et al. International consensus statement on an update of the classification criteria for definite antiphospholipid syndrome (APS). J Thromb Haemost 2006;4:295–306.

and IgM phospholipids, respectively, and standard G unit and standard M unit for anti-β_2-glycoprotein I IgG and IgM, respectively. For anticardiolipin antibodies and anti-β_2-glycoprotein I antibodies, a result must be greater than the 99th percentile to be considered positive. Repeat levels should be obtained at least 12 weeks later for confirmation.

The pathophysiology of the syndrome involves antiphospholipid antibodies binding to endothelial or platelet phospholipids, which then triggers platelet activation and aggregation and ultimately thrombosis. In patients with APS and recurrent pregnancy loss who have had a prior thrombotic event, such as the described patient, prophylactic anticoagulation with heparin and low-dose aspirin would be recommended throughout the pregnancy and for 6 weeks postpartum. Nonpregnant APS patients should avoid estrogen-containing contraceptives.

Antiphospholipid syndrome. Practice Bulletin No. 132. American College of Obstetricians and Gynecologists. Obstet Gynecol 2012;120:1514–21.

Giannakopoulos B, Krilis SA. The pathogenesis of the antiphospholipid syndrome. N Engl J Med 2013;368:1033–44.

Miyakis S, Lockshin MD, Atsumi T, Branch DW, Brey RL, Cervera R, et al. International consensus statement on an update of the classification criteria for definite antiphospholipid syndrome (APS). J Thromb Haemost 2006;4:295–306.

4
Endometrial polyp

A 30-year-old woman, gravida 2, para 2, comes to your office with bleeding between menstrual periods. Her Pap test 3 months ago was normal and speculum examination in your office identifies a normal cervix and vaginal mucosa. The urinary human chorionic gonadotropin test result is negative. Saline sonohysterography reveals a 1-cm endometrial polyp. She is interested in future fertility. The treatment option that is most likely to improve her symptoms is

* (A) dilation and curettage
* (B) operative hysteroscopy
* (C) endometrial ablation
* (D) periodic oral progestin withdrawal

Endometrial polyps are benign, focal overgrowths of endometrial tissue that are covered by epithelium and contain variable amounts of glands, stroma, and blood vessels. They may commonly cause intermenstrual bleeding and are diagnosed easily by two-dimensional ultrasonography or saline-infused sonohysterography. Because polyps also may be asymptomatic, their causative role in abnormal menstrual bleeding has been questioned. The prevalence of polyps in women who report abnormal uterine bleeding ranges from 13% to 50%. Approximately 10% of asymptomatic premenopausal women were found to have small endometrial polyps.

The likelihood of an endometrial polyp being malignant is small, even with symptomatic irregular bleeding. The lifetime risk of developing adenocarcinoma of the endometrium is approximately 2–3%. Somewhat higher rates are observed in obese women with unopposed estrogen, such as patients with polycystic ovary syndrome. Oral progestin would induce cyclic menses in this population but would not cause eradication of the endometrial polyp. Endometrial biopsy is helpful in evaluation of irregular bleeding secondary to endometrial hyperplasia; however, in the described patient, an endometrial polyp has been confirmed with sonohysterography. A dilation and curettage without hysteroscopy may have the potential to miss resection of the pathology. Endometrial ablation is contraindicated if future fertility is desired.

Hysteroscopic resection of endometrial polyps usually is performed in order to exclude atypical or malignant endometrial changes, to relieve abnormal uterine bleeding, or to potentially improve fertility. The effect that an endometrial polyp has on fertility is unclear. The benefit of surgical polypectomy in regard to fertility also is debated. Theorized mechanisms by which endometrial polyps could adversely affect reproductive performance include irregular intraendometrial bleeding; creation of an inflammatory endometrial response similar to an intrauterine device; an obstructive defect that inhibits sperm transport; a physical surface area effect that prevents exposure of the embryo to the endometrium; and an endocrine surface area effect in which increased endometrium surface area results in increased secretion of glycodelin, which has been shown to inhibit sperm binding to the zona pellucida. When evaluated retrospectively, surgical treatments to remove polyps have been noted to result in improvement in bleeding symptoms and in high satisfaction rates.

Progestin withdrawal triggers release of lysosomal enzymes in the endometrial stromal cells, which promote enzymatic digestion of all the components of the glands, stroma, and connective tissue of the endometrial functionalis. This cyclical process of autodigestion occurs monthly after progesterone production stops with involution of the corpus luteum.

Lieng M, Istre O, Qvigstad E. Treatment of endometrial polyps: a systematic review. Acta Obstet Gynecol Scand 2010;89:992–1002.

Perez-Medina T, Bajo-Arenas J, Salazar F, Redondo T, Sanfrutos L, Alvarez P, et al. Endometrial polyps and their implication in the pregnancy rates of patients undergoing intrauterine insemination: a prospective, randomized study. Hum Reprod 2005;20:1632–5.

Silberstein T, Saphier O, van Voorhis BJ, Plosker SM. Endometrial polyps in reproductive-age fertile and infertile women. Isr Med Assoc J 2006;8:192–5.

Tjarks M, Van Voorhis BJ. Treatment of endometrial polyps. Obstet Gynecol 2000;96:886–9.

5
Evaluation of male factor infertility

A 38-year-old nulligravid woman requests your help to achieve pregnancy. Her male partner is noted to have azoospermia on semen analysis, but with normal volume; the result is confirmed on repeat testing. His general physical examination does not reveal any additional abnormalities. The next step in evaluation of the male partner should include assessment of

    (A) white blood cells in the seminal fluid
*   (B) fructose in the semen
    (C) postejaculatory urinalysis for sperm
    (D) antisperm antibodies

In addition to the male partner's history, semen analysis is the mainstay of the male evaluation. Although it should not be considered a test of fertility, an abnormal semen analysis suggests that the probability of achieving fertility is lower than normal. It should be performed after 2–3 days of abstinence. The semen should be collected by masturbation and without the use of lubricants. Couples may opt to use intercourse for collection; if this is their preference, a special collection condom should be provided. The sample should be transported to the laboratory within 1 hour of collection while kept at body temperature (specimen cup should be placed close to the skin during transport). The laboratory then assesses the specimen for time needed to liquefy, sperm count, sperm motility, sperm morphology, and number of round cells, which may signify the presence of either white blood cells or immature sperm.

Table 5-1 shows normal semen parameters according to the World Health Organization. In cases in which an abnormality in the semen parameters is detected, the semen analysis should be repeated 2–3 months later, given that spermatogenesis takes 60–80 days to complete. Thus, a semen analysis collected today reflects biologic influences 60–80 days ago.

Azoospermia (no sperm detected in the specimen) can be the result of either obstructive or nonobstructive causes. In cases where azoospermia is encountered, a physical examination should be performed to rule out obstructive problems such as congenital absence of the vas deferens, which may be bilateral and is commonly seen in individuals affected by cystic fibrosis gene mutations. Fructose, the energy source in sperm, is absent in cases of obstructive azoospermia. The next step in management of the described male partner is to evaluate for fructose in the semen.

In addition to fructose testing, evaluation of gonadotropins, prolactin, and sex hormones needs to be done in order to evaluate for testicular failure (ie, high follicle-stimulating hormone and luteinizing hormone levels with a low testosterone level) or hypogonadotropic hypogonadism (ie, low levels of both hormones with or without high prolactin, and low testosterone level). Genetic testing with chromosomal analysis and Y chromosome microdeletion analysis also should be performed because the chance of encountering a genetic abnormality is approximately 15% in men with azoospermia. These tests also should be considered in severe oligospermia where the risk of genetic abnormalities is approximately 5%.

Fructose testing will differentiate between obstructive and nonobstructive causes of azoospermia. In addition, results can be obtained rapidly during a semen analysis. Cystic fibrosis testing will be of value if obstructive azoospermia is suspected and the physical examination reveals absent vas deferens. White blood cells may affect sperm function but would not be the cause of azoospermia. Antisperm antibodies are a cause of low motility but not azoospermia. Postejaculatory urinalysis for sperm should not be performed except in cases of presence of fructose in semen, which indicates a nonobstructive etiology.

TABLE 5-1. Normal Semen Parameters by World Health Organization Five Criteria

Parameter	Value
Volume	1.5 mL
Concentration	15 million/mL
Progressive motility	32%
Total motility	40%
Normal forms (morphology)	4% by Kruger strict criteria

Cooper TG, Noonan E, von Eckardstein S, Auger J, Baker HW, Behre HM, et al. World Health Organization reference values for human semen characteristics. Hum Reprod Update 2010; 16:231–45.

6

Primary ovarian insufficiency

A 29-year-old nulligravid woman is concerned about her fertility and wants a second opinion. She has noticed spacing out of her menstrual intervals to every 3 months along with some vasomotor symptoms and insomnia. She tells you that her younger sister was diagnosed with "premature menopause," but she is not sure of the specifics. Her primary care provider obtained a day-3 follicle-stimulating hormone (FSH) level of 44.3 mIU/mL and a 46,XX karyotype. A repeat day-3 FSH was 56.1 mIU/mL and estradiol level was 16 pg/mL. She was told by her primary care provider that she has entered menopause and can no longer conceive. You inform her that the best next step to explain the findings is to screen for

* (A) cystic fibrosis
* (B) fragile X syndrome
* (C) Niemann–Pick disease
* (D) Gaucher disease

The diagnosis of secondary amenorrhea as a result of ovarian insufficiency is made when a patient younger than 40 years has amenorrhea for 3 or more months with laboratory findings of an elevated FSH level in the menopausal range (greater than 40 mIU/mL on two separate occasions weeks apart). Any patient in this age group who experiences secondary amenorrhea should be evaluated, especially if she has symptoms of estrogen deficiency such as hot flushes or vaginal dryness.

The causes of primary ovarian insufficiency include genetic abnormalities of the X chromosome (eg, 45,X and fragile X premutations), enzyme defects such as 17α-hydroxylase deficiency, aromatase deficiency, or galactosemia. Other etiologies include chemotherapy, radiation, viral infection, autoimmune disorders, and unknown causes.

This patient has primary ovarian insufficiency and there is reason to suspect that her younger sister also is affected, which would point to a genetic cause. Her chromosomal analysis is 46,XX. A karyotype was obtained in order to rule out the presence of a Y chromosome. Because of the potential risk of developing germ cell tumors, the presence of a Y chromosome is an indication to remove abnormal intra-abdominal gonads.

Premature ovarian failure was first noted to be associated with heterozygotic carriers of the fragile X premutation in the early 1990s. This syndrome is now known as primary ovarian insufficiency. Affected individuals will have expansion of the CGG tandem repeat on the X chromosome in a region that has been named the fragile X mental retardation 1 (*FMR1*) gene. Approximately 16–24% of women with this premutation will go on to develop primary ovarian insufficiency. In contrast, women with primary ovarian insufficiency have a 3% risk of fragile X premutation with no family history and up to 12% when one or more siblings are affected. Women with infertility due to primary ovarian insufficiency who have a normal karyotype should be tested for the fragile X premutation. The best next step to explain this patient's results is to screen for fragile X syndrome.

Cystic fibrosis is an autosomal recessive disorder that leads to the production of thickened secretions that can affect pulmonary, gastrointestinal, and reproductive functions in individuals who are homozygous for the disease. Niemann–Pick disease describes a group of diseases that are inherited in an autosomal recessive manner. Affected patients typically have deficient acid sphingomyelinase activity or defective transport of low-density lipoprotein cholesterol. Gaucher disease patients have defective acid beta glucosidase activity. Each of these conditions is an autosomal recessive disorder and could be seen in two siblings. However, cystic fibrosis, Niemann–Pick disease, and Gaucher disease do not cause primary ovarian insufficiency.

Current evaluation of amenorrhea. Practice Committee of the American Society for Reproductive Medicine. Fertil Steril 2006;86(suppl): S148–55.

Testing and interpreting measures of ovarian reserve: a committee opinion. Practice Committee of the American Society for Reproductive Medicine. Fertil Steril 2012;98:1407–15.

7

Migraine and use of oral contraceptives

A 31-year-old woman has a history of frequent migraines without aura around the time of her menses. She has read that an oral contraceptive (OC) might help control these headaches. She reports that she experiences throbbing, centrally located pain; numbness of her fingers; and nausea that occur 1–3 days before and during menses. The symptom that she experiences that will influence the choice of OC is

  (A) throbbing, centrally located pain
  (B) migraine without aura
* (C) numbness of fingers
  (D) nausea

Intermittent headaches are not a contraindication to the use of OCs unless the patient has migraine headaches preceded by an aura that often includes the presence of associated focal neurologic signs. A migraine headache preceded by an aura increases twofold the risk of a stroke in an otherwise healthy individual.

The presence of an isolated headache can be due to a number of factors. Some headaches can be associated with the fall of estrogen and progesterone levels 1–3 days before menses (in the late luteal phase), which can lead to menstrual-related headache, sometimes referred to as catamenial headache.

Other headaches can be classified as migraine. The headache is usually the most prominent feature of a number of associated neurologic and systemic symptoms that compose a migraine attack. Migraine attacks are common and occur in approximately 43% of women at some time in their life. Patients with a migraine headache often have a throbbing, centrally located pain that begins slowly and worsens over 1–2 hours.

Migraines associated with an aura are designated "classic migraines," whereas migraines without an aura are known as "common migraines." The aura that precedes a classic migraine is a perceptual disturbance, often a repetitive telltale sensation that occurs for a few minutes or up to an hour before the headache. The aura may persist for the duration of the migraine, often leaving the patient disoriented and confused. Migraine attacks with aura often are preceded by nausea and occasional vomiting. Affected individuals also demonstrate a marked sensitivity to light, smell, touch, or sound. The aura may manifest as the perception of a strange light or an unpleasant smell, or it may begin with confusing thoughts or experiences. It is not uncommon for an affected individual to have multiple different auras before their headache, but many have the same aura symptom(s) before their migraine attack.

Migraines may be precipitated by stress, alcohol, and foods that are rich in tyramine and tryptophan such as red wine, chocolate, and ripe cheeses. Migraines are usually cyclical in frequency, but can be completely absent for a period. Women with a history of migraine headaches without an aura are not at increased risk of stroke unless they have other major risk factors, such as tobacco smoking, hypertension, diabetes mellitus, or age greater than 35 years. Headaches also may be exacerbated after initiation of OCs.

Women with migraines and an associated preceding aura often have associated focal neurologic symptoms. These symptoms are the most severe and worrisome. Focal neurologic deficits can affect the functions of the brain, spinal cord, and central nervous system, which leads to alterations in movement (paralysis, weakness, tremor), sensation (paresthesia, numbness), speech, vision, and hearing. In the described patient, the symptom that will influence the choice of OC is the reported numbness of her fingers.

Women with a history of migraine headaches with aura should use a non-estrogen-mediated contraceptive method, such as the progestin-only OC, an intrauterine device, a progestin injection or implant, or barrier methods to prevent unintended conception.

Other management options for menstrual migraine include initiation of nonsteroidal antiinflammatory drugs begun 7 days before the onset of menses and continued throughout the patient's menses. Triptan medications, such as sumatriptan succinate or naratriptan hydrochloride, often are used in conjunction with nonsteroidal antiinflammatory drugs. The triptan medication is started 2–3 days before menses and continued for a total treatment period of 5–6 days.

Bonnema RA, McNamara MC, Spencer AL. Contraception choices in women with underlying medical conditions. Am Fam Physician 2010; 82:621–8.

Charles A. Advances in the basic and clinical science of migraine. Ann Neurol 2009;65:491–8.

Understanding and using the U.S. Selected Practice Recommendations for Contraceptive Use, 2013. Committee Opinion No. 577. American College of Obstetricians and Gynecologists. Obstet Gynecol 2013;122: 1132–3.

U.S. Medical Eligibility Criteria for Contraceptive Use, 2010. Centers for Disease Control and Prevention. MMWR Recomm Rep 2010;59 (RR-4):1–86.

U.S. Selected Practice Recommendations for Contraceptive Use, 2013: adapted from the World Health Organization selected practice recommendations for contraceptive use, 2nd edition. Division of Reproductive Health, National Center for Chronic Disease Prevention and Health Promotion, Centers for Disease Control and Prevention. MMWR Recomm Rep 2013;62(RR-05):1–60.

Use of hormonal contraception in women with coexisting medical conditions. ACOG Practice Bulletin No. 73. American College of Obstetricians and Gynecologists. Obstet Gynecol 2006;107:1453–72.

8

Assisted reproductive technology and ovarian hyperstimulation syndrome

A 36-year-old patient with primary infertility undergoes in vitro fertilization (IVF). She develops abdominal pain, bloating, and oliguria after her egg retrieval. The risk factor most commonly associated with these complications is

    (A) advanced maternal age
    (B) diminished ovarian reserve
    (C) obesity
*   (D) polycystic ovaries

Ovarian hyperstimulation syndrome is a complication of assisted reproductive techniques. This disease state is characterized by enlarged ovaries and acute fluid shifts from the intravascular space into the extravascular space (peritoneal cavity and thoracic cavity). These therapies can include IVF and, less frequently, clomiphene citrate (low risk) and aromatase inhibitors, such as tamoxifen citrate and letrozole. Ovarian hyperstimulation syndrome is caused by increased vascular permeability through ovarian hypersecretion of vascular endothelial growth factor (VEGF) activating VEGF receptor 2 (VEGF-2); VEGF levels are increased after human chorionic gonadotropin (hCG) is administered in assisted reproductive technology (ART) cycles to help induce ovulation and final maturation of the growing oocytes. As levels of VEGF increase after hCG administration, it stimulates vascular permeability by interacting with the VEGF-2 receptor.

Massive extravasation caused by ovarian hyperstimulation syndrome can lead to hemoconcentration with reduced end-organ perfusion, especially to the kidneys. Additionally, third spacing into the pleural space can lead to respiratory distress syndrome. Vascular changes associated with ovarian hyperstimulation syndrome ultimately lead to an increased risk of thromboembolic events in these patients.

The primary symptoms of ovarian hyperstimulation syndrome are abdominal fullness and shortness of breath. The incidence of moderate-to-severe ovarian hyperstimulation syndrome can vary between 5% and 8% of all total IVF cycles. Mild symptoms typically comprise lower abdominal pain, nausea, vomiting, and diarrhea. Moderate-to-severe symptoms include persistent and worsening ascites. Serious symptoms consist of rapid weight gain, tense ascites, hemodynamic instability from orthostatic hypotension, tachypnea, and progressive oliguria. Severe symptoms, seen in up to 1.4% of all women who undergo ART cycles, can be life threatening (Table 8-1). Laboratory findings associated with serious

TABLE 8-1. Classification of Ovarian Hyperstimulation Syndrome Symptoms

Classification of Ovarian Hyperstimulation Syndrome Symptoms	Clinical Features	Laboratory Values
Mild	Abdominal discomfort/distention Mild nausea/vomiting Diarrhea Enlarged ovaries	
Moderate	Mild features plus Ultrasonographic evidence of ascites	Elevated hematocrit greater than 41% Elevated leukocytes greater than 15,000/mm^3

Severe	Mild and moderate features plus
	  Clinical evidence of ascites
	  Hydrothorax
	  Severe dyspnea
	  Oliguria/anuria
	  Intractable nausea/vomiting
	  Tense ascites
	  Low blood pressure/central venous pressure
	  Rapid weight gain
	  Syncope
	  Severe abdominal pain
		Hematocrit greater than 51%
		Elevated leukocytes greater than 25,000/mm^3
		Creatinine clearance less than 50 mL/min
		Creatinine greater than 1.6 mL/min
		Sodium less than 135 mEq/L
		Potassium greater than 5 mEq/L
		Elevated liver enzymes

Critical	Mild, moderate, and severe features plus
	  Anuria/acute renal failure
	  Arrhythmia
	  Thromboembolism
	  Pericardial effusion
	  Massive hydrothorax
	  Arterial thrombosis
	  Acute respiratory distress syndrome
	  Sepsis	

ovarian hyperstimulation syndrome include hemoconcentration, leukocytosis, hyponatremia, hyperkalemia, elevated liver enzymes, and increased serum creatinine.

Risk factors for ovarian hyperstimulation syndrome include high antral follicle counts, polycystic ovary syndrome, young age, and a previous history of ovarian hyperstimulation syndrome. Women of advanced maternal age are less likely to have numerous antral follicles; thus the described patient is less likely to be at risk of hyperstimulation. The risk of ovarian hyperstimulation syndrome increases with the number of developing ovarian follicles. Women with poor stimulation are less likely to hyperstimulate. Women with polycystic ovaries have a high number of potential eggs recruited for growth over a treatment cycle in response to the use of exogenous gonadotropins. Therefore, they are at higher risk of hyperstimulation.

Prevention of ovarian hyperstimulation syndrome is the hallmark of good IVF treatment. This is done by optimal selection of stimulation protocol; withholding the hCG trigger; cancellation of the ART cycle; freezing all embryos; and not performing an embryo transfer, thus preventing pregnancy. Pregnancy will only make ovarian hyperstimulation syndrome worse.

Treatment for patients with ovarian hyperstimulation syndrome is largely supportive. However, patients with severe ovarian hyperstimulation syndrome can have fatal complications; they require immediate treatment to help maintain circulatory volume and electrolyte balance. Paracentesis for patients with severe pain or abdominal distention can help to improve patient comfort.

Treatment options include cabergoline, a dopamine agonist that has been demonstrated to reverse VEGF-2-dependent increased vascular permeability. Clinically, this medication can significantly reduce the incidence of moderate ovarian hyperstimulation syndrome, as well as pelvic fluid accumulation and hemoconcentration. Implantation and ongoing pregnancy–live-birth rates of IVF patients at risk of ovarian hyperstimulation syndrome appear not to be affected by cabergoline administration. Cabergoline administration is warranted in women at risk of ovarian hyperstimulation syndrome, and such women should be offered this medication to help reduce the incidence of ovarian hyperstimulation syndrome.

Patients who undergo ART are at risk of adnexal torsion. The risk is believed to be less than 1%. Patients who have undergone an ART cycle have enlargement of the ovaries. These patients are at an 11-fold increased risk of adnexal torsion. This is especially true for patients whose treatment was complicated by ovarian hyperstimulation syndrome. Delayed diagnosis of ovarian torsion can lead to eventual ischemic injury of the ovary and loss of the affected ovary.

Polycystic ovaries are the risk factor most commonly associated with the described patient's symptoms. Advanced maternal age, diminished ovarian reserve, and obesity are not the factors most commonly associated with these complications.

Gomez R, Soares SR, Busso C, Garcia-Velasco JA, Simon C, Pellicer A. Physiology and pathology of ovarian hyperstimulation syndrome. Semin Reprod Med 2010;28:448–57.

Humaidan P, Quartarolo J, Papanikolaou EG. Preventing ovarian hyperstimulation syndrome: guidance for the clinician. Fertil Steril 2010;94:389–400.

Maxwell KN, Cholst IN, Rosenwaks Z. The incidence of both serious and minor complications in young women undergoing oocyte donation. Fertil Steril 2008;90:2165–71.

Ovarian hyperstimulation syndrome. Practice Committee of American Society for Reproductive Medicine. Fertil Steril 2008;90:S188–S193.

9

Patient with gastric bypass

A 34-year-old nulligravid woman comes to your office 18 months after Roux-en-Y gastric bypass surgery. Her body mass index is 28 (calculated as weight in kilograms divided by height in meters squared). She now has regular cycles and her hypertension has resolved. She is a vegetarian. Before attempting pregnancy, the most appropriate immediate management for this patient is

  (A) begin timed intercourse
* (B) evaluate for nutritional deficiencies
  (C) pelvic ultrasonography
  (D) wait 6 more months before attempting pregnancy

Bariatric surgery has become an alternative surgical choice for morbidly obese patients who cannot lose weight through nonsurgical approaches. A patient with a body mass index of 40 or higher (or 35 or higher with comorbidities) is a candidate for bariatric surgery. The two most common bariatric surgical procedures are Roux-en-Y gastric bypass (restrictive and malabsorptive) and gastric banding (restrictive). Because these procedures are becoming more common, understanding their management and complications is vital for preconception and conception care in patients of reproductive age.

Quality of life in general improves after bariatric surgery. This is important because it sets the stage for a healthier pregnancy for the mother and the fetus. Bariatric surgery can lead to a general improvement in health and a decrease specifically in gestational diabetes mellitus, hypertension, preeclampsia, sleep apnea, gastric reflux, urinary incontinence, and degenerative joint disease. Children of obese mothers are at risk of childhood obesity. Surgical and nonsurgical preconception weight loss help break this cycle.

This patient now has regular cycles and should attempt pregnancy on her own with timed intercourse or ovulation predictor kits. Weight loss 2 years after bariatric surgery ranges between 20 kg and 40 kg. More than 70% of anovulatory patients will resume regular menses after losing 20 kg. Because the described patient is a vegetarian with recent bariatric surgery, a nutritional status evaluation is indicated before she attempts pregnancy. When her nutritional status is optimized, she then can attempt pregnancy.

Nutritional deficiencies can occur after bariatric surgery. This patient had Roux-en-Y gastric bypass surgery. Nutritional deficiencies mostly are seen with malabsorptive procedures because part of the intestine is bypassed. Serious nutritional deficiencies include vitamin D, vitamin B_{12}, vitamin B_1 (thiamine), magnesium, folic acid, calcium, and ferritin deficiencies. Patients most at risk of nutritional deficits include vegetarians such as the described patient. In a vegetarian, an evaluation for micronutrient deficiencies is warranted. Preconception evaluation and consultation with a nutritionist should be considered for a patient who has had bariatric surgery.

Complications of bariatric surgery carried out during pregnancy include intestinal obstruction, gastrointestinal hemorrhage, hernias, anastomotic leaks, and band complications. Abdominal pain, nausea, and vomiting in a pregnant patient who has undergone bariatric surgery need to be evaluated carefully. Early consultation with a bariatric surgeon is warranted.

Pelvic ultrasonography would not be the most appropriate immediate need for this patient. However, pelvic ultrasonography should be part of the overall evaluation of any infertile patient.

The first year after bariatric surgery is when most weight loss occurs (catabolic phase). In the second year, the rate of loss decreases and stabilizes. The American College of Obstetricians and Gynecologists advises bariatric surgery patients to delay pregnancy for 12–18 months after surgery, during the rapid weight-loss phase. Waiting to conceive may help the patient avoid nutritional deficiencies and intrauterine growth restriction. Postponing pregnancy an additional 6 months is not warranted. Studies have compared pregnancy outcomes among patients in the first year after bariatric surgery. In patients who became pregnant in the first year after bariatric surgery, no increases were observed in pregnancy complications or comorbidities, such as hypertension, diabetes mellitus, and congenital malformations.

Bariatric surgery and pregnancy. ACOG Practice Bulletin No. 105. American College of Obstetricians and Gynecologists. Obstet Gynecol 2009;113:1405–13.

Merhi ZO. Impact of bariatric surgery on female reproduction. Fertil Steril 2009;92:1501–8.

Moran LJ, Norman RJ. The effect of bariatric surgery on female reproductive function. J Clin Endocrinol Metab 2012;97:4352–4.

Sheiner E, Edri A, Balaban E, Levi I, Aricha-Tamir B. Pregnancy outcome of patients who conceive during or after the first year following bariatric surgery. Am J Obstet Gynecol 2011;204:50.e1–6.

10

Macroadenoma and galactorrhea

A 28-year-old nulligravid woman comes to your clinic after 6 months of amenorrhea. Her menstrual irregularity started initially as irregular cycles 1 year ago, followed by amenorrhea. On evaluation, she is not pregnant, but her prolactin level is elevated to 85 ng/mL. She has a normal thyroid-stimulating hormone level. Repeat testing shows a persistently high prolactin elevation of 91 ng/mL. Magnetic resonance imaging reveals the image shown in Figure 10-1. The serum test level most likely to be elevated in further evaluations is

(A) follicle-stimulating hormone (FSH)
* (B) insulin-like growth factor 1 (IGF-1)
(C) antimüllerian hormone
(D) morning cortisol

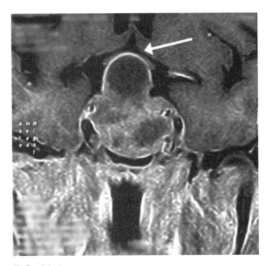

FIG. 10-1

Secondary amenorrhea is defined as a lack of three consecutive menstrual cycles or cessation of bleeding for 6 months. In addition to pregnancy, amenorrhea may be a subsidiary factor to a number of etiologies, such as hypothalamic–pituitary causes, ovarian failure, anovulation secondary to polycystic ovary syndrome, and anatomic abnormalities that affect the outflow tract.

Hypothalamic–pituitary tumors and masses can lead to hypogonadotropism and amenorrhea. Prolactin-producing tumors (most commonly adenomas) are a common type of pituitary adenoma that leads to hyperprolactinemia. Hyperprolactinemia can have other numerous causes, such as pituitary stalk compression–transection, hypothalamic lesions, medications (eg, dopamine receptor blockers, tricyclic antidepressants, selective serotonin reuptake inhibitors, and alpha-methyldopa), breast stimulation, chest wall trauma or injury, hypothyroidism, and renal failure (from decreased prolactin clearance). Prolactin-secreting adenomas, or prolactinomas, are the most common type of secreting pituitary adenoma. Prolactin secretion is primarily under inhibitory control by dopamine. Normal prolactin levels are less than 20 ng/mL in nonpregnant women, and the upper limit of normal varies by assay.

Most patients with prolactinomas will have a microadenoma and, therefore, their signs and symptoms will be secondary to the hyperprolactinemia alone. However, patients with macroadenomas may have neurologic signs and symptoms from the compressive effect of the macroadenoma on surrounding structures, with symptoms such as headaches and visual abnormalities. In the presence of a pituitary macroadenoma, it is important to evaluate for causes of pituitary adenomas other than prolactinomas, such as growth hormone (GH)-secreting adenomas (via an IGF-1 evaluation), thyrotropin-secreting adenomas, corticotropin-secreting adenomas, and gonadotropin-secreting adenomas (via alpha subunit). In the presence of pituitary macroadenomas, it is wise to check thyroid, adrenal, and IGF-1 levels as well as the alpha subunit. Growth hormone-producing adenomas are the next most common type of pituitary adenoma, following prolactinomas. In cases of GH-producing adenoma, IGF-1 rather than GH levels are evaluated because GH levels are difficult to measure and variable in their response. The IGF-1 level acts as a surrogate marker.

In women, hyperprolactinemia can cause galactorrhea (in approximately one third of cases), oligomenorrhea or amenorrhea, and osteopenia. The development of oligomenorrhea and amenorrhea are likely caused by the disruption of normal gonadotropin-releasing hormone pulsatility by the reflex increase in central dopamine.

Loss of gonadotropin-releasing hormone pulsatility results in abnormal luteinizing hormone and FSH secretion with resultant anovulation and oligomenorrhea or amenorrhea. The long-term lack of estradiol production may result in osteopenia or osteoporosis.

Hyperprolactinemia can be diagnosed in the presence of an elevated serum prolactin level. Thyroid function evaluation and a pregnancy test (to rule out pregnancy as a physiologic cause of hyperprolactinemia) should be performed. Magnetic resonance imaging of the brain should be obtained in any patient with a significantly elevated prolactin level. However, some practitioners advocate imaging for all patients with hyperprolactinemia because mildly elevated prolactin levels may be due to the pituitary stalk compression by a non-prolactin-secreting pituitary macroadenoma or a craniopharyngioma.

Prolactinomas may be treated medically or surgically. Medical management with the use of dopamine receptor agonist is the primary approach for treatment of prolactinomas, including macroprolactinomas, as long as the patient is free from neurologic symptoms that require surgical intervention. Surgical intervention may be used if the response to medical management of a macroadenoma is not adequate.

For this patient, the serum test most likely to be elevated is IGF-1, given that GH-producing adenomas are the second most common type of pituitary macroadenoma. Although FSH and morning cortisol levels can be evaluated, they are less likely to reveal the diagnosis because FSH-secreting and adrenocorticotropic hormone-secreting macroadenomas are rare. Measurement of the antimüllerian hormone level is not relevant to the evaluation of hyperprolactinemia but is helpful in the evaluation of ovarian reserve.

Beshay VE, Beshay JE, Halvorson LM. Pituitary tumors: diagnosis, management, and implications for reproduction. Semin Reprod Med 2007;25:388–401.

Davis JR. Prolactin and reproductive medicine. Curr Opin Obstet Gynecol 2004;16:331–7.

Shibli-Rahhal A, Schlechte J. Hyperprolactinemia and infertility. Endocrinol Metab Clin North Am 2011;40:837–46.

11

In vitro fertilization and intracytoplasmic sperm

A 36-year-old nulligravid woman with primary infertility comes to your office for evaluation. She has been attempting pregnancy for 12 months and has had regular menstrual cycles. Hysterosalpingography is normal and basal antral follicle count is 15. Ovarian reserve testing reveals a follicle-stimulating hormone level of 6.5 mIU/mL and an antimüllerian hormone level of 2.5 ng/mL. Semen analysis concentration is 3 million/mL with 15% motility and 1% strict morphology. Repeat semen analysis is similar. Male karyotype, Y chromosome microdeletion, total testosterone level, and follicle-stimulating hormone level are all normal. You consider recommending in vitro fertilization (IVF) with intracytoplasmic sperm injection (ICSI). You counsel the patient that a complication unique to ICSI is

* (A) imprinting disorders
* (B) multiple pregnancy
* (C) ovarian hyperstimulation syndrome
* (D) decreased fertilization rates

The described patient has a negative infertility evaluation. Her partner's initial and repeat semen analyses show oligoasthenoteratozoospermia. Based on World Health Organization guidelines, a normal sperm concentration has greater than 15×10^6 spermatozoa/mL, greater than 40% motility, and greater than 4% normal forms by strict Kruger morphology. The semen analysis is the most important initial step in the evaluation of male factor infertility. Once initial screening reveals a low concentration of sperm (5–10 million/mL), a reproductive urologic consultation is warranted. The initial evaluation will include a medical history and physical examination in addition to a paternal karyotype, Y chromosome analysis, transrectal ultrasonography, and an endocrine evaluation.

Genetic evaluation of the patient's partner is indicated before performing ICSI with his sperm. Approximately

10–15% of men who are azoospermic or oligospermic will have deletions in the Y chromosome. Regions on the long arm of the Y chromosome contain genes for spermatogenesis. There are three regions referred to as azoospermia factor (*AZF*) regions: 1) *AZFa* (proximal), 2) *AZFb* (central), and 3) *AZFc* (distal). A gene deletion in the entire region of *AZFa* or *AZFb* usually results in azoospermia. Testicular biopsies with these deletions often result in no sperm being available for ICSI. Men with a deletion in *AZFc* will have either oligospermia (sperm in their ejaculate) or azoospermia. In contrast to men with *AZFa* and *AZFb* deletions, men with azoospermia and an *AZFc* deletion will have sperm on testicular extraction for ICSI. The male partner needs to be aware that his male offspring may inherit his Y chromosome abnormality and also may prove to be oligospermic if they have a Y chromosome deletion.

Approximately 5% of men with severe oligospermia (less than 5 million/mL) will have a karyotypic abnormality. In the infertile male with sex chromosome aneuploidy, two thirds will have Klinefelter syndrome (47,XXY). Offspring of men with Klinefelter syndrome are at an increased risk of having a chromosomal or genetic defect. Thus, the male partner should have a karyotype before IVF with ICSI.

In vitro fertilization with ICSI is indicated in this couple based on the male partner's abnormal sperm sample. In vitro fertilization involves sequential steps, including controlled ovarian hyperstimulation with gonadotropins, oocyte vaginal retrieval under ultrasonographic guidance, fertilization in the laboratory (by means of conventional insemination or ICSI), and transcervical embryo transfer. Box 11-1 lists the indications for IVF.

Intracytoplasmic sperm injection is a procedure in which a single sperm is immobilized and drawn into a pipette. Using a micromanipulator, an oocyte is stabilized, a pipette traverses the zona and oolemma, and the sperm is injected into the ooplasm. This manipulation activates the oocyte, achieving fertilization rates comparable with conventional IVF. Box 11-2 lists the indications for ICSI.

There may be an association between the use of ICSI and Beckwith–Wiedemann syndrome. Beckwith–Wiedemann syndrome is an imprinting disorder that is a fetal overgrowth and neoplasia syndrome (macrosomia, macroglossia, midline abdominal wall defects, predisposition to embryonal cancer) with the genetic error on chromosome 11. Beckwith–Wiedemann syndrome involves stable alterations in DNA itself with no sequence changes. This is referred to as an epigenetic change. The use of ICSI may affect the epigenetics of early embryogenesis. Most Beckwith–Wiedemann syndrome cases are due to loss or failure of methylation. Disorders of imprinting are rare, with an incidence of 1 in 10,000–30,000 persons. In children born after IVF/ICSI, the rate of Beckwith–Wiedemann syndrome cases

---

BOX 11-1

Indications for In Vitro Fertilization

- Male factor infertility
- Tubal factor infertility
- Endometriosis and pelvic adhesive disease
- Ovulatory dysfunction; patients hyper-responsive to gonadotropins
- Unexplained infertility when superovulation fails
- Diminished ovarian reserve
- Fertility preservation
- Preimplantation genetic diagnosis and aneuploidy screening

BOX 11-2

Indications for Intracytoplasmic Sperm Injection

- Moderate-to-severe male factor infertility
- Surgically retrieved sperm
- Preimplantation genetic diagnosis and aneuploidy screening cases
- Previous poor or failed fertilization
- Cryopreserved oocytes

---

may reach 1 in 2,700 births. Another imprinting disorder that has been linked to ICSI that is even less common is Angelman syndrome. In counseling the described patient, you should tell her that the complication unique to ICSI is the risk of imprinting disorders.

The goal of IVF is to have one healthy child at a time. The risk factor for multiple pregnancy with IVF is the transfer of more than one embryo, not ICSI per se. The American Society for Reproductive Medicine has recommendations for the number of embryos to be transferred in an IVF cycle based on patient age, embryo quality, and reproductive history. Historically, IVF centers have compensated for lower implantation rates by transferring more embryos, which sometimes resulted in high-order multiples. In the last 10 years, IVF pregnancy rates have increased significantly because of improved embryo transfer technique, advanced culture media systems, and well-managed oocyte stimulation protocols. Elective single-embryo transfer is becoming more prevalent as patients are more carefully chosen in order to reduce the risk of multiple pregnancy. More recently, embryo biopsy for aneuploidy screening will allow transfer of a single euploid embryo and dramatically reduce the chance of a multiple pregnancy.

Ovarian hyperstimulation syndrome is a risk of IVF, not ICSI, and is associated with gonadotropin stimulation during an IVF cycle. The pathophysiology of ovarian hyperstimulation syndrome is increased capillary permeability with intravascular fluid moving into third space compartments. Mildly affected patients often experience bloating, shortness of breath, nausea, and decreased urination. One of the main mediators is increased vascular endothelial growth factor, which is driven by high estradiol levels and human chorionic gonadotropins. Risk factors include an elevated anti-müllerian hormone level, previous history of ovarian hyperstimulation syndrome, polycystic ovary syndrome, young age, high antral follicle count, rapidly increasing estradiol levels, and estrogen levels above 2,500 pg/mL. The key to prevent ovarian hyperstimulation syndrome is to recognize risk factors and choose an ovulation induction protocol and gonadotropin dose carefully.

The use of ICSI increases fertilization rates in properly selected patients. Patients who have had poor or failed fertilization in a prior cycle have an indication for ICSI in subsequent IVF cycles. Thus, ICSI can restore fertilization rates by overcoming either an oocyte or sperm function problem.

Diagnostic evaluation of the infertile male: a committee opinion. Practice Committee of American Society for Reproductive Medicine [published erratum appears in Fertil Steril 2013;99:951]. Fertil Steril 2012;98:294–301.

Forman EJ, Hong KH, Treff NR, Scott RT. Comprehensive chromosome screening and embryo selection: moving toward single euploid blastocyst transfer. Semin Reprod Med 2012;30:236–42.

Multiple gestation associated with infertility therapy: an American Society for Reproductive Medicine Practice Committee opinion. Practice Committee of American Society for Reproductive Medicine. Fertil Steril 2012;97:825–34.

Ovarian hyperstimulation syndrome. Practice Committee of American Society for Reproductive Medicine. Fertil Steril 2008;90:S188–93.

Vermeiden JP, Bernardus RE. Are imprinting disorders more prevalent after human in vitro fertilization or intracytoplasmic sperm injection? Fertil Steril 2013;99:642–51.

12

Hydrosalpinges and in vitro fertilization–embryo transfer

A 29-year-old nulligravid woman with regular menstrual cycles comes to your office after 18 months of infertility. Her husband had a normal semen analysis. Hysterosalpingography 1 year ago showed large bilateral hydrosalpinges and distal obstruction (Appendix B). Before proceeding with in vitro fertilization (IVF), the best next step in management is

    (A) saline sonohysterography
    (B) repeat hysterosalpingography
*   (C) bilateral salpingectomy
    (D) neosalpingostomy of dilated tubes
    (E) antibiotic therapy alone

Hydrosalpinges, large dilated fallopian tubes that are damaged intrinsically and no longer function properly, may be observed by means of ultrasonography, hysterosalpingography, or both. They contain fluid that is believed to retrograde spill back within the uterus and lead to decreased implantation of transferred embryos. The hydrosalpinx fluid may interfere mechanically with embryonic apposition or reduce endometrial receptivity. In the presence of hydrosalpinges, the expression of endometrial receptivity markers (integrin α, β, and γ [eg, integrin]) is reduced. This fluid is believed to be toxic to embryos.

A Cochrane review and an American Society for Reproductive Medicine practice committee guideline have noted that pregnancy rates are decreased by approximately 50% in patients with hydrosalpinges. The Cochrane review demonstrated improved pregnancy rates with laparoscopic salpingectomy for hydrosalpinges before IVF. Pregnancy rates with IVF return to normal levels after surgically excising the damaged tubes. Hysteroscopic closure of hydrosalpinges using an implantable contraceptive device has been theorized to yield similar results. A recent pooled analysis shows good efficacy, but evidence from a randomized controlled trial is lacking.

The described patient is not a candidate for reassurance because hydrosalpinges are a marker for permanent tubal disease. Hydrosalpinges will not resolve spontaneously and repeat imaging with hysterosalpingography or saline sonohysterography is not necessary. Therefore, the hydrosalpinges must be surgically addressed. Neosalpingostomy of dilated tubes is not the treatment of choice because it is unlikely to result in a successful intrauterine pregnancy. Bilateral salpingectomy is the best choice for the described patient. Antibiotic therapy alone would not

Arora P, Arora RS, Cahill D. Essure® for management of hydrosalpinx prior to in vitro fertilization—a systematic review and pooled analysis. BJOG 2014;121:527–36.

Bildirici I, Bukulmez O, Ensari A, Yarali H, Gurgan T. A prospective evaluation of the effect of salpingectomy on endometrial receptivity in cases of women with communicating hydrosalpinges. Hum Reprod 2001;16:2422–6.

Camus E, Poncelet C, Goffinet F, Wainer B, Merlet F, Nisand I, et al. Pregnancy rates after in-vitro fertilization in cases of tubal infertility with and without hydrosalpinx: a meta-analysis of published comparative studies. Hum Reprod 1999;14:1243–9.

Johnson N, van Voorst S, Sowter MC, Strandell A, Mol BWJ. Surgical treatment for tubal disease in women due to undergo in vitro fertilisation. Cochrane Database of Systematic Reviews 2010, Issue 1. Art. No.: CD002125. DOI: 10.1002/14651858.CD002125.pub3.

Kontoravdis A, Makrakis E, Pantos K, Botsis D, Deligeoroglou E, Creatsas G. Proximal tubal occlusion and salpingectomy result in similar improvement in in vitro fertilization outcome in patients with hydrosalpinx. Fertil Steril 2006;86:1642–9.

Strandell A, Lindhard A, Waldenstrom U, Thorburn J. Hydrosalpinx and IVF outcome: cumulative results after salpingectomy in a randomized controlled trial. Hum Reprod 2001;16:2403–10.

13

Ovarian cancer in a *BRCA1*-positive patient

A 33-year-old nulligravid woman comes to your office for her annual well-woman examination. She tells you that her mother is battling breast cancer. She is of Ashkenazi Jewish heritage and reports early onset of breast and ovarian cancer throughout her mother's family. The patient and her mother have tested positive for the *BRCA1* gene mutation. She currently is taking oral contraceptives (OCs) and is afraid that she is increasing her chances of developing cancer. The contraceptive that offers the most noncontraceptive health benefits for this patient is

   (A) progestin-only OC
* (B) combination estrogen–progestin OC
   (C) medroxyprogesterone acetate
   (D) intrauterine device

Breast and ovarian cancer are among the leading causes of cancer death in the United States. Most patients with breast and ovarian cancer have sporadic cancer formation. Only a small percentage of all cases of breast and ovarian cancer (approximately 4–7%) are considered to be familial. During their lifetime, *BRCA1* mutation carriers have an approximate 18–60% chance of developing ovarian cancer and a 54–85% chance of developing breast cancer.

Controversy continues about the role of OCs and birth control. Good evidence exists that OCs have a protective effect against ovarian cancer. The described patient is relatively young and nulliparous. Women who use OCs have been found to have approximately a 46% reduction in risk of ovarian cancer compared with women who have never taken OCs. The risk reduction appears to be associated with longer length of use, and the protective effect has been demonstrated to be as long as 30 years after discontinuation of use. The risk reduction does not appear to be related to high-, moderate-, or low-dose contraceptives. All formulations of OCs seem to confer ovarian cancer protection to users.

Oral contraceptives can reduce the risk of ovarian cancer by approximately 50% in *BRCA1* mutation carriers. The protective effects of OCs appear to be similar in women without a *BRCA1* mutation compared with women who are *BRCA1* mutation carriers. The breast cancer risk for *BRCA1* mutation carriers and the use of OCs is subject to debate. Studies have demonstrated conflicting results for and against increased risk of breast cancer. Mastectomy and bilateral salpingo-oophorectomy confer the highest risk reduction of breast and ovarian cancer in *BRCA1* mutation carriers. However, the described patient has not completed childbearing. Until then, OCs offer a significant risk reduction of ovarian cancer even in this patient population.

Medroxyprogesterone acetate has only been shown to decrease the risk of endometrial cancer. The levonorgestrel intrauterine device can be used to help treat malignancy of the endometrium, although varying degrees of treatment effectiveness have been described. Progestin therapy is particularly successful to help maintain fertility in patients who have been diagnosed with endometrial cancer and who wish to maintain fertility.

14

Endometriosis and infertility

A 25-year-old nulligravid woman has been attempting to become pregnant for 2 years. She was asymptomatic when she was taking combination oral contraceptives. Since she stopped taking oral contraceptives, she has experienced progressively worsening dysmenorrhea, dyspareunia, and pelvic pain, which significantly interfere with her daily life. Bimanual examination reveals evidence of right adnexal fullness and tenderness. Transvaginal ultrasonography shows a normal uterus and a 5-cm right ovarian cyst with a homogeneous "ground glass" appearance suggestive of an endometrioma. The best surgical treatment to improve her fertility and decrease her pain is laparoscopy with

(A) cauterization of cyst wall
(B) right oophorectomy
(C) drainage of endometrioma
* (D) resection of endometrioma cyst wall

Endometriosis is characterized by lesions of endometrial-like tissue outside the uterus; the condition is associated with concurrent inflammation, pelvic pain, and infertility. The mechanisms responsible for endometriosis are believed to be retrograde menstruation, celomic metaplasia, and lymphatic spread in immunologically and genetically susceptible individuals. It is recognized that the stage and extent of disease may not correlate with its symptoms, reproductive outcome, or recurrence risk. Although the cause of endometriosis is unknown, it is likely to be multifactorial in origin, including genetic factors with possible epigenetic influences promoted through environmental exposures. Endometriosis is associated with pelvic–abdominal pain and infertility. Other symptoms include dysmenorrhea, dyspareunia, heavy menstrual bleeding, nonmenstrual pelvic pain, pain at ovulation, dyschezia, and dysuria. Endometriosis-related infertility is associated with the severity of disease, and although the exact mechanism for infertility is unknown, it is related to impaired tubo-ovarian function, inflammation, ovarian endometriomas, reduced egg quality, and implantation.

The standard for making the diagnosis of endometriosis is histologic confirmation because the false-positive rate with laparoscopic visualization alone may approach 50%, especially in women with minimal or mild endometriosis. Laparoscopy also allows for staging based on the 1997 revised American Society for Reproductive Medicine scoring system.

Because of a lack of data from sufficient randomized controlled clinical trials, the best way to treat endometriosis at different clinical stages is unclear. The World Endometriosis Society sponsored the Montpellier Consortium Consensus conference to bring together representatives of multinational societies to reach a consensus on the management of endometriosis. A process was developed to score the current evidence to make recommendations for future care and treatment. Whenever possible, laparoscopic surgery should always be performed rather than laparotomy. In cases of severe endometriosis, gynecologic surgeons may consider the option of limiting surgical excision at an initial operation and refer the patient to a surgeon experienced in endometriosis surgery. Studies suggest that the first surgical intervention delivers the greatest benefit compared with subsequent surgical procedures, with pain improvement at 6 months of approximately 83% for the first excisional procedure versus 53% for subsequent surgeries. Excessive numbers of repeat laparoscopic procedures should be avoided.

The Montpellier conference emphasized consideration of the ovarian reserve before surgery in women who are experiencing infertility. Growing evidence suggests that surgical treatment of endometriomas contributes to reduced ovarian reserve. Laparoscopic surgical removal of stage I and II endometriosis improves fertility. It is unclear whether excision is better than ablation; however, it is recommended to excise lesions where possible, especially in cases of deep endometriosis associated with pain. To date, no controlled clinical trials have addressed whether surgery improves fertility for stage III and IV endometriosis.

Adjunct therapy with gonadotropin-releasing hormone agonist after laparoscopic surgery has not been shown to benefit fertility. Use of gonadotropin-releasing hormone agonist after surgery is not recommended because it delays pregnancy at a time when fertility has been improved by surgery. Recent evidence suggests that laparoscopic excision (cystectomy) should be performed whenever possible for endometriomas greater than 4 cm in diameter because it improves fertility more than ablation (drainage and coagulation). However, the surgeon needs to take great care to identify the tissue planes and to carefully dissect the endometrioma to avoid removing normal ovarian tissue, which will affect the ovarian reserve. Reducing the amount of electrosurgical hemostasis to control bleeding is recommended, as is the use of suturing. The best surgical treatment to improve the described patient's fertility and to decrease her pain is laparoscopy with resection of the endometrioma cyst wall.

Oophorectomy should be avoided because there is normal ovarian tissue that can be preserved and that will improve the patient's future fertility. Likewise, excessive use of cautery, although effective for ablating endometriosis, leads to excessive destruction of normal ovarian tissue, reducing ovarian reserve. Insufficient data exist at this time in regard to the value of robotic surgery in endometriosis care. However, limited recent studies have suggested that robotic surgery appears to be at least comparable with conventional laparoscopic surgery but with higher cost and prolonged operative times.

Hart RJ, Hickey M, Maouris P, Buckett W. Excisional surgery versus ablative surgery for ovarian endometriomata. Cochrane Database of Systematic Reviews 2008, Issue 2. Art. No.: CD004992. DOI: 10.1002/14651858.CD004992.pub3.

Johnson NP, Hummelshoj L. Consensus on current management of endometriosis. World Endometriosis Society Montpellier Consortium. Hum Reprod 2013;28:1552–68.

Somigliana E, Berlanda N, Benaglia L, Vigano P, Vercellini P, Fedele L. Surgical excision of endometriomas and ovarian reserve: a systematic review on serum antimüllerian hormone level modifications. Fertil Steril 2012;98:1531–8.

15

Reproductive function, nutritional status, and protein levels

A 17-year-old female gymnast comes to your office with secondary amenorrhea. She is 1.63 m (64 in.) tall and weighs 41 kg (91 lb). She reports exercise in excess of 21 hours per week. She is virginal and has a negative pregnancy test. Her pelvic examination is normal. She has a follicle-stimulating hormone level of 3.0 and an undetectable luteinizing hormone level. Negative energy balance in this patient causes low serum levels of

* (A) leptin
  (B) ghrelin
  (C) cortisol
  (D) corticotropin-releasing hormone
  (E) adrenocorticotropic hormone

A relationship exists between nutritional status and reproductive function. Profound caloric restriction can lead to poor reproductive function. Leptin is a 16kDa protein that acts as a satiety signal. Leptin exhibits a diurnal pattern that corresponds to consumption of meals. Women with eating disorders and low caloric intake have low circulating levels of leptin (Fig. 15-1).

Leptin-deficient animals exhibit impaired gonadotropin-releasing hormone secretion and are anovulatory. Restoration of leptin levels restores ovulatory status. Leptin also has been associated with the maintenance of luteinizing hormone pulsatility and the preovulatory surge associated with ovulation. Leptin levels in female athletes were observed over a 20-week period and noted to be decreased over the 20-week study period. Women with exercise-induced amenorrhea have lower leptin levels than ovulatory females. In this study, leptin levels of women with exercise-induced amenorrhea were compared with those with ovulatory controls, and, when controlling for body fat levels among the two groups,

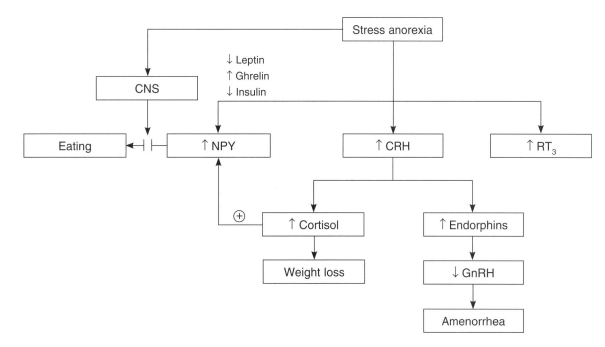

FIG. 15-1. Relationship of various hormone levels in women and stress anorexia, eating, weight loss, and amenorrhea. Abbreviations: CNS, central nervous system; CRH, corticotropin-releasing hormone; GnRH, gonadotropin-releasing hormone; NPY, neuropeptide Y; RT_3, reverse triiodothyronine. (Speroff L, Fritz MA. Clinical gynecologic endocrinology and infertility. 8th ed. Philadelphia [PA]: Lippincott Williams and Wilkins; 2011. p. 875.)

little difference was observed between the two patient populations. This suggests that factors other than exercise are important in the modulation of leptin levels.

Ghrelin is a 28-amino acid peptide that provides energy balance. It can cross the blood–brain barrier and can modulate the vagus nerve and nucleus tractus solitarius at the level of the brain. Ghrelin levels are decreased by food intake and obesity. Its production is increased in fasting and caloric-restrictive states. High ghrelin levels are seen in women with anorexia, bulimia, and exercise-induced amenorrhea.

Thin, amenorrheic individuals with caloric restriction can have a nutritional state of anorexia. These patients often exhibit hypercortisolism in response to elevated corticotropin-releasing hormone and adrenocorticotropic hormone secretion. The most likely laboratory finding to explain this patient's secondary amenorrhea is low levels of serum leptin.

Chou SH, Chamberland JP, Liu X, Matarese G, Gao C, Stefanakis R, et al. Leptin is an effective treatment for hypothalamic amenorrhea. Proc Natl Acad Sci U S A 2011;108:6585–90.

De Souza MJ, Leidy HJ, O'Donnell E, Lasley B, Williams NI. Fasting ghrelin levels in physically active women: relationship with menstrual disturbances and metabolic hormones. J Clin Endocrinol Metab 2004;89:3536–42.

Lawson EA, Donoho D, Miller KK, Misra M, Meenaghan E, Lydecker J, et al. Hypercortisolemia is associated with severity of bone loss and depression in hypothalamic amenorrhea and anorexia nervosa. J Clin Endocrinol Metab 2009;94:4710–6.

16
Unexplained infertility

A 32-year-old nulligravid woman comes to your office with her partner after failing to achieve a pregnancy despite unprotected intercourse for the past 18 months. They are both healthy and take no medications. She has regular monthly menstrual cycles. You recommend that she proceed with hysterosalpingography and ovarian reserve testing, both of which are normal. Her partner's semen analysis is also normal. Given these findings, the most appropriate cost-effective first step is

(A) expectant management for 6 months
(B) clomiphene citrate for 3 months
(C) intrauterine insemination (IUI) alone for 3 months
* (D) clomiphene and intrauterine insemination for 3 months
(E) in vitro fertilization (IVF)

Approximately 25% of couples that are evaluated for fertility concerns do not have an identifiable diagnosis and instead have "unexplained infertility." The basic infertility evaluation occurs after 6–12 months of attempted conception. The evaluation includes confirmation of ovulatory status, testing to evaluate tubal patency and the uterine cavity, semen analysis, and testing of ovarian reserve. In the past, other tests such as a postcoital test, laparoscopy, and endometrial dating were performed, but such steps have been shown to be of no assistance in terms of improving pregnancy rates. The natural decline in fertility due to increasing age clearly contributes to challenges with conception, especially because the average age of first-time mothers increased from age 21.4 years in 1970 to age 25 years in 2006.

In couples with otherwise unexplained infertility, the monthly fecundity rate without treatment varies between 1.8% and 3.8%, which is much lower than couples without fertility challenges (20–25% per month). The monthly fecundity rate tends to be lower for couples in which the woman is older or the length of infertility is longer. Treatment plans should be tailored to each couple, and if the proposed plan does not result in a pregnancy after three to four cycles, alternative treatments should be considered. Lifestyle changes should be part of the conversation, with the aim to eliminate cigarette smoking, achieve a normal body mass index, and reduce caffeine intake. These lifestyle changes can be beneficial for couples who are attempting conception on their own or with assistance. Although many couples with unexplained infertility may conceive on their own if they are given enough time, the low likelihood of conception combined with their steadily declining fecundity due to age warrants that such patients be offered treatment rather than expectant management.

Treatments, such as clomiphene, are used commonly for conditions associated with anovulation. Women with unexplained infertility are already ovulating on a regular basis. Consideration should be given to such medications to achieve superovulation. Clomiphene and other medications have the potential to produce two to three mature follicles during a given cycle. Many studies have evaluated the benefit of clomiphene with intercourse in cases of unexplained infertility and have found little benefit compared with no treatment. Given the lack of benefit and the increased risk of twins and, on rare occasion, higher-order multiples, care should be taken before routinely using clomiphene with intercourse.

Intrauterine insemination involves washing semen with sterile placement of a concentrated volume of motile sperm into the uterine cavity at the approximate time of ovulation. In a recent trial to compare IUI in natural cycles with expectant management, no appreciable improvement in pregnancy rates was observed with IUI alone.

The use of clomiphene combined with IUI to achieve superovulation has been studied and found to have some benefit. The Fast Track and Standard Treatment trial evaluated approximately 500 couples with unexplained infertility who were treated with clomiphene in combination with IUI. During 1,294 cycles, the monthly fecundity was 9.5%, a significant increase compared with the expected 2–4% monthly fecundity with expectant management. Overall, given the low cost and relative simplicity of using clomiphene together with IUI, this is the most cost-effective first step for the treatment of couples with unexplained infertility.

Treatments such as IVF can prove to be very successful in women with unexplained infertility, depending on the woman's age. In 2011, the live-birth rate in the United States for couples younger than 35 years with unexplained infertility and who used IVF was 42.5%. Although IVF is clearly a more effective treatment per cycle than other options, it is usually not the first-line option for couples with unexplained infertility because of its significant cost and relatively complex treatment plan.

Bhattacharya S, Harrild K, Mollison J, Wordsworth S, Tay C, Harrold A, et al. Clomifene citrate or unstimulated intrauterine insemination compared with expectant management for unexplained infertility: pragmatic randomised controlled trial. BMJ 2008;337:a716.

Centers for Disease Control and Prevention. Assisted reproductive technology. 2011 National summary report. Section 2: ART cycles using fresh, nondonor eggs or embryos (part B). Atlanta (GA): CDC; 2014. Available at: http://www.cdc.gov/art/ART2011/section2b.htm. Retrieved March 20, 2014.

Fisch P, Casper RF, Brown SE, Wrixon W, Collins JA, Reid RL, et al. Unexplained infertility: evaluation of treatment with clomiphene citrate and human chorionic gonadotropin. Fertil Steril 1989;51:828–33.

Reindollar RH, Regan MM, Neumann PJ, Levine BS, Thornton KL, Alper MM, et al. A randomized clinical trial to evaluate optimal treatment for unexplained infertility: the fast track and standard treatment (FASTT) trial. Fertil Steril 2010;94:888–99.

Speroff L, Fritz MA. Female infertility. In: Speroff L, Fritz MA, editors. Clinical gynecologic endocrinology and infertility. 8th ed. Philadelphia (PA): Lippincott Williams and Wilkins; 2011. p. 1185–90.

17

Obesity and contraceptive choices

A 29-year-old obese woman with a body mass index of 32 (calculated as weight in kilograms divided by height in meters squared) has just undergone an elective abortion and is considering her contraceptive options. The best reversible contraceptive choice for this woman compared with a nonobese woman is

(A) combination oral contraceptives (OCs)
(B) contraceptive patch
(C) injectable progestin
* (D) long-acting reversible contraceptive

In the United States, obesity is increasing; it is estimated that approximately 61% of adult women are overweight or obese. Obesity is a major contributor to many medical problems, including type 2 diabetes mellitus, coronary artery disease, stroke, and pulmonary dysfunction. These health conditions often are magnified when a woman becomes pregnant. Pregnancy places the patient at risk of developing gestational diabetes, preeclampsia, a venous thromboembolic event, or a delivery complication. Pregnancies complicated by maternal obesity also are associated with an increased prevalence of macrosomic infants, stillbirths, and neonatal deaths. Given these increased health risks for pregnant obese women, it is important to recognize differences in contraceptive methods practiced by obese women compared with nonobese women.

In a recent retrospective review of 7,262 charts of women who had a first-trimester termination of a pregnancy, women were counseled after their abortion on the use of contraceptive options, such as combination hormonal contraceptives (OCs, ring, and patch), the intrauterine device (IUD), the progestin implant, injectable progestin, condoms, and laparoscopic tubal ligation. After performing a multivariable analysis adjusting for age, education, race, and tobacco use, overweight and obese women were more likely to select the IUD, combination hormonal ring, or tubal ligation and were less likely to choose injectable progestin (Table 17-1). No relation was found between being overweight or obese and the choice of condoms, OCs, or the progestin implant after a first-trimester abortion. These findings suggest that after counseling following an elective termination of pregnancy, overweight and obese women select different contraception options than nonobese women. Knowledge of these predispositions may help the health care provider to foster improved patient adherence to their postabortion contraceptive method.

Use of long-acting reversible contraception, such as an IUD or a progestin implant, may be a better choice in an obese individual than estrogen-containing contraceptive methods, such as combination OCs, the patch, or the ring. Obese individuals have an inherent increased baseline risk of thrombosis, which may be further exacerbated by estrogen-containing contraceptive methods. Although a recent study found no difference in the unintended pregnancy rate of overweight and obese users of OCs, the patch, and the ring, these methods may increase morbidity because of occult coexistent conditions, such as hypertension and diabetes, which may have escaped initial detection. Although injectable progestin use would be safer, its association with increased weight gain may be responsible for overweight and obese women adversely selecting it after an abortion.

TABLE 17-1. Contraceptive Method Chosen After Elective Abortion

Method	Overweight	Obese
Intrauterine device	OR =1.3 95% CI, 1.28–1.32 $P<.01$	OR =1.6 95% CI, 1.59–1.61 $P<.01$
Combination contraceptive ring	OR =1.4 95% CI, 1.28–1.52 $P=.04$	OR =1.6 95% CI, 1.57–1.63 $P<.01$
Laparoscopic tubal ligation	OR =1.5 95% CI, 1.44–1.62 $P=.03$	OR =2.9 95% CI, 2.79–3.01 $P<.01$
Injectable progestin	OR =0.70 95% CI, 0.59–0.81 $P=.02$	OR =0.52 95% CI, 0.48–0.56 $P<.01$

Abbreviations: CI, confidence interval; OR, odds ratio.

Scott-Ram R, Chor J, Bhogireddy V, Keith L, Patel A. Contraceptive choices of overweight and obese women in a publically funded hospital: possible clinical implications. Contraception 2012;86:122–6.

McNicholas C, Zhao Q, Secura G, Allsworth JE, Madden T, Peipert JF. Contraceptive failures in overweight and obese combined hormonal contraceptive users. Obstet Gynecol 2013;121:585–92.

18
Heavy menstrual bleeding

A 48-year-old multiparous woman comes to your office with a multiple-year history of heavy menses that adversely affects her daily activities. Her examination is negative. Ultrasonography reveals posterior uterine thickening consistent with adenomyosis. You categorize her bleeding as abnormal uterine bleeding (AUB) with adenomyosis. She requests nonsurgical, nonhormonal treatment. The nonhormonal treatment with the highest likelihood of decreasing her menstrual flow and improving her quality of life is

    (A) ibuprofen
    (B) mefenamic acid
* (C) tranexamic acid
    (D) black cohosh
    (E) ethamsylate

Heavy menstrual bleeding is an important cause of morbidity in premenopausal women. In an effort to simplify the nomenclature and move away from such terms as menorrhagia, menometrorrhagia, and polymenorrhea, the International Federation of Gynecology and Obstetrics developed a classification system for causes of AUB that classifies the etiologies of AUB into two categories:

1. Structural causes, ie, polyp, adenomyosis, leiomyomata, malignancy (PALM)
2. Coagulopathy, ovulatory dysfunction, endometrial, iatrogenic, and "not yet classified" (COEIN)

The two sets of causes constitute the PALM–COEIN classification system (Appendix C).

Although hormonal preparations and surgery are effective therapies, other medical therapies have established efficacy and are preferred by some patients. Nonsteroidal antiinflammatory drugs (NSAIDs) reduce prostaglandin levels, which often are elevated in women with AUB. Studies have shown that NSAIDs have a beneficial effect on dysmenorrhea. A recent Cochrane review analyzed the data from 17 randomized controlled trials that examined the effects of NSAIDs on the reduction of AUB. As a group, NSAIDs were more effective than placebo at reducing AUB, but all were less effective than tranexamic acid, danazol, and the levonorgestrel intrauterine device. Although NSAIDs were shown to be as effective as oral contraceptives and ethamsylate, most studies were underpowered. In terms of efficacy in reducing AUB, no difference has been shown between NSAIDs such as naproxen and mefenamic acid.

Tranexamic acid, a synthetic derivative of the amino acid lysine, is an antifibrinolytic agent that acts by binding to plasminogen and blocking the interaction with fibrin, thereby preventing the dissolution of the fibrin clot. In multiple large randomized trials, tranexamic acid significantly reduced blood loss compared with placebo in a variety of surgical procedures, including joint replacements, cardiopulmonary bypass, and prostatectomy. In many instances, tranexamic acid reduced the number of blood transfusions associated with surgery, and reduced all-cause mortality and death due to bleeding in trauma patients. In Europe, the use of tranexamic acid in gynecology has become commonplace.

A recent randomized controlled trial tested the effectiveness of tranexamic acid in women with AUB. Patients were prescribed tranexamic acid (n=115) or placebo (n=78) for the first 5 days of menses for 6 treatment months after quantification of menstrual flow was determined at baseline. Women who received tranexamic acid had significantly greater reduction in AUB (40% decrease) compared with placebo (8% decrease), had significantly improved quality-of-life scores, and had no difference in adverse events compared with placebo. Even though tranexamic acid is an antifibrinolytic agent and there appears to be no increased risk of thromboembolic events with combined use of tranexamic acid and oral contraceptives, their concomitant use is contraindicated.

Tranexamic acid appears to be well tolerated and effective. Other studies with much larger cohorts of women showed 80% satisfaction with tranexamic acid therapy at 3 months. In terms of the described patient, the treatment that has the highest likelihood of markedly decreasing her AUB and improving her quality of life is tranexamic acid.

Ethamsylate is a hemostatic agent used in the treatment of heavy menstrual bleeding, although its effectiveness appears to be less than that of tranexamic acid. Black cohosh has limited effectiveness for the treatment of AUB.

Diagnosis of abnormal uterine bleeding in reproductive-aged women. Practice Bulletin No. 128. American College of Obstetricians and Gynecologists. Obstet Gynecol 2012;120:197–206.

Lethaby A, Duckitt K, Farquhar C. Non-steroidal anti-inflammatory drugs for heavy menstrual bleeding. Cochrane Database of Systematic Reviews 2013, Issue 1. Art. No.: CD000400. DOI: 10.1002/14651858.CD000400.pub3.

Lukes AS, Moore KA, Muse KN, Gersten JK, Hecht BR, Edlund M, et al. Tranexamic acid treatment for heavy menstrual bleeding: a randomized controlled trial. Obstet Gynecol 2010;116:865–75.

McCormack PL. Tranexamic acid: a review of its use in the treatment of hyperfibrinolysis. Drugs 2012;72:585–617.

19

Contraception for a patient with venous thromboembolism

A 25-year-old woman develops a pulmonary embolism during her first pregnancy and is found to be heterozygous for a factor V Leiden mutation. After 6 months of postpartum therapy with warfarin sodium, anticoagulation therapy is discontinued. She is planning on becoming pregnant in the next 3 years. The most appropriate contraceptive for her is

   (A) combination hormonal vaginal contraceptive ring
* (B) progestin-only contraceptive
   (C) combination oral contraceptive (OC)
   (D) combination contraceptive transdermal patch

The risk of venous thromboembolism in women correlates with aging, obesity, smoking, genetic thrombophilias, autoimmune disorders, pregnancy, and estrogen exposure. The annual incidence of venous thromboembolism is approximately 4 in 10,000 nonhormone users in their twenties, increasing to approximately 1 in 1,000 nonhormone users at age 45 years. This incidence has increased with time, most likely because of increasing body mass index in the population. Obesity and smoking independently increase the odds of venous thromboembolism approximately twofold to threefold. Genetic thrombophilias are inherited as autosomal recessive disorders and include factor V Leiden mutation, prothrombin *G20201A* mutation, and mutations leading to decreased protein C, protein S, and antithrombin. Other conditions include hyperhomocysteinemia, acquired activated protein C resistance, and elevated factor VIIIc. Of these genetic thrombophilias, factor V Leiden mutation is the most common type. Most of these conditions increase the risk of venous thromboembolism in reproductive-aged women who do not use hormones by threefold to fivefold except for prothrombin *G20201A* mutation, which has the least risk. Compound heterozygous and homozygous variants of these conditions carry a greater risk than single mutations. Patients with significant titers of anticardiolipin antibodies and lupus anticoagulant also have increased risk of venous thromboembolism.

Estrogens and the synthetic ethinyl estradiol in combination OCs primarily increase the risk of venous thromboembolism by altering hemostatic factor production by the liver. Procoagulant changes include increased prothrombin fragments, D-dimers, soluble fibrin, fibrinogen, factor VII, factor VIII, Ristocetin cofactor activity, and fibrinogen degradation. Additionally, there is a decrease in antithrombin, protein S, and plasminogen activator inhibitor 1, which also increases coagulation. The estrogen-induced changes in coagulation are dose dependent. In the absence of estrogen, progesterone and progestins have negligible effects on hemostatic factors.

Combination OCs increase the risk of venous thromboembolism by approximately threefold, ie, to approximately seven venous thromboembolism events in 10,000 women per year. The risk is dramatically higher in women with a thrombophilia, making it a contraindication for the use of combination OCs. For example, the risk of a venous thromboembolism in a woman with factor V Leiden heterozygote increases from threefold in a woman who does not use combination OCs (compared with individuals without mutations) to 15-fold in a woman who uses combination OCs.

The type of progestin in combination OCs changes the risk of venous thromboembolism. Third-generation progestins (norgestimate, desogestrel) and particularly the progestin drospirenone in combination OCs may have higher venous thromboembolism risks compared with the first-generation progestin, norethindrone, and the second-generation progestin, levonorgestrel. The increased risk correlates with the increase in sex hormone-binding globulin (SHBG) associated with combination OCs that contain these progestins. In addition, SHBG levels correlate with the degree of activated protein C resistance. Combination OCs with norgestimate and desogestrel compared with levonorgestrel result in higher circulating prothrombin fragments, factor VII, and decreased protein C.

Originally it was thought that nonoral administration of combination contraceptives, including use of the patch and the vaginal ring, would yield less risk because the increase in hemostatic factors was expected to be less. No clinical study to date has proven this hypothesis. The contraceptive patch containing ethinyl estradiol and norelgestromin has a higher risk of venous thromboembolism compared with combination OCs containing the similar progestin, norgestimate. The reason may be related to higher circulating levels of ethinyl estradiol from the patch. Based on limited data, the vaginal ring releasing ethinyl estradiol and etonogestrel (an active metabolite of desogestrel) may have a higher venous thromboembolism risk compared with desogestrel-containing combination OCs. To date, this product has been less well studied than other combination OCs.

Progestin-only contraceptive products available in the United States include norethindrone OCs, the etonogestrel implant, the levonorgestrel intrauterine device (IUD), and depot medroxyprogesterone acetate (DMPA). Studies show no increase in venous thromboembolism risk with norethindrone OCs or the levonorgestrel IUD. The etonogestrel implant has not been adequately studied, but hemostasis factors are not changed with use of the implant, suggesting a low likelihood of venous thromboembolism risk. Traditionally, use of DMPA was not thought to impose a venous thromboembolism risk because DMPA does not have substantial effects on coagulation factors. This idea has been challenged by more recent data that show an increase in venous thromboembolism in women who use DMPA compared with women who use nonhormonal contraceptive methods. Current information would suggest that the least risky hormonal contraceptives for women at high risk of venous thromboembolism are progestin-only contraceptives, such as norethindrone OCs, the levonorgestrel IUD, and the etonogestrel implant. Although any IUD would be safe and effective, her desire for a pregnancy within 3 years makes this approach less appropriate.

Plu-Bureau G, Maitrot-Mantelet L, Hugon-Rodin J, Canonico M. Hormonal contraceptives and venous thromboembolism: an epidemiological update. Best Pract Res Clin Endocrinol Metab 2013;27:25–34.

Raps M, Helmerhorst F, Fleischer K, Thomassen S, Rosendaal F, Rosing J, et al. Sex hormone-binding globulin as a marker for the thrombotic risk of hormonal contraceptives. J Thromb Haemost 2012;10:992–7.

Wu O, Robertson L, Langhorne P, Twaddle S, Lowe GD, Clark P, et al. Oral contraceptives, hormone replacement therapy, thrombophilias and risk of venous thromboembolism: a systematic review. The Thrombosis: Risk and Economic Assessment of Thrombophilia Screening (TREATS) Study. J Thromb Haemost 2005;94:17–25.

20

Premature thelarche

A 2-year-old girl is brought to your clinic by her mother. The girl has a 3-month history of breast growth without vaginal bleeding. Physical examination shows that, compared with girls her age, she is in the 75th percentile for height and the 90th percentile for weight. In addition, she has Tanner stage 3 breast development and Tanner stage 1 pubic hair development. The most appropriate combination of tests to assess her problem is

(A) gonadotropin-releasing hormone (GnRH) agonist stimulation test and pelvic ultrasonography
(B) follicle-stimulating hormone (FSH), luteinizing hormone (LH), and prolactin levels
(C) dehydroepiandrosterone sulfate and testosterone levels
* (D) estradiol and thyroid-stimulating hormone (TSH) levels plus bone age
(E) growth hormone stimulation test and vaginoscopy

Thelarche is the estradiol-dependent growth of breast tissue with stages originally described by Marshall and Tanner (Fig. 20-1). Breast development is typically the initial physical sign of puberty and is dependent on nutritional status and adiposity. Compared with the mid-1800s, pubertal development now occurs approximately 4 years earlier. In addition, significant ethnic differences in the onset of pubertal development have been observed, ie, on average, African American girls experience thelarche at ages 8.9–9.5 years, Mexican American girls at age 9.8 years, and white girls at ages 10–10.4 years. Other classifications of breast development are available based on areolar diameter, pigmentation, and contour.

Traditionally, *premature thelarche* is defined as breast development before age 8 years. Yet, given the mean difference in thelarche among girls of different ethnicities, some experts suggest that premature thelarche be redefined as breast development before age 6 years in African American girls and age 7 years in white girls. This criterion would hold true as long as other factors are not present, such as rapid progression of puberty, ie, bone age greater than 2 years ahead of chronologic age and predicted height of less than 150 cm or 2 standard deviations or more below the genetic target height; central nervous system findings, eg, headaches, seizures, or other neurologic symptoms; and pubertal progression that affects the emotional health of the family or the girl.

Premature thelarche most commonly presents in two age categories: 1) from age 6 months to 2 years and 2) age 5–7 years. Approximately 60% of cases are in the younger age group. Approximately 50% will have regression within 4 years, 30–35% will persist without progression, and 10–15% will develop precocious puberty. No criterion or test is available that can predict which girls with premature thelarche will subsequently develop precocious puberty. The initial presenting symptom is breast development, most commonly Tanner stage 2, which may be asymmetrical or unilateral. There should be no signs or symptoms of sexual hair development or uterine bleeding. The differential diagnoses include initial presentation of precocious puberty, iatrogenic causation from exogenous steroids, and the rare possibility of an estrogen-producing tumor.

Initial evaluation of premature thelarche should include a medical history, review of growth records, and physical examination. Initial diagnostic testing includes blood estradiol and TSH levels and bone age. Hypothyroidism may be associated with premature puberty and functional ovarian cysts. For the described patient, blood estradiol and TSH levels plus bone age would be the most appropriate combination of tests at this time.

Girls with premature thelarche have normal to slightly increased growth velocity and normal to slightly advanced bone age, but less than 2 standard deviations. If these variables are normal, close follow-up every 4–6 months is reasonable. Any change in signs or symptoms or abnormal laboratory findings may require further testing for early precocious puberty, including baseline FSH and LH levels and a GnRH stimulation test. Normal levels of estradiol, FSH, and LH do not exclude early precocious puberty and require a GnRH stimulation test. Premature thelarche is characterized by an FSH-dominant response, whereas central precocious puberty is characterized by an LH-dominant response with GnRH stimulation. In the United States, GnRH is no longer available and stimulation testing is performed with the use of a GnRH agonist. Pelvic ultrasonography also may result in findings that discriminate premature thelarche from central precocious

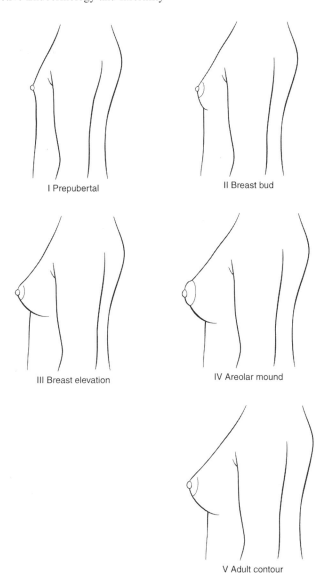

FIG. 20-1. Tanner staging of breast development. (Modified from Speroff L, Glass RH, Kase NG. Clinical gynecologic endocrinology and infertility. 6th ed. Baltimore: Lippincott Williams & Wilkins, 1999. p. 398–9.)

puberty. In premature thelarche, no increase in uterine size is observed and the ovaries lack follicular development. In central precocious puberty, increased uterine dimensions for age and follicles may be seen. Androgen testing, such as dehydroepiandrosterone sulfate and testosterone level testing, is not needed in the absence of sexual hair development. Growth hormone stimulation testing would be considered in cases of short stature. Vaginoscopy is indicated for the evaluation of unexplained vaginal bleeding in a child.

No treatment is necessary with a diagnosis of premature thelarche, but close follow-up is mandatory given that there is no way to predict which children will subsequently develop precocious puberty. Initially, children should be seen at 4–6-month intervals. Girls with premature thelarche have a normal age of menarche and normal final predicted height.

The etiology of premature thelarche remains unclear. Suggested hypotheses include increased sensitivity of the breast to estrogen, increased estrogen production from an ovarian cyst, exposure to environmental estrogen-like compounds, and transient activation of the hypothalamic–pituitary–gonadal axis. Although the estradiol level is normal in girls with premature thelarche using standard clinical assays, most assays do not have the necessary sensitivity to detect a difference. Studies using ultrasensitive recombinant bioassay support higher estradiol levels, making a change in breast sensitivity less likely. Ovarian cysts are not observed with premature thelarche. Epidemics of premature thelarche reported in the past outside the United States support the hypothesis that estrogen-like environmental compounds might be responsible for such outbreaks of mass premature thelarche. It is unlikely that an environmental etiology is responsible for most sporadic cases. The classic finding of elevated FSH response to GnRH is consistent with an early partial activation of the hypothalamic–pituitary–gonadal axis, but the cause of this phenomenon remains to be determined.

de Vries L, Guz-Mark A, Lazar L, Reches A, Phillip M. Premature thelarche: age at presentation affects clinical course but not clinical characteristics or risk to progress to precocious puberty. J Pediatr 2010;156:466–71.

Kaplowitz PB, Oberfield SE. Reexamination of the age limit for defining when puberty is precocious in girls in the United States: implications for evaluation and treatment. Drug and Therapeutics and Executive Committees of the Lawson Wilkins Pediatric Endocrine Society. Pediatrics 1999;104:936–41.

Klein KO, Mericq V, Brown-Dawson JM, Larmore KA, Cabezas P, Cortinez A. Estrogen levels in girls with premature thelarche compared with normal prepubertal girls as determined by an ultrasensitive recombinant cell bioassay. J Pediatr 1999;134:190–2.

21
Recurrent pregnancy loss

A 25-year-old woman seeks your help after three consecutive first-trimester miscarriages. As part of your evaluation, you send her blood and her partner's blood for chromosome testing. If a translocation is detected, the most likely scenario is

    (A) patient—reciprocal
* (B) patient—robertsonian
    (C) partner—reciprocal
    (D) partner—robertsonian

Recurrent pregnancy loss is defined as two or more failed clinical pregnancies (ie, pregnancies visualized by ultrasonography). Evaluation of pregnancy loss should be initiated after two first-trimester pregnancy losses. The most common contributors to recurrent pregnancy loss are maternal anatomic abnormalities, maternal antiphospholipid syndrome, and cytogenetic abnormalities in either partner. Cytogenetic abnormalities make up approximately 5% of etiologies for recurrent pregnancy loss. If recurrent pregnancy loss is diagnosed, blood should be collected from both parents and sent for karyotyping.

Reciprocal and robertsonian translocations can be identified in couples with recurrent pregnancy loss. Reciprocal translocation, the most common chromosomal rearrangement in humans, refers to the exchange of chromosome material between nonhomologous chromosomes. Robertsonian translocation refers to whole arm exchange between acrocentric chromosomes (chromosomes 13, 14, 15, 21, and 22). Carriers of reciprocal translocations can produce balanced and unbalanced gametes. Robertsonian translocation carriers tend to produce more unbalanced gametes. Translocations are more common in women than in men. It is therefore more likely to detect a robertsonian translocation in the female partner if a chromosomal abnormality exists. However, this should not be a reason for excluding the male partner from evaluation.

Cytogenetic evaluation of the products of conception can be performed for the evaluation as well as the psychologic benefit of the couple. This approach has several pitfalls. The material may fail to grow in a culture, leading to a lack of diagnosis, and there is also a potential for maternal contamination. Even if a cytogenetic abnormality in the product of conception is identified, this does not exclude other causes of recurrent pregnancy loss, which should still be considered. In addition, approximately one half of all first-trimester pregnancy losses may be associated with sporadic aneuploidies (trisomies or monosomies). Aneuploidies do not tend to be repetitive and also may be related to maternal age, but maternal age may not be the only factor leading to these aneuploidies. Treatment for couples with parental cytogenetic abnormalities should start with genetic counseling. Couples also may opt for preimplantation genetic screening with transfer of unaffected embryos.

Evaluation and treatment of recurrent pregnancy loss: a committee opinion. Practice Committee of the American Society for Reproductive Medicine. Fertil Steril 2012;98:1103–11.

Hassold T, Hunt P. Maternal age and chromosomally abnormal pregnancies: what we know and what we wish we knew. Curr Opin Pediatr 2009;21:703–8.

Hirshfeld-Cytron J, Sugiura-Ogasawara M, Stephenson MD. Management of recurrent pregnancy loss associated with a parental carrier of a reciprocal translocation: a systematic review. Semin Reprod Med 2011;29:470–81.

Stephenson MD, Awartani KA, Robinson WP. Cytogenetic analysis of miscarriages from couples with recurrent miscarriage: a case–control study. Hum Reprod 2002;17:446–51.

22

Contraception for a patient with *BRCA* gene mutation

A 22-year-old nulligravid woman is a *BRCA2* mutation carrier. She consults you about her contraceptive options. She desires future children and is interested in reliable, reversible birth control. You inform her that oral contraceptives (OCs) decrease the risk of cancer of the endometrium and cancer of the

(A) cervix
(B) breast
* (C) ovary
(D) colon

In recent decades, awareness of a genetic link to breast cancer has become better understood by the scientific community, and today screening tests for *BRCA1* and *BRCA2* mutations are frequently performed. Approximately 0.04–0.2% of U.S. women and 2% of women of Ashkenazi Jewish descent are estimated to be *BRCA* mutation carriers. Reproductive factors may influence the risk of developing breast and ovarian cancer in women who carry *BRCA* mutations. Menarche before age 12 years, low parity, younger age at first childbirth, and breastfeeding all have been shown to modulate the risk of breast cancer. Many mutation carriers will opt for risk-reduction surgery, such as bilateral prophylactic mastectomy and salpingo-oophorectomy, when they have completed their reproductive goals. As *BRCA* mutation screening becomes more common, more women may be identified as carriers at a younger age. Given the national trend toward older age at the time of the first full-term pregnancy, *BRCA* mutation carriers may struggle with decisions about contraception options. Many of these women will opt for nonhormonal options such as the copper intrauterine device or barrier contraception. However, when considering hormonal contraceptive options in *BRCA* mutation carriers, it is important to weigh the benefits and risks.

An increased risk of cervical cancer has been demonstrated among OC users in the general population. Some studies have shown an increased risk of cervical cancer in *BRCA* mutation carriers, although this association has not been confirmed conclusively. Women who are *BRCA* mutation carriers should be screened for cervical cancer in a manner similar to that for the general population, with Pap tests and human papillomavirus testing. Data do not demonstrate that use of OCs modulates the risk of cervical cancer in *BRCA* mutation carriers.

Many studies have attempted to evaluate the effect of OC use on breast cancer development in *BRCA* mutation carriers; these studies have shown conflicting results. Most of the studies are retrospective with potential recall bias or small sample size. Most such studies show a small increased risk of breast cancer, although the risk does not appear to be greatly increased or statistically significant but rather is comparable with the effect in OC users who were not carriers of *BRCA* mutations. A recent meta-analysis found that new formulations of OCs are not associated with an increased risk of breast cancer among *BRCA* mutation carriers compared with older formulations used before 1975. Given the uncertainty of a mildly increased risk of breast cancer, mutation carriers should consider the substantially decreased risk of other cancers when weighing the risk–benefit ratio of OC use.

Several investigations have demonstrated that OCs are protective against ovarian cancer in the general population. A large meta-analysis of 23,257 cases and 87,303 controls confirmed these findings and found a significant reduction in ovarian cancer in women who had ever used OCs. The risk of ovarian cancer decreased by 20% for every 5 years of OC use. This risk reduction has been confirmed in *BRCA* mutation carriers. Another large meta-analysis of *BRCA1* and *BRCA2* mutation carriers who did or did not use OCs found a significant reduction in ovarian cancer risk among OC users (summary relative risk, 0.5; 95% confidence interval, 0.33–0.75). This benefit became more significant for every additional 10 years of OC use in mutation carriers.

Data remain limited and inconsistent in regard to the risk of colon cancer in *BRCA* mutation carriers. Guidelines for colon cancer screening for mutation carriers remain the same as those for the general population. The use of OCs does not alter the risk of colon cancer in *BRCA* mutation carriers.

Cibula D, Gompel A, Mueck AO, La Vecchia C, Hannaford PC, Skouby SO, et al. Hormonal contraception and risk of cancer. Hum Reprod Update 2010;16:631–50.

Hereditary breast and ovarian cancer syndrome. ACOG Practice Bulletin No. 103. American College of Obstetricians and Gynecologists and Society of Gynecologic Oncologists. Obstet Gynecol 2009;113:957–66.

Iodice S, Barile M, Rotmensz N, Feroce I, Bonanni B, Radice P, et al. Oral contraceptive use and breast or ovarian cancer risk in *BRCA1/2* carriers: a meta-analysis. Eur J Cancer 2010;46:2275–84.

23
Müllerian anomalies

A 19-year-old woman with müllerian agenesis would like to become sexually active. Upon physical examination, you find she has Tanner stage V breast and pubic hair development and a 2-cm blind-ending vagina. The best next step in management to create a functional vagina for her is

* (A) neovagina using a skin graft
* (B) neovagina using large bowel
* (C) laparoscopic vaginal expansion
* * (D) sequential vaginal dilators
* (E) vaginal estrogen cream

Müllerian dysgenesis refers to patients with the absence of a vagina and variations in the presence or absence of a uterus. The term "müllerian agenesis" implies complete absence of all müllerian structures, and typically a patient will present with different degrees of development. Müllerian agenesis occurs in approximately 1 in 4,000 births and also is known as Mayer–Rokitansky–Küster–Hauser syndrome. Müllerian agenesis occurs in an otherwise phenotypically normal female with amenorrhea and a normal 46,XX karyotype. Characteristically, in addition to vaginal agenesis, the uterus is hypoplastic or absent and, if present, the fallopian tubes are rudimentary. The ovaries are normal in structure and function because they derive from a different embryologic source. Associated anomalies may include renal aplasia, spinal anomalies, anorectal malformations, and deafness.

When treating a patient with these anomalies, the correct diagnosis and assessment of the structures and length of the vagina are important. Evaluations of associated anomalies and the urinary tract are essential. Assessment can be accomplished without surgical intervention by using ultrasonography or magnetic resonance imaging. Given the varying degrees of the anomalies, proper preparation, follow-up, and emotional assessment are also critical. It is important to emphasize to the patient and her family that she has reproductive and sexual options. Fertility treatments need to be discussed, including the use of a gestation carrier.

Several treatment options are currently available to develop a functional vagina for sexual intercourse. However, the concept of developing a new vagina requires significant emotional support and should be attempted only when the patient is ready to start sexual intimacy and is emotionally mature. Patients who have been monitored after developing a neovagina have reported some degree of dyspareunia at penetration but otherwise report satisfying sexual activity.

Several surgical interventions are available, the most common being the creation of a neovagina with a skin graft (McIndoe operation). It is a highly successful procedure in motivated patients but is still a major operation entailing hospitalization, anesthesia, and creation of the graft. It can result in postoperative pain. If the graft is not used with certain frequency it can constrict, and reoperations are fairly difficult for the patient to handle. Variations of the McIndoe operation are available with the utilization of a loop of the small or large bowel, buccal mucosa, amnion, and an absorbable adhesion barrier to create the neovagina. However, each procedure carries similar risks to those of the original procedure and would certainly not be the first choice today. Vaginal estrogen creams may help as a lubricant to introduce dilators, and may increase tissue growth, healing, and pliability of the tissues, but use of a cream alone will not help develop a neovagina.

Another technique is the Vecchietti operation, which uses a traction device that is attached and wired laparoscopically to the vaginal "dimple." Progressive traction is placed, allowing for invagination. The technique requires regular use of dilators and coitus. The Vecchietti operation has been reported with high success rates; however, it still requires a laparoscopic procedure with its inherent complications, anesthesia, and several days of hospitalization in order to increase the traction on the device.

The most successful and least invasive procedure to achieve a functional vagina is the use of sequential vaginal dilators with pressure applied by the patient in 20-minute intervals over several months. The pressure can be applied by the patient (Frank method) or by sitting on a bicycle seat that has been previously fitted with dilators that are changed to progressively enlarge the vagina to the desired size (Ingram method). These methods are under the control of the patient and can be done at home without the need for surgery or anesthesia. Sequential vaginal dilators have been demonstrated to be effective and to provide patients with a satisfying intimate life within a matter of months.

Brucker SY, Rall K, Campo R, Oppelt P, Isaacson K. Treatment of congenital malformations. Semin Reprod Med 2011;29:101–12.

Frank RT. The formation of an artificial vagina without an operation. Am J Obstet Gynecol 1938;35:1053–5.

Ingram JM. The bicycle seat stool in the treatment of vaginal agenesis and stenosis: a preliminary report. Am J Obstet Gynecol 1981;140: 867–73.

Nakhal RS, Creighton SM. Management of vaginal agenesis. J Pediatr Adolesc Gynecol 2012;25:352–7.

24
Hysteroscopic complications

A 33-year-old woman visits your office with heavy menstrual bleeding. Her evaluation includes saline infusion ultrasonography showing a 2-cm mass consistent with a type 0 submucosal myoma. A hysteroscopic myomectomy is recommended. The most common complication with hysteroscopic resection of a type 0 submucosal myoma using glycine as a distention media is

* (A) uterine perforation
  (B) cervical laceration
  (C) excessive bleeding
  (D) hyponatremia
  (E) fluid overload

Operative hysteroscopy has become common in gynecology for many procedures, including endometrial ablation, submucosal leiomyomata removal, lysis of intrauterine adhesions, polypectomy, incision of uterine septum, tubal sterilization, tubal cannulation for infertility, and removal of intrauterine devices. With improvements in equipment, many procedures can be performed in an office or an outpatient surgery center with local anesthesia or conscious sedation.

The necessary equipment includes a hysteroscope, light source, camera and monitor, distention medium, and instruments. Hysteroscopes are available in varying diameters for diagnostic (2.7 mm) or operative (7–10 mm) use and as flexible or rigid scopes. Rigid hysteroscopes are available with 0-, 12-, and 30-degree lenses. Operative hysteroscopes typically have a single operating channel for placement of semirigid or rigid instruments. A single inflow channel may be used with diagnostic scopes, whereas dual inflow and outflow channels for continuous flow are found in operative scopes.

Distention media for diagnostic hysteroscopy include carbon dioxide (CO_2) and normal saline (NS) or lactated ringers, although CO_2 is used less often because of increased operative time, shoulder pain, and vasovagal reactions. Media for operative hysteroscopy include NS, lactated ringers, nonelectrolyte solutions of 1.5% glycine, 3% sorbitol, 5% mannitol, and 5% dextrose. Because blood would not disperse in the solution, high viscosity, nonelectrolyte dextran 70 was used commonly in the past to improve image quality. Dextran 70, however, is used infrequently now because of the higher risk of anaphylaxis, pulmonary edema, and disseminated intravascular coagulation compared with other agents.

Instruments for hysteroscopy include operative scissors, graspers, biopsy forceps, electrocautery devices, and morcellators. Electrocautery devices include unipolar loops for cutting and ablation, which require a nonelectrolyte solution as mentioned previously. Bipolar devices for cutting and ablation also are available that can be used with NS or lactated ringers. Morcellator systems are available with a company-specific hysteroscope. The morcellating attachment consists of two hollow tubes. The inner tube rotates and oscillates within the outer tube to remove tissue through a distal opening in the outer tube. This device is used for hysteroscopic polypectomy and myomectomy.

Anesthesia for hysteroscopy depends upon the degree of needed cervical dilation, the degree of needed uterine distention, and the length of the case. Diagnostic hysteroscopy with a flexible scope typically requires no anesthesia. Polypectomy, IUD removal, and tubal sterilization may be performed with a paracervical block with or without conscious sedation. More lengthy procedures, such as extensive lysis of adhesions, septum incision, and myomectomy usually require monitored anesthesia care. General anesthesia may be needed in cases of concomitant laparoscopy or simultaneous transabdominal ultrasonography performed for hysteroscopic lysis of adhesions, septum incision, myomectomy, and tubal cannulation.

Hysteroscopy is a relatively safe procedure with a complication rate of 0.95–3%. The most frequent complication is uterine perforation, with an incidence of 0.7–3%. Perforation is most common with hysteroscopic lysis of adhesions followed by myomectomy and uterine septum incision. Perforation is recognized by a sudden decrease in or loss of distention, increased bleeding, or visualization of bowel or omentum. A small fundal perforation with a sound, small dilator or hysteroscope may be observed for clinical signs of intraperitoneal bleeding. Lateral perforations and perforations that occur with sharp or electrocautery instruments are more likely to result in bleeding or bowel injury and may require laparoscopic evaluation.

Excessive fluid absorption occurs in 0.2–0.76% of surgical cases. Factors that increase fluid absorption include type of procedure (eg, myomectomy, lysis of adhesions, or septum incision), intrauterine pressure, and length of surgery. Defects created in the endometrium, such as hitting the fundus during cervical dilation, may increase fluid absorption. The main risk with isotonic solutions is fluid overload, which can cause pulmonary edema. Generally, the surgery should be concluded with a total fluid deficit of 2,500 mL, remembering to account for intravenous and hysteroscopic fluids. Nonelectrolyte media add an additional risk of hyponatremia. For this reason, the fluid deficit should be limited to no more than 1,500 mL. As the deficit approaches 750 mL or greater, intraoperative blood sodium levels should be considered and the procedure stopped with the onset of hyponatremia. Fluid absorption can be minimized by maintaining a continuous fluid flow (not having the outflow closed) and keeping the intrauterine pressure below the mean arterial pressure. Intracervical injection of a dilute vasopressin solution at 12-, 3-, 6-, and 9-o'clock also will decrease fluid absorption. Inadvertent intravascular injection of vasopressin may result in severe hypertension and bradycardia.

Cervical laceration is most likely caused by a single-tooth tenaculum with strong countertraction during cervical dilation. Depending on the depth and bleeding, this may be controlled with application of silver nitrate or require suture repair. Pretreatment with vaginal misoprostol before surgery may ease cervical dilation in difficult cases. Adverse effects may include nausea, diarrhea, and abdominal pain.

Bleeding during hysteroscopy may be due to a uterine perforation, cervical laceration, or interruption of a myometrial vessel. Heavy bleeding in the absence of a coagulation or platelet disorder is uncommon. Bleeding without evidence of perforation or laceration is best controlled by placement of an intrauterine balloon. This may be accomplished with a specific triangular stent or a Foley catheter.

Air embolism is a rare but often fatal hysteroscopic complication. Studies show that air enters the vasculature during most hysteroscopic procedures. Thus, the difficulty is not the mere presence of air but the volume of air that enters. Methods to decrease air entering the uterus include minimizing the degree of Trendelenburg position, flushing all tubing with media, and minimizing repetitive instrument insertions through the cervix that can cause piston-like transmission of air into the uterus.

Cooper NA, Khan KS, Clark TJ. Local anaesthesia for pain control during outpatient hysteroscopy: systematic review and meta-analysis. BMJ 2010;340:c1130.

Jansen FW, Vredevoogd CB, van Ulzen K, Hermans J, Trimbos JB, Trimbos-Kemper TC. Complications of hysteroscopy: a prospective, multicenter study. Obstet Gynecol 2000;96:266–70.

25
Physiology of the menstrual cycle

Hormone therapy prescribed for a variety of conditions may mimic physiologic menstrual events. The administration of cyclic progesterone for a patient with polycystic ovary syndrome mimics

    (A) human chorionic gonadotropin (hCG) rescue of the corpus luteum
    (B) hormone-mediated proliferation of the endometrium
*   (C) involution of the corpus luteum
    (D) pulsatile luteinizing hormone from the anterior pituitary

Menstruation is a result of the profound tissue remodeling that occurs each month in women of reproductive age. Endometrial tissue responds to the sex steroid hormones produced in the follicular and luteal phases of the ovarian cycle. Estradiol, produced by the ovaries at approximately days 4–5 of the cycle, induces growth and proliferation of the endometrium. The epithelial and stromal cells undergo mitoses and multiply.

After withdrawal of steroid hormone support, specifically involution of the corpus luteum, and decline in progesterone, the functionalis layer of the endometrium undergoes extensive changes, resulting in complete tissue breakdown. The underlying basalis layer contains the progenitor cells that regenerate the functionalis layer in each cycle. Figure 25-1 illustrates the normal menstrual cycle.

To reproduce this cyclical exposure and withdrawal from progesterone, a short course of oral progestin may be given. Completion of the course of progestin results in decreasing serum levels, mimicking involution of the progesterone-producing corpus luteum. Response from the endometrium involves tissue breakdown and menses. If pregnancy takes place, hCG is responsible for maintaining the corpus luteum and progesterone production.

The administration of cyclic progesterone for a patient with polycystic ovary syndrome is not mimicked by hCG rescue of the corpus luteum. Neither the action of hormone-mediated proliferation of the endometrium nor that of pulsatile luteinizing hormone from the anterior pituitary mimics the administration of cyclic progesterone in such a patient.

Beshay VE, Carr B. The normal menstrual cycle and the control of ovulation. Available at http://www.endotext.org/. Retrieved December 4, 2013.

Matteson KA, Rahn DD, Wheeler TL 2nd, Casiano E, Siddiqui NY, Harvie HS, et al. Nonsurgical management of heavy menstrual bleeding: a systematic review. Society of Gynecologic Surgeons Systematic Review Group. Obstet Gynecol 2013;121:632–43.

Nair AR, Taylor HS. The mechanism of menstruation. In: Amenorrhea: a case-based clinical guide. Contemporary endocrinology. New York (NY): Humana Press; 2010. p. 21–34.

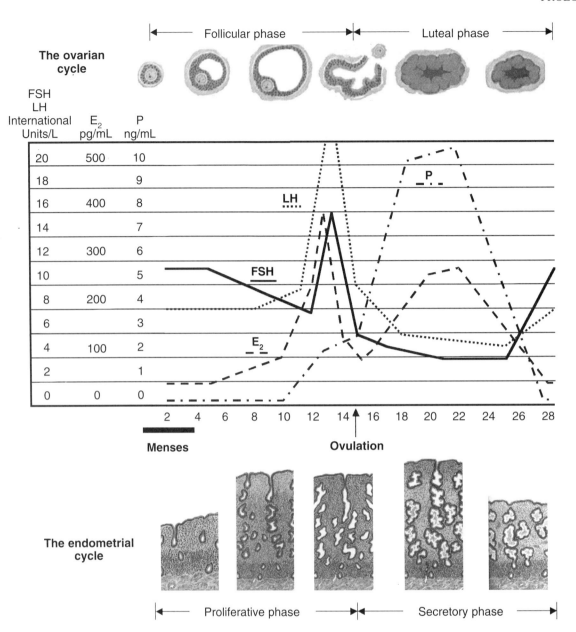

FIG. 25-1. The normal menstrual cycle. Abbreviations: E_2, estradiol; FSH, follicle-stimulating hormone; LH, luteinizing hormone; P, progesterone. (Fritz MA. Evaluation of the female: ovulation. Cedars MI, editor. Infertility: practical pathways in obstetrics and gynecology. New York (NY): The McGraw-Hill Companies, Inc.; 2005.)

26
Pregnancy termination

A 23-year-old primigravid woman requests a second-trimester pregnancy termination for a fetal anomaly and asks about her options in terms of medical therapy. The medication that is most efficacious in combination with misoprostol, and which doubles the chance of complete expulsion by 15 hours, is

    (A) etonogestrel
    (B) medroxyprogesterone acetate
    (C) norethindrone
*   (D) mifepristone
    (E) dienogest

Abortions in the second trimester (ie, at 13–24 weeks) can be performed either surgically by means of dilation and evacuation (D&E) or medically by use of misoprostol, a synthetic orally active E_1 prostaglandin. Because of a shortage in the number of individuals who are skilled in performing D&E procedures and an unequal distribution of these health care providers throughout the country, some women do not have a surgical option and must use medical therapy for their second-trimester abortions. The operating health care provider's prior experience and technique in performing D&E procedures are vital. Although second-trimester abortion procedures account for only 10–15% of all abortions worldwide, they are associated with two thirds of the major complications.

One way to shorten the induction time (first misoprostol dose to abortion) is to pretreat these women with mifepristone 12–48 hours before use of the misoprostol. Mifepristone is an antiprogestin that was approved by the U.S. Food and Drug Administration for the termination of first-trimester pregnancies in September 2000. Mifepristone binds to the progesterone receptor with a five-times-greater affinity than progesterone. Mifepristone is a corticosteroid receptor antagonist and has a three-times-greater affinity for the corticosteroid receptor than dexamethasone. Mifepristone has less affinity (only 25%) for the testosterone receptor and does not bind to the estrogen or mineralocorticoid receptors. In blocking the progesterone receptor, mifepristone is thought to increase uterine sensitivity to misoprostol and is associated with cervical softening and pliability. In first-trimester medical abortions, mifepristone pretreatment promotes endometrial instability and sloughing. In a randomized, placebo-controlled, double-blind trial of mifepristone use before misoprostol use, women had a statistically significant shortening of their induction time to abortion with mifepristone pretreatment. The medical protocol consisted of ingestion of either oral mifepristone or a placebo at home 1 day before a 15-hour hospital admission. In the hospital, women used buccal misoprostol every 3 hours up to five doses or until the fetus and placenta were expelled. Pretreatment with mifepristone resulted in more than twice the chance of complete uterine evacuation within 15 hours (79.8% versus 36.9%, relative risk, 2.16; 95% confidence interval, 1.70–2.75).

In a recent systematic review of the mifepristone to misoprostol dosing interval and its effect on induction abortion times, the authors concluded that shortening the mifepristone–misoprostol interval to 12–24 hours did not compromise the safety or efficacy of the second-trimester medical termination procedure and markedly reduced the total abortion time.

Etonogestrel, medroxyprogesterone acetate, norethindrone, and dienogest are all progestins and pretreatment with any of them would not improve the efficacy of misoprostol alone for medical therapy for a second-trimester abortion. Etonogestrel (3-ketodesogestrel) is the progestin that resides within the implant used for contraception. Medroxyprogesterone acetate and norethindrone are orally active synthetic progestins. Dienogest is an orally active progestin utilized in a recently introduced oral contraceptive pill.

Ngoc NT, Shochet T, Raghavan S, Blum J, Nga NT, Minh NT, et al. Mifepristone and misoprostol compared with misoprostol alone for second-trimester abortion: a randomized controlled trial. Obstet Gynecol 2011;118:601–8.

Shaw KA, Topp NJ, Shaw JG, Blumenthal PD. Mifepristone-misoprostol dosing interval and effect on induction abortion times: a systematic review. Obstet Gynecol 2013;121:1335–47.

27
Müllerian dysgenesis

A 25-year-old woman with primary amenorrhea visits your office to discuss her future reproductive options. She has müllerian dysgenesis with a functional vagina through progressive dilation. She is concerned about her future reproductive potential and wishes to have a safe approach that maximizes her genetic participation. The best option is

    (A) zygote intrafallopian transfer (ZIFT)
*   (B) in vitro fertilization (IVF) and use of a gestational carrier
    (C) donor oocyte and use of a gestational carrier
    (D) adoption

Müllerian dysgenesis, also referred to as müllerian agenesis or Mayer–Rokitansky–Küster–Hauser syndrome, occurs in 1 in 4,000–10,000 females and is characterized by a 46,XX karyotype and normal ovaries that produce the normal array of female hormones. It is the second most common etiology of primary amenorrhea, with gonadal dysgenesis being the first and androgen insensitivity the third. Müllerian dysgenesis occurs in patients in the early teenage years or adolescence with primary amenorrhea. On physical examination, patients with this condition have normal height, breast development, external genitalia, and pubic and axillary hair. The vagina is absent or may be a short pouch without a cervix.

The patient has successfully created a functional vagina through progressive dilation. The nonsurgical approach to vaginal creation is the first-line therapy. Up to 90% of motivated patients are able to dilate and create a functional vagina. Counseling is very important to the success of these patients. Most literature reports normal sexual desire, arousal, and orgasm when these patients are compared with controls. A subset of women with müllerian dysgenesis are not successful at self-dilation or do not wish to self-dilate. For such women, surgical creation of a neovagina usually is performed when patients are ready for vaginal intercourse. The most common surgical approach is the McIndoe operation: creation of a neovagina using a skin graft. Postoperatively, a dilator is used to keep the space patent. Other operations also are available for neovaginal creation. The choice depends on the experience of the operating surgeon.

Even though patients have normal ovaries, one or both ovaries may be located at the pelvic brim, in the inguinal canal, or may be hypoplastic. The location is important for the patient who desires in vitro fertilization and will dictate the need for abdominal or vaginal oocyte retrieval. Most women with müllerian dysgenesis will have a nonfunctioning rudimentary uterus. Up to 7% of patients will have a functional endometrium present in the rudimentary uterine structure. Cyclical pelvic pain may occur initially when functional endometrium is present, at times necessitating surgical removal to relieve pain and decrease the potential for endometriotic implants.

An uncommon presentation of primary amenorrhea is cervical agenesis in which the patient has a normal midline uterus with no cervix. One solution to this problem is to carry out ZIFT. Pregnancy has been reported using ZIFT with laparoscopic embryo transfer into the fallopian tube. Because the described patient does not have a midline uterus, ZIFT is not an option.

In vitro fertilization with a gestational carrier is the preferred modality for patients with müllerian dysgenesis to have their own genetic offspring. Offspring are likely to have normal müllerian anatomy at birth. Practices may have mentors or patients with müllerian dysgenesis who can talk to other patients with the condition and serve as role models. Local and national organizations offer patient referral, support groups, and information (see www.ghr.nlm.nih.gov/condition/mayer-rokitansky-kuster-hauser-syndrome).

The described patient has questions about her future reproductive potential. Proper compassionate counseling is the most important goal for this consultation. Gestational surrogates can be a family member (sister, cousin) or an unrelated carrier. Ovulation induction is followed by oocyte retrieval. Ovaries can be lateral and near the pelvic brim, making transvaginal oocyte retrieval difficult. Some patients may require laparoscopic or transabdominal oocyte retrieval. Embryo transfer will occur in a hormonally primed surrogate carrier. Pregnancy rates for IVF are excellent and are related to the chronologic age of the woman's oocyte.

Using donor oocytes is not warranted at this time. This patient is aged 25 years and is actively interested in her reproductive options. When she is ready to build a family, she can use her own oocytes. Donor oocytes may be needed in the rare scenario in which a patient presents with müllerian dysgenesis in her 40s or has ovaries that are not amenable to oocyte retrieval.

Adoption is an option for patients with müllerian dysgenesis who are not interested in IVF with a surrogate carrier or for whom IVF is not an option. For this patient, it is a future option that needs to be discussed. Adoption is a valid tool for family building.

Edmonds DK, Rose GL, Lipton MG, Quek J. Mayer–Rokitansky–Kuster–Hauser syndrome: a review of 245 consecutive cases managed by a multidisciplinary approach with vaginal dilators. Fertil Steril 2012; 97:686–90.

Müllerian agenesis: diagnosis, management, and treatment. Committee Opinion No. 562. American College of Obstetricians and Gynecologists. Obstet Gynecol 2013;121:1134–7.

Reichman DE, Laufer MR. Mayer–Rokitansky–Kuster–Hauser syndrome: fertility counseling and treatment. Fertil Steril 2010;94:1941–3.

Thijssen RF, Hollanders JM, Willemsen WN, van der Heyden PM, van Dongen PW, Rolland R. Successful pregnancy after ZIFT in a patient with congenital cervical atresia. Obstet Gynecol 1990;76:902–4.

28

Perimenopausal changes

A 48-year-old woman with irregular menstrual cycles comes to your office to discuss the perimenopausal transition. You inform her that the finding most characteristic of the early menopausal transition is

    (A) vasomotor symptoms
    (B) urogenital atrophy
    (C) amenorrhea
*   (D) persistently irregular menstrual cycles

Perimenopause is defined as the interval of a woman's life before actual menopause until 12 months after the final menstrual period. Because the timing of the final menstrual period is established retrospectively, it is difficult to prospectively estimate when the perimenopausal stage will end. The hallmark of this phase is the occurrence of intermittent anovulation that leads to the clinical sign of irregular vaginal bleeding. The vaginal bleeding may have no pattern of frequency, varying from infrequent menses every 12 weeks to daily vaginal bleeding and spotting. Traditionally, the World Health Organization has defined the beginning of perimenopause as a break in a woman's regular menstrual cycles, but not more than 3 months without a period. When a woman has 3–11 months of amenorrhea, she is considered to be in the late perimenopausal stage.

In 2001, the Stages of Reproductive Aging Workshop recommended terminology and established a staging system for reproductive aging. That became the criterion standard for characterizing reproductive aging through menopause. In 2011, at the follow-up workshop, the Stages of Reproductive Aging Workshop + 10, modifications were made to the 2001 staging system and the perimenopausal interval was divided into early and late phases of the menopausal transition (Table 28-1).

The hallmark of the early menopausal transition, Stage –2, part of the perimenopause stage, is the presence of *increased variability in menstrual cycle length*, defined as a persistent difference of 7 days or more in the length of consecutive cycles. *Persistence* was defined as recurrence within 10 cycles of the first variable length cycle. The early menopausal transition also is characterized by elevated but variable early follicular phase follicle-stimulating hormone (FSH) levels, low antimüllerian hormone levels, and a low antral follicle count. The overall duration of the early menopausal transition is thought to be variable among women.

The late menopause transition, Stage –1, part of the perimenopause stage, is marked by the occurrence of amenorrhea of 60 days or longer. Menstrual cycles are characterized by increased variability in cycle length, extreme fluctuations in hormone levels, and increased prevalence of anovulation. During a random blood draw, FSH levels usually will be greater than 25 international units/L, but if associated with an elevated estradiol level, FSH levels sometimes may be lower than 25 international units/L. Associated vasomotor symptoms are likely to occur during the late menopausal transition. The overall duration of the late menopausal transition is believed to last 1–3 years. The late menopausal transition ends at the

TABLE 28-1. The Stages of Reproductive Aging Workshop + 10 Staging System for Reproductive Aging in Women

Stage	Menarche →						Final Menstrual Period (0) →			
	-5	-4	-3b	-3a	-2	-1	+1a	+1b	+1c	+2
	Reproductive				Menopausal Transition		Postmenopause			
Terminology	Early	Peak	Late		Early	Late	Early		Late	
					Perimenopause					
Duration		Variable			Variable	1–3 years	2 years (1 + 1)		3–6 years	Remaining lifespan

Principal Criteria

Menstrual cycle	Variable to regular	Regular	Regular	Subtle changes in flow/length	Variable length Persistent 7 days or greater difference in length of consecutive cycles	Interval of amenorrhea of 60 or more days				

Supportive Criteria

Endocrine										
FSH			Low	Variable*	↑Variable*	↑25 international units/L or more[†]	↑Variable		Stabilizes	
AMH			Low	Low	Low	Low	Low		Very low	
Inhibin B			Low	Low	Low	Low	Low		Very low	
Antral follicle count			Low	Low	Low	Low	Very low		Very low	

Descriptive Characteristics

Symptoms						Vasomotor symptoms *likely*	Vasomotor symptoms *most likely*			*Increasing symptoms of urogenital atrophy*

Abbreviations: AMH, antimüllerian hormone; FSH, follicle-stimulating hormone.
*Blood draw on cycle days 2–5. ↑ = elevated.
[†]Approximate expected level based on assays using current international pituitary standard.
Harlow SD, Gass M, Hall JE, Lobo R, Maki P, Rebar RW, et al. Executive summary of the Stages of Reproductive Aging Workshop + 10: addressing the unfinished agenda of staging reproductive aging. STRAW + 10 Collaborative Group. J Clin Endocrinol Metab 2012;97:1159–68.

final menstrual period and is dated retrospectively. The perimenopause stage extends into the early postmenopausal stage and is defined to end 1 year after the final menstrual period, ie, Stage +1a, part of the perimenopause stage.

Given the aforementioned definitions of perimenopause, the presence of vasomotor symptoms and an interval of amenorrhea of 60 days or longer would be consistent with the changes seen in the late menopausal transition. The presence of urogenital atrophy is thought not to occur during the perimenopausal transition, but in the late postmenopausal stage (Stage +2).

Harlow SD, Gass M, Hall JE, Lobo R, Maki P, Rebar RW, et al. Executive summary of the Stages of Reproductive Aging Workshop + 10: addressing the unfinished agenda of staging reproductive aging. STRAW + 10 Collaborative Group. J Clin Endocrinol Metab 2012;97:1159–68.

Soules MR, Sherman S, Parrott E, Rebar R, Santoro N, Utian W, et al. Executive summary: Stages of Reproductive Aging Workshop (STRAW). Fertil Steril 2001;76:874–8.

29

Donor egg in vitro fertilization

A 47-year-old nulligravid woman is interested in attempting pregnancy using anonymous donor eggs with in vitro fertilization (IVF). She is an obese woman with a long-standing history of poorly controlled hypertension. She does not smoke. You explain to her the process of selecting an egg donor and the medical procedures associated with uterine preparation. Next you discuss with her the risks of pregnancy in older women. The most important preconception test that she should undergo before considering pregnancy is

* (A) echocardiography
   (B) confirmation of rubella immunity
   (C) mammography
   (D) diabetes mellitus screening
   (E) Pap test

In recent decades, women who are past the age at which natural conception is generally possible have been able to consider IVF with an egg donor. The method is available to

- women with primary ovarian insufficiency before age 40 years
- women in their 40s who have decreased fecundity
- women who have loss of fertility potential (eg, loss of fertility caused by oophorectomy or cancer therapy)

For such women, becoming pregnant after donor egg IVF does not pose any significant additional risk compared with a spontaneous conception. Egg donation disproportionately occurs in women older than 40 years, and that alone often results in their pregnancies being classified as high risk. In addition, the frequency of multiple gestations is high with the use of donor egg cycles. In 2010, approximately 38% of pregnancies that resulted from donor egg IVF were multiple gestations, most of which were twins. Other maternal and fetal complications can occur during pregnancy in older women, even with singleton pregnancies. Such complications include preterm labor; gestational hypertension; diabetes mellitus; preeclampsia; hemolysis, elevated liver enzymes, and low platelet count (HELLP) syndrome; abnormal placentation; and premature rupture of membranes. One study of 74 women aged 45–56 years who received donor eggs found that the incidence of pregnancy complications was 38%.

Ideally, all women who consider conception, either with assistance or on their own, should have a preconception evaluation with their health care provider. A preconception visit is an opportunity to address behavioral and medical conditions that may affect the safety of pregnancy for the mother and child. Many women do not see a health care provider before conception. However, infertile women who seek assistance in conception in particular should receive a thorough preconception evaluation with consideration of use of preconception folic acid. For an older patient who wishes to receive donor eggs, this counseling could be provided by a maternal–fetal medicine specialist who can review the added risks associated with the woman's age.

It is recommended that women with certain known health problems have an appropriate evaluation before consideration of pregnancy. Women with chronic hypertension should confirm that they are taking medication that is safe for their pregnancy and will control their blood pressure before conception. Women who have a history of poorly controlled hypertension, as in the described patient, should be evaluated for serious sequelae that can affect the pregnancy. Echocardiography is the most essential preconception test for this patient because it can identify left ventricular hypertrophy that could cause life-threatening cardiomyopathy during pregnancy.

Rubella, also known as German measles, is a childhood disease that is now rare in much of the world. Most individuals in the United States received rubella vaccination as children, but for some individuals, the vaccine loses efficacy as they age. The rubella vaccine is a live, attenuated vaccine and is contraindicated during pregnancy, so testing for immunity and subsequent vaccination must occur during the preconception evaluation. Although confirmation of rubella immunity is important before attempting pregnancy, it is not the most important preconception test for this patient.

Screening for breast and cervical cancer is essential for all women in their 40s, whether they are attempting pregnancy or not. Before beginning an egg donation cycle, it is appropriate to make sure the patient has had her routine Pap test and mammography; any concerning test results are more easily pursued before pregnancy. This should be a routine part of a preconception visit and is not unique to this specific case.

Given that maternal hyperglycemia can present a detrimental fetal risk, screening women at risk of diabetes mellitus before pregnancy is essential. The described patient is not at an especially high risk of diabetes; she should have routine screening as part of her preconception evaluation, but diabetes screening is not the most essential preconception test for her.

American College of Obstetricians and Gynecologists. Hypertension in pregnancy. Washington, DC: American College of Obstetricians and Gynecologists; 2013. Available at: http://www.acog.org/Resources_And_Publications/Task_Force_and_Work_Group_Reports/Hypertension_in_Pregnancy. Retrieved March 20, 2014.

Centers for Disease Control and Prevention. Assisted reproductive technology. 2011 National summary report. Section 2: ART cycles using fresh, nondonor eggs or embryos (part B). Atlanta (GA): CDC; 2014. Available at: http://www.cdc.gov/art/ART2011/section2b.htm. Retrieved March 20, 2014.

Sauer MV, Paulson RJ, Lobo RA. Oocyte donation to women of advanced reproductive age: pregnancy results and obstetrical outcomes in patients 45 years and older. Hum Reprod 1996;11:2540–3.

30
Labial adhesions in children

A 5-year-old girl has had difficulty voiding over the past 6 weeks. Genital examination reveals that the labia are fused with a white band of adhesions. The urethra and clitoris are not visualized, but you observe a small pinpoint opening at the introitus. The most appropriate management is

    (A) clindamycin hydrochloride cream
    (B) clobetasol propionate cream
* (C) estrogen cream
    (D) inpatient surgical separation of the labia
    (E) office surgical separation of the labia

Labial adhesions, also termed labial agglutination, labial fusion, or vulvar synechiae, are a relatively uncommon condition. The condition occurs in up to 1.8% of female prepubertal patients, with a peak incidence at approximately 13–23 months of age. Although most patients are asymptomatic, some experience complications such as local inflammation, dysuria, urinary tract infection, and obstruction. Labial adhesions also can occur with chronic urinary tract infection, chronic inflammation such as vulvovaginitis, genital trauma, and hypoestrogenism. All patients with labial adhesions should therefore be screened for urinary tract infections. During labial trauma, denuding of the upper squamous epithelial layer of labial mucosa leads to formation of connective tissue bridges between the healing labia because of vulvovaginitis and poor local hygiene. Approximately 80% of labial adhesions resolve spontaneously within 1 year, and persistence after the onset of puberty is very rare. It is important to note that the bulk of medical literature on this subject consists of case series and other observational studies.

Management options range from reassurance for asymptomatic patients to surgical treatment for severe cases and those resistant to conservative treatment. The finding of labial adhesions in pubertal girls also may suggest sexual abuse and warrants a careful disclosure interview and examination by skilled personnel to determine if there is evidence of sexual interference.

Locally applied estrogen cream changes the vaginal epithelium from a thin and atrophic environment to a thicker environment rich in glycogen with more acidic vaginal secretions that prevent inflammation and infection. For symptomatic patients, conservative treatment with estrogen cream is recommended as the first-line treatment. Treatment may be required for several weeks to achieve separation of the labia minora. Controversy exists regarding treatment in patients with asymptomatic labial adhesions, given that most cases will resolve spontaneously within 1 year of puberty.

Patients with thin adhesions can be treated with topical estrogen but patients with dense, fibrous adhesions are less likely to respond to topical estrogen therapy alone. If conservative therapy with topical estrogen has failed or if a girl presents with urinary retention, additional measures are necessary. Manual or surgical separation is then indicated and should be performed under anesthesia to limit potential emotional and psychologic trauma from the procedure.

As an antibiotic cream, clindamycin is primarily used for the treatment of bacterial vaginosis, which would not be expected in a pediatric patient except in the case of sexual abuse. Topical clobetasol cream is used to treat lichen sclerosus and historically has been thought to be ineffective in the treatment of labial adhesions. The authors of one small retrospective study reported modest success with topical betamethasone valerate. They found estrogen and betamethasone to be equally efficacious. However, the small number of study subjects limits the strength of their conclusions. For now, estrogen cream remains the first-line treatment for labial adhesions.

Berkoff MC, Zolotor AJ, Makoroff KL, Thackeray JD, Shapiro RA, Runyan DK. Has this prepubertal girl been sexually abused? JAMA 2008;300:2779–92.

Eroglu E, Yip M, Oktar T, Kayiran SM, Mocan H. How should we treat prepubertal labial adhesions? Retrospective comparison of topical treatments: estrogen only, betamethasone only, and combination estrogen and betamethasone. J Pediatr Adolesc Gynecol 2011;24:389–91.

Kumar RK, Sonika A, Charu C, Sunesh K, Neena M. Labial adhesions in pubertal girls. Arch Gynecol Obstet 2006;273:243–5.

Tebruegge M, Misra I, Nerminathan V. Is the topical application of oestrogen cream an effective intervention in girls suffering from labial adhesions? Arch Dis Child 2007;92:268–71.

31
Risks and benefits of hormone therapy

The Women's Health Initiative (WHI) study was the largest study of its kind to examine hormone therapy in menopausal women. The study included two randomized clinical trials: a study of the use of estrogen–progestin in women with a uterus and a study of the use of estrogen alone in women without a uterus. The study showed that estrogen–progestin has no benefit in terms of protection against the primary outcome of

    (A) dementia
* (B) cardiovascular disease
    (C) colon cancer
    (D) hip fracture

The WHI study was designed to assess the effect of estrogen–progestin on cardiovascular events in healthy postmenopausal women. A total of 161,809 women aged 50–79 years were enrolled in the study. The study was planned to run for 8 years, but it was stopped after 5.2 years when the number of new breast cancer cases reached a preset safety threshold for harm.

The WHI study had two randomized controlled trials of postmenopausal estrogen–progestin: 1) daily estrogen–progestin and 2) unopposed estrogen in hysterectomized women. The estrogen–progestin arm was discontinued first because of a statistically significant increase in invasive breast cancer and a trend toward increased cardiovascular events. The estrogen-only arm was discontinued after 6.8 years in a controversial decision. Unlike the combined estrogen–progestin arm, none of the preset boundaries had been crossed, although there was an increase in stroke risk as well as a trend toward increased risk of dementia or cognitive impairment. Interim statistical analysis also had concluded that the primary endpoint, coronary heart disease reduction, would not have been realized even if the study had continued to its planned endpoint.

The WHI investigators reported an overall increase in stroke in the estrogen–progestin arm. This large and well-designed randomized clinical trial demonstrated that daily estrogen–progestin does not have any cardioprotective benefits against cardiovascular disease. In fact, there was an increase in cardiovascular events in this arm of the study. The findings of the WHI study were consistent with the vast majority of smaller clinical trials that preceded it. However, the cardiovascular results were an exception. The WHI study showed that estrogen–progestin had no benefit in terms of protection against the incidence of cardiovascular disease.

It has been postulated that the role of estrogen–progestin varies with the timing of initiation of therapy and the proximity of use to the onset of menopause. The current recommendations from regulatory agencies and professional associations, such as the American College of Obstetricians and Gynecologists, advise the use of hormone therapy only for moderate-to-severe hot flushes. Osteoporosis may be treated with estrogens when other treatment modalities have failed or are poorly tolerated.

The risks of estrogen–progestin in younger menopausal women are reassuring. Table 31-1 illustrates that in women aged 50–54 years, the individuals most likely to benefit from receiving estrogen–progestin, this therapy carries a small increase in absolute risk, ie, 1.56 net adverse events per 1,000 women-years. By comparison, for estrogen-only therapy the net increase was 0.08 events per 1,000 women-years. Estrogen-only therapy and estrogen–progestin therapy are associated with a 20% increase in stroke. For women aged 50–54 years, pulmonary embolism, coronary heart disease, and breast cancer were increased only in the estrogen–progestin arm of the WHI study.

Anderson GL, Limacher M, Assaf AR, Bassford T, Beresford SA, Black H, et al. Effects of conjugated equine estrogen in postmenopausal women with hysterectomy: the Women's Health Initiative randomized controlled trial. Women's Health Initiative Steering Committee. JAMA 2004;291:1701–12.

Grady D. Clinical practice. Management of menopausal symptoms. N Engl J Med 2006;355:2338–47.

Postmenopausal estrogen therapy: route of administration and risk of venous thromboembolism. Committee Opinion No. 556. American College of Obstetricians and Gynecologists. Obstet Gynecol 2013;121: 887–90.

Rossouw JE, Anderson GL, Prentice RL, LaCroix AZ, Kooperberg C, Stefanick ML, et al. Risks and benefits of estrogen plus progestin in healthy postmenopausal women: principal results From the Women's Health Initiative randomized controlled trial. Writing Group for the Women's Health Initiative Investigators. JAMA 2002;288:321–33.

Wassertheil-Smoller S, Hendrix SL, Limacher M, Heiss G, Kooperberg C, Baird A, et al. Effect of estrogen plus progestin on stroke in postmenopausal women: the Women's Health Initiative: a randomized trial. WHI Investigators. JAMA 2003;289:2673–84.

TABLE 31-1. Relative Risks of Disease Outcomes from the Women's Health Initiative Trials and Estimates of Absolute Differences in Risk Among Women Aged 50–54 Years*

Outcome	Estrogen Plus Progestin[†] RR (95% CI)	Absolute Difference in Risk[‡]	Estrogen Only[§] RR (95% CI)	Absolute Difference in Risk[‡]
Coronary heart disease	1.29 (1.02–1.63)	0.26	0.91 (0.75–1.12)	–
Stroke	1.41 (1.07–1.85)	0.20	1.39 (1.10–1.77)	0.20
Pulmonary embolism	2.13 (1.39–3.25)	0.45	1.34 (0.87–2.06)	–
Invasive breast cancer	1.26 (1.00–1.59)	0.93	0.77 (0.59–1.01)	–
Colon cancer	0.63 (0.43–0.92)	–0.18	1.08 (0.75–1.55)	–
Hip fracture	0.66 (0.45–0.98)	–0.10	0.61 (0.41–0.91)	–0.12
Net adverse outcomes per 1,000 women per year		1.56		0.08

Abbreviations: CI, confidence interval; RR, relative risk.

*A dash denotes that the RR was not statistically different.

[†]In this trial, 16,608 postmenopausal women without hysterectomy were randomly assigned to receive 0.625 mg of conjugated estrogen plus 2.5 mg of medroxyprogesterone acetate per day or an identical placebo and were followed for an average of 5.2 years. (Rossouw JE, Anderson GL, Prentice RL, LaCroix AZ, Kooperberg C, Stefanick ML, et al. Risks and benefits of estrogen plus progestin in healthy postmenopausal women: principal results from the Women's Health Initiative randomized controlled trial. Writing Group for the Women's Health Initiative Investigators. JAMA 2002;288:321-33 and Shumaker SA, Legault C, Rapp SR, Thal L, Wallace RB, Ockene JK, et al. Estrogen plus progestin and the incidence of dementia and mild cognitive impairment in postmenopausal women: the Women's Health Initiative Memory Study: a randomized controlled trial. JAMA 2003;289:2651-62.)

[‡]The absolute difference in risk equals the rate per 1,000 women per year among women from 50 years to 54 years of age who were treated with hormones, minus the rate in untreated women of the same age. Absolute risks of disease in untreated women are based on rates of confirmed outcomes (except pulmonary embolism, which was self-reported) among 12,381 women in the Women's Health Initiative Observational Study, who were followed for 95.8 months. (Women's Health Initiative's Scientific Resources website, http://www.whiscience.org/data/data_outcomes.php accessed June 11, 2009.) Absolute risk among hormone-treated women was calculated by multiplying the relative risk for each outcome from the Women's Health Initiative randomized trials by the absolute risk among untreated women. Overall relative risks from the Women's Health Initiative randomized trials are used rather than age-specific relative risks, because there were no statistically significant differences in relative risks according to age. Absolute differences in risk are calculated only for relative risks that were significantly different (with $\alpha < .05$) from 1.0.

[§]In this trial, 10,739 postmenopausal women with hysterectomy were randomly assigned to receive 0.625 mg of conjugated estrogen per day or an identical placebo and were followed for an average of 6.8 years. (Anderson GL, Limacher M, Assaf AR, Bassford T, Beresford SA, Black H, et al. Effects of conjugated equine estrogen in postmenopausal women with hysterectomy: the Women's Health Initiative randomized controlled trial. Women's Health Initiative Steering Committee. JAMA 2004;291:1701-12 and Shumaker SA, Legault C, Kuller L, Rapp SR, Thal L, Lane DS, et al. Conjugated equine estrogens and incidence of probable dementia and mild cognitive impairment in postmenopausal women: Women's Health Initiative Memory Study. JAMA 2004;291:2947-58.)

Grady D. Management of menopausal symptoms. N Engl J Med 2006;355:2338-47. Copyright 2006 by the Massachusetts Medical Society.

32
Fertility preservation in a patient undergoing cancer treatment

A 27-year-old nulligravid woman is referred to your office because she has a recent diagnosis of breast cancer. She and her partner desire future childbearing. The treatment most likely to preserve their fertility is

* (A) embryo banking
* (B) oocyte banking
* (C) ovarian tissue freezing
* (D) gonadal suppression with gonadotropin-releasing hormone (GnRH) agonist
* (E) ovarian transposition

In vitro fertilization is a process in which exogenous gonadotropins are administered to a patient, eggs are surgically extracted and fertilized using either partner sperm or donor sperm, and the resultant embryos are frozen for future use. The technique has been used successfully for couples undergoing infertility treatment. Until recently, embryo cryopreservation was the main recognized treatment for reproductive-aged cancer patients whose treatment may render them sterile.

Before 2012, oocyte cryopreservation was considered experimental. In 2012, the experimental designation was lifted by the American Society for Reproductive Medicine, which reports that there are limited data on thawed cryopreserved oocytes. However, the limited data suggest that there is no increased risk of chromosomal abnormalities, birth defects, or developmental deficits in offspring born with the use of this technology.

Modifications in the cryopreservation techniques for oocytes have improved the survival of cells. The vitrification freezing technique allows the oocyte to be cryopreserved in a glass-like state. The technology also is available to those who do not have a partner or to individuals who are unwilling to cryopreserve embryos.

Oocyte cryopreservation is not as successful as embryo cryopreservation. Large observational clinical studies demonstrate that implantation and pregnancy rates may be lower when using frozen oocytes compared with frozen embryos. Embryos are multicellular and can survive the freeze and thaw process well. Although egg freezing is no longer considered experimental, ongoing research is still necessary to assess the long-term effects of such treatments. At this time, embryo cryopreservation offers the best treatment to preserve this patient's fertility.

Ovarian transposition is a surgical technique to move ovaries outside the field of radiation in cancer patients who require radiation treatment to the pelvis. The described patient does not require pelvic irradiation, and therefore relocation of her ovaries offers her no fertility preservation.

In vitro maturation is a technique to mature oocytes in vitro that are either retrieved from unstimulated ovaries or from ovarian tissue. This technique has shown early promise; however, the culture system associated with this technique is complex and still requires validation through published evidence-based research. At this time, it is not a reliable fertility preservation option for cancer patients.

Ovarian tissue freezing is a technique in which the cortex of an ovary is surgically stripped from the ovary. The obtained ovarian tissue is subsequently frozen for future use. The technique comprises the freezing of ovarian tissue and reimplantation of the tissue to the patient once she is cancer free or in remission. The main concern in regard to this technique is possible reintroduction into the body of cancer-containing cells. An additional issue is that only a limited series of studies have been published to demonstrate that ovarian tissue freezing is a reliable, reproducible, and successful technique.

Researchers have theorized that gonadal suppression with GnRH agonist preserves hormonal function and fertility. However, randomized control trials have shown mixed results. As a consequence, GnRH agonist would not be the patient's best choice for reliable fertility preservation.

Badawy A, Elnashar A, El-Ashry M, Shahat M. Gonadotropin-releasing hormone agonists for prevention of chemotherapy-induced ovarian damage: prospective randomized study. Fertil Steril 2009;91:694–7.

Mature oocyte cryopreservation: a guideline. Practice Committees of the American Society for Reproductive Medicine and the Society for Assisted Reproductive Technology. Fertil Steril 2013;99:37–43.

Sverrisdottir A, Nystedt M, Johansson H, Fornander T. Adjuvant goserelin and ovarian preservation in chemotherapy treated patients with early breast cancer: results from a randomized trial. Breast Cancer Res Treat 2009;117:561–7.

33
Turner syndrome

A 22-year-old nulligravid woman with secondary amenorrhea visits your clinic. Her follicle-stimulating hormone level is 45 international units/L. She is 1.6 m (63 in.) tall and experienced normal pubertal development with spontaneous onset of menstrual bleeding at age 13 years. The most likely karyotype finding is

   (A) 46,X,del(Xp)
   (B) 45,X/46,XY
   (C) 45,X
* (D) 45,X/46,XX/47,XXX

Turner syndrome was originally a clinical syndrome of phenotypic findings described by Henry Turner in 1938 before the discovery that its genetic basis is due to the loss of a X chromosome, which results in a 45,X karyotype. The classic physical findings of Turner syndrome are short stature of less than 1.47 m (58 in.), lack of secondary sexual development, webbed neck, and cubitus valgus. The incidence of Turner syndrome is approximately 1 in 2,500 live female births. Approximately 50% of Turner patients have a pure 45,X cell line without detectable mosaicism; the remaining 50% of patients have a wide variety of mosaicism.

The presentation of Turner syndrome is variable and is not uniformly correlated with the presence or degree of mosaicism. The most important characteristics of Turner syndrome include the cardiac and renal anomalies. The cardiac defects include coarctation of the aorta, bicuspid aortic valve, and dilated aortic root. Dissection with or without rupture of the aorta can occur in any Turner syndrome patient at any time in her life without identifiable risk factors. Turner syndrome patients with hypertension or a dilated aortic root are at greater risk of aortic dissection. Patients with Turner syndrome who conceive spontaneously or through oocyte donation are at greater risk of maternal morbidity and mortality because of the increased cardiovascular demands of pregnancy. Furthermore, after pregnancy, such individuals remain at increased risk of cardiovascular events. The maternal mortality rate for Turner syndrome patients who conceive is estimated to be 2%, which represents a 150-fold increase compared with the general population. Box 33-1 lists phenotypic findings in Turner syndrome patients.

Besides short stature, most Turner syndrome patients have absent or delayed puberty, but the presentation can be quite variable. Even in 45,X Turner syndrome individuals, 11% can have puberty with spontaneous onset of menses at the appropriate time. The occurrence of pubertal changes with menses appears to increase with the presence of mosaicism. Approximately 34% of 45,X/46,XX and 68% of 45,X/46,XX/47,XXX Turner syndrome individuals have spontaneous onset of menses. In one review, Turner syndrome patients with a 46,X,del(Xp) karyotype did not initiate puberty. It appears that deletion of the short arm of one of the X chromosomes is sufficient to lead to the Turner syndrome phenotype. Typically, the karyotype in a Turner syndrome patient is only determined from the DNA isolated from peripheral venous blood leukocytes. The karyotype also can be obtained from other tissue such as fibroblasts from a muscle biopsy or ovarian cells from an ovarian tissue biopsy. Karyotypes have been found to differ depending on the source of the tissue.

BOX 33-1

Phenotypic Findings in Turner Syndrome Patients

- Short stature less than 1.47 m (58 in.), sexual infantilism, webbed neck, and cubitus valgus
- Cardiac defects (eg, coarctation of the aorta, bicuspid aortic valve, and dilated aortic root)
- Renal anomalies (eg, horseshoe kidney and double or cleft renal pelvis)
- Edema of the hands and feet
- Low posterior hairline
- Nail dysplasia or hypoplasia
- Rotated ears
- Small mandible
- Multiple pigmented nevi
- Broad shield chest leading to widely spaced nipples
- Short fourth metacarpal
- High arched palate

Bondy CA. Care of girls and women with Turner syndrome: a guideline of the Turner Syndrome Study Group. Turner Syndrome Study Group. J Clin Endocrinol Metab 2007;92:10–25.

Other 45,X Turner syndrome mosaic karyotypes have some characteristic findings. The most common Turner syndrome mosaic that is seen is a woman with a 45,X cell line with an additional isochromosome of the X cell line, 46,X,i(Xq). An isochromosome forms when there is loss of the two short arms of the X chromosome and fusing of the two remaining long arms of the X chromosome. This 45,X/46,X,i(Xq) mosaic has been associated with a higher risk of Hashimoto thyroiditis and hypothyroidism. Turner syndrome 45,X mosaic patients with a 46,XY cell line can occur in 5–10% of Turner syndrome individuals. Women with a Y chromosome cell line are at increased risk of gonadal transformation into a gonadoblastoma and subsequent malignant transformation into a dysgerminoma. A prophylactic gonadectomy should be performed if any Y chromosome is present to prevent the formation of these potential neoplastic processes. Turner syndrome patients are at increased risk of many future abnormalities according to the Turner Syndrome Society (www.turnersyndrome.org). Comprehensive recommendations for ongoing monitoring of patients with Turner syndrome have been published recently (Box 33-2).

In approximately 75% of 45,X patients with Turner syndrome, the lone X chromosome is maternal in origin. This is likely due to the instability of the Y chromosome that is lost during male meiosis, leading to a genetically defective sperm that fertilizes the oocyte. The exception to this tendency is a patient with 45,X/46,X,i(Xq) Turner syndrome in whom approximately 50% of the i(Xq) material is paternal and the remaining material is maternal.

BOX 33-2

Recommendations for Ongoing Monitoring of Patients with Turner Syndrome

All ages
Cardiological evaluation as indicated
Blood pressure annually
Ear, nose, and throat examination and audiology every 1–5 years

Girls under age 5 years
Social skills at age 4–5 years

School age
Liver and thyroid screening annually
Celiac screen every 2–5 years
Educational and social progress annually
Dental and orthodontic as needed

Older girls and adults
Fasting lipids and blood sugar annually
Liver and thyroid screening annually
Celiac screen as indicated
Age-appropriate evaluation of pubertal development and psychosexual adjustment

Bondy CA. Care of girls and women with Turner syndrome: a guideline of the Turner Syndrome Study Group. Turner Syndrome Study Group. J Clin Endocrinol Metab 2007;92:10–25.

Table 33-1 shows the effects of X chromosome parental origin of the Turner syndrome phenotype.

TABLE 33-1. Effects of X Chromosome Parental Origin on Turner Syndrome Phenotype

Stature*	X^P (23)	X^M (56)	P value	
Height (SE), cm	147.4 (1.8)	147.1 (1.1)	0.89	
% Predicted height†	89.3 (0.7)	89.4 (0.9)	0.98	
% Maternal height	90.0 (0.8)	90.1 (0.8)	0.99	
Developmental anomalies	X^P (50)	X^M (133)	P value	CI‡
Renal anomaly§	12/50 (24%)	33/133 (25%)	0.92	–0.18, 0.15
Webbed neck	11/50 (22%)	47/133 (35%)	0.11	–0.30, 0.03
Abnormal aortic valve‖	12/50 (24%)	33/128¶ (26%)	0.83	–0.19, 0.14
Any cardiovascular defect#	26/50 (52%)	71/128¶ (24%)	0.69	–0.22, 0.13

Abbreviations: CI, confidence interval; SE, standard error.

*Stature was evaluated in participants aged less than 15 years that had reached final height and had never been treated with growth hormone. Developmental anomalies were evaluated in all subjects aged 7 years and older.

†Predicted height based on average of maternal and (paternal height: 13 cm).

‡CI, confidence interval on the difference (X^P minus X^M) of proportions.

§Renal anomalies included single or horseshoe kidney and duplication of collecting system.

‖Includes bicuspid and partially fused aortic valves.

¶Inadequate cardiovascular studies on five X^M subjects.

#Cardiovascular defects include abnormal aortic valve, aortic coarctation, partial anomalous pulmonary venous connection, left superior vena cava, and elongated transverse aortic arch.

With kind permission from Springer Science+Business Media. Bondy CA, Matura LA, Wooten N, Troendle J, Zinn AR, Bakalov VK. The physical phenotype of girls and women with Turner syndrome is not X-imprinted. Hum Genet 2007;121:469–74.

Genetic abnormalities are likely in the described patient who has secondary amenorrhea and an elevated follicle-stimulating hormone level suggestive of premature ovarian failure. Because she had appropriate pubertal changes with spontaneous menses, a deletion of the short arm of the X chromosome, ie, 46,X,del(Xp), is unlikely. Patients with a 46,X,del(Xp) karyotype have not been reported to undergo puberty. She is also taller than patients with a pure 45,X karyotype, suggesting a mosaic karyotype. In comparing a 45,X/46,XX/47,XXX karyotype with the 45,X/46,XY karyotype, the 45,X/46,XX/47,XXX karyotype is more likely to be associated with a history of pubertal change and spontaneous menses.

Bondy CA. Care of girls and women with Turner syndrome: a guideline of the Turner Syndrome Study Group. Turner Syndrome Study Group. J Clin Endocrinol Metab 2007;92:10–25.

Bondy CA, Matura LA, Wooten N, Troendle J, Zinn AR, Bakalov VK. The physical phenotype of girls and women with Turner syndrome is not X-imprinted. Hum Genet 2007;121:469–74.

Karnis MF. Fertility, pregnancy, and medical management of Turner syndrome in the reproductive years. Fertil Steril 2012;98:787–91.

Zhong Q, Layman LC. Genetic considerations in the patient with Turner syndrome—45,X with or without mosaicism. Fertil Steril 2012; 98:775–9.

34

Postmenopausal uterine bleeding and hormone therapy

A 63-year-old woman, gravida 2, para 2, has been taking hormone therapy (HT) for 11 years. She tells you that she has been experiencing vaginal spotting in recent weeks. She has not been sexually active for more than 10 years and her pelvic examination shows no signs of infection. Transvaginal ultrasonography indicates no evidence of polyps, leiomyomas, or ovarian enlargement. The endometrial stripe is 3 mm. Your next step in management is

* (A) observation
* (B) saline sonohysterography
* (C) endometrial sampling
* (D) hysteroscopy
* (E) dilation and curettage

Endometrial abnormalities in asymptomatic women are rare. Patients who have postmenopausal bleeding often will have abnormalities of the endometrium that are either benign or malignant. The clinical approach to postmenopausal bleeding requires prompt evaluation for endometrial carcinoma. Bleeding that occurs in patients who are taking HT may be caused by poor adherence, inadequate dosing, or poor absorption of the medication. Clinicians should use endometrial biopsy or transvaginal ultrasonography to assess patients with postmenopausal bleeding.

Cancer of the endometrium is the most common gynecologic malignancy among U.S. women. Although vaginal bleeding is the presenting sign in more than 90% of postmenopausal patients with endometrial carcinoma, the vast majority of cases of postmenopausal bleeding are due to atrophic changes of the vagina or endometrium. The incidence of endometrial cancer in patients who present with postmenopausal bleeding is approximately 1–14% depending on age and risk factors. The probability of malignancy is greatly reduced in patients with an endometrial thickness of less than or equal to 4 mm. In patients with an endometrial thickness of more than 4 mm, endometrial sampling is warranted to rule out malignancy. If the endometrial thickness is less than or equal to 4 mm, endometrial sampling is not required. Observation would be the appropriate management for the described patient at this time. Saline sonohysterography, endometrial sampling, hysteroscopy, and dilation and curettage are all more invasive and are not clinically indicated.

Endometrial thickness is measured as the maximum anterior–posterior distance of the endometrial stripe on a long-axis transvaginal view of the uterus. Adequate visualization of the endometrium by means of transvaginal ultrasonography may not be possible in all patients because of anatomic changes such as prior surgery, obesity, or myomas. Transvaginal ultrasonography can be useful in the triage of patients who underwent endometrial sampling and had insufficient tissue for diagnosis. No further evaluation is needed if transvaginal ultrasonography demonstrates an endometrial lining of 4 mm or less, as in the described patient. However, a clinician should pursue additional evaluation if bleeding recurs or if the endometrium cannot be adequately visualized by transvaginal ultrasonography.

Office endometrial sampling is very accurate in the detection of endometrial carcinoma in women with postmenopausal bleeding and has a sensitivity of 99.6%. It is a minimally invasive alternative to dilation and curettage or hysteroscopy. Endometrial sampling would not be indicated in the asymptomatic patient. Clinicians may wish to identify patients at high risk of chronic estrogen exposure because of various conditions, such as obesity, dysfunctional uterine bleeding, anovulation, polycystic ovary syndrome, hepatic disease, diabetes mellitus, and hypothyroidism. Endometrial sampling also may be carried out before the initiation of HT.

Approximately 40% of postmenopausal women who start continuous, combined estrogen–progestin HT regimens will experience bleeding during the first 4–6 months of treatment. This percentage decreases to 10–20% after 1 year. Therefore, observation over time is an effective strategy for some patients. Any woman who receives long-term HT who has achieved amenorrhea and then presents with abnormal uterine bleeding should be evaluated with ultrasonography followed by uterine sampling if the endometrium is greater than 4 mm.

If the patient has recurrent bleeding despite repeated HT dose adjustments, a structural uterine abnormality may be responsible. Up to 27% of cases of abnormal uterine bleeding while receiving HT can be attributed to leiomyomas, polyps, adenomyosis, endometrial hyperplasia, or endometrial cancer. Sonohysterography can be a valuable tool to evaluate polyps and leiomyomas. Drug interactions, coagulation defects, and concomitant use of tamoxifen citrate also have been implicated in postmenopausal bleeding.

Observation is the most appropriate step for this patient because her lining is less than 4 mm. Her bleeding is most likely due to atrophic changes. Patients receiving combination HT will obtain protective effects from the progestin exposure, but there is still a risk of developing endometrial cancer. For example, the Women's Health Initiative study evaluated 16,608 women with an intact uterus who were randomized to receive estrogen plus progestin compared with placebo and monitored for a mean of 62 months. There were 22 cases of endometrial cancer in the estrogen and progestin group compared with 25 cases in the placebo group. Although the hazard ratio showed a reduction in the number of cases of endometrial cancer in the treated group, the differences were not statistically significant.

Moroney JW, Zahn CM. Common gynecologic problems in geriatric-aged women. Clin Obstet Gynecol 2007;50:687–708.

The role of transvaginal ultrasonography in the evaluation of postmenopausal bleeding. ACOG Committee Opinion No. 440. American College of Obstetricians and Gynecologists. Obstet Gynecol 2009:114:409–11.

Rossouw JE, Anderson GL, Prentice RL, LaCroix AZ, Kooperberg C, Stefanick ML, et al. Risks and benefits of estrogen plus progestin in healthy postmenopausal women: principal results from the Women's Health Initiative randomized controlled trial. Writing Group for the Women's Health Initiative Investigators. JAMA 2002;288:321–33.

Speroff L, Fritz MA. Postmenopausal hormone therapy. In: Speroff L, Fritz MA. Clinical gynecologic endocrinology and infertility. 8th ed. Philadelphia (PA): Lippincott Williams and Wilkins; 2011. p. 794–6.

35

Intrauterine microinsert and pregnancy

A 31-year-old woman underwent tubal sterilization by means of a hysteroscopically placed permanent contraceptive microinsert 4 years ago. She is in a new relationship and is considering in vitro fertilization–embryo transfer (IVF–ET) to conceive. The likelihood of endometrial tissue completely covering the microinsert (encapsulation) with no visible outer coils seen at the time of hysteroscopy is at least

* (A) 25%
  (B) 40%
  (C) 55%
  (D) 70%
  (E) 85%

Female sterilization with the use of a hysteroscopically placed permanent contraceptive microinsert was approved by the U.S. Food and Drug Administration in 2002. After undergoing the procedure, some women have desired to conceive and many have considered using IVF–ET to resolve their tubal factor infertility. By design and variations in the placement technique of the microinserts, the number of outer coils of the microinsert that extend into the uterine cavity might vary from one individual to another. The manufacturer recommends that 3–8 expanded outer coils protrude into the uterine cavity after initial placement. Two potential areas of concern for a woman who is considering an embryo transfer with the microinsert in place are 1) a possible decreased implantation rate per embryos transferred and 2) the possible disruption of the normal development of a fetus if any remnant of the outer coils protrudes into the uterine cavity. To address these concerns, the manufacturer has stated in the package insert that "Effects, including risks, of . . . [the] inserts on IVF have not been evaluated. Risks to the patient, fetus and continuation of pregnancy are unknown."

To quantify these concerns, investigators in one study performed a second-look hysteroscopy procedure in 22 patients (20 with uterine bleeding and 2 pre-IVF). They demonstrated that 25% of the women (4 of 16 patients) had complete tissue encapsulation of the microinsert 13–43 months after microinsert placement. *Complete tissue encapsulation* was defined as endometrial tissue ingrowth covering the microinsert. It was hypothesized that the endometrial tissue around the tubal ostium used the trailing coils of the microinsert as a scaffolding structure to gradually cover over the microinsert and exclude it from the uterine cavity. The investigators also demonstrated that over time the trailing coils protruding into the uterine cavity shortened. Given the fact that these individuals were observed for a variable time interval of 13–43 months, the estimate that at least 25% of the women would demonstrate complete tissue encapsulation 4 years (48 months) later would appear to be a conservative one.

In the time since the initial introduction of the contraceptive microinsert system, clinicians have extended the indications for the system to include its off-label use to occlude a hydrosalpinx proximally as an alternative to a salpingectomy procedure before an embryo transfer. The hysteroscopic occlusion may be preferable to laparoscopic tubal ligation or salpingectomy because of either the patient's prior history of abdominal surgeries or obesity, as well as to minimize operative anesthesia time. Prior studies have demonstrated an improved implantation and delivery rate after surgical removal of proximal tubal occlusion of a hydrosalpinx.

In a small prospective, single-arm, clinical study of 20 women with unilateral or bilateral hydrosalpinx visible by transvaginal ultrasonography, microinserts were placed in all patients in an ambulatory setting without complications. Clinicians inserted the microinserts with a mean number of three outer coils protruding into the uterine cavity (range: 1–4 outer coils). This deeper placement of the microinsert rather than the 3–8 expanded outer coils recommended by the manufacturer was done to minimize a potential effect of the microinsert on embryo implantation and development. Proximal tubal occlusion was confirmed 3 months after placement by hysterosalpingography in 95% of patients (19 out of 20) and 96% of treated fallopian tubes (26 out of 27). After the hysterosalpingography, 45 embryo transfer procedures were performed in 19 patients, resulting in 18 pregnancies with 12 live births, 6 miscarriages, and 1 premature delivery with eventual death of the infant for a 63% cumulative live-birth rate per patient and a 27% cumulative live-birth rate per embryo transfer. To further evaluate the safety and efficacy of this procedure, the authors are conducting a multicenter randomized controlled trial to compare microinsert placement with salpingectomy for the treatment of a hydrosalpinx before IVF–ET.

36
Leiomyoma in infertility

A nulliparous 40-year-old woman comes to your office with a 1-year history of heavy menstrual bleeding and a normal hemoglobin level. She has no pain or additional symptoms. Office ultrasonography reveals a thickened endometrium measuring 22 mm. Follow-up sonohysterography is suggestive of the presence of a 1.5-cm submucosal leiomyoma type 2 based on the leiomyoma subclassification system (Fig. 36-1), a subset of the International Federation of Gynecology and Obstetrics classification system for causes of abnormal uterine bleeding in nongravid women of reproductive age (Appendix C). She would like to preserve her future fertility and requests nonsurgical options for management. The most effective long-term treatment option for her is

* (A) levonorgestrel intrauterine device (IUD)
  (B) gonadotropin-releasing hormone (GnRH) agonist
  (C) mifepristone
  (D) medroxyprogesterone acetate
  (E) letrozole

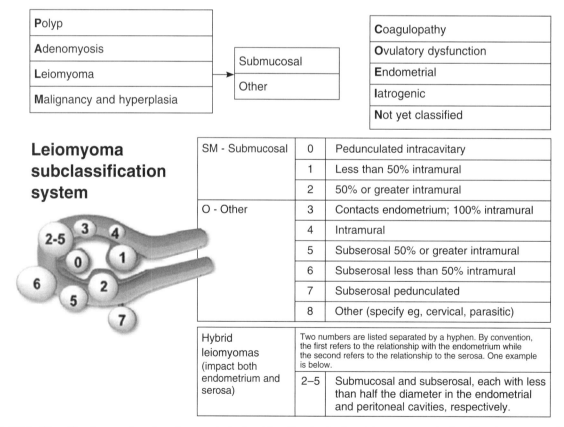

FIG. 36-1. Classification system for abnormal uterine bleeding including leiomyoma subclassification system in nongravid women of reproductive age. (Munro MG. Uterine leiomyomas, current concepts: pathogenesis, impact on reproductive health, and medical, procedural, and surgical management. Obstet Gynecol Clin North Am 2011;38: 703–31 copyright 2011 with permission from Elsevier. Adapted from Munro MG. Abnormal uterine bleeding. New York: Cambridge University Press, 2010.)

Uterine leiomyomas are common uterine neoplasms found in approximately 70–80% of premenopausal women. It has been estimated that leiomyomas account for 30–40% of the approximately 600,000 hysterectomies performed annually in the United States. Growth of leiomyomas is largely dependent on female gonadal steroids, especially estrogens and progesterone, and after menopause they can spontaneously regress. Women with leiomyomas experience symptoms such as infertility, abnormal uterine bleeding, and pelvic pain or pressure; however, most cases of leiomyomas are asymptomatic. Although there has been an increase in the number of minimally invasive procedures to surgically address these neoplasms, the use of appropriate nonsurgical therapy should be discussed with the patient as a viable alternative.

The use of GnRH agonist induces amenorrhea secondary to suppression of gonadotropin release and halted folliculogenesis, creating a hypoestrogenic state that leads to size reduction. This may be helpful for short-term courses to reduce the size of the leiomyoma before surgery and to be able to convert an otherwise difficult or open case into a minimally invasive one. It has been shown that GnRH agonist results in a reduction of the leiomyoma and total uterine volume by a mean of approximately 50% by 12 weeks of treatment. However, the effect is temporary because the leiomyoma will return to baseline level within a few months after the cessation of therapy. Therefore, GnRH agonist therapy is not a long-term solution to leiomyomas. Adverse effects of GnRH agonist therapy associated with hypoestrogenemia include vasomotor symptoms, vaginal atrophy, and osteopenia if the therapy lasts longer than 6 months.

The use of mifepristone, a selective progesterone receptor modulator, has been shown to reduce or even eliminate abnormal uterine bleeding associated with leiomyomas. Some studies have shown that endometrial hyperplasia is an adverse effect of the use of mifepristone. Large-scale clinical trials are necessary to further elucidate the value of this approach.

Insufficient evidence exists in regard to the use of medroxyprogesterone acetate, a systemic progestin, for women with leiomyomas. Systemic progestins may temporarily stabilize the endometrium, but are ultimately ineffective as a treatment for leiomyomas.

Aromatase inhibitors, such as letrozole, inhibit the physiological conversion of androgens to estrogens in the ovary and peripheral tissues. The relative hypoestrogenemia induced by these medications can reduce leiomyoma volume without vasomotor symptoms. Long-term use of such medications in premenopausal women can induce folliculogenesis and ovarian cysts. To date, there have been no large-scale trials of aromatase inhibitors to treat leiomyomas.

The best option to treat the described patient's leiomyoma is a levonorgestrel IUD. The levonorgestrel IUD delivers 20 micrograms per day of active hormone into the uterine cavity over a period of 5 years with minimal systemic absorption. It is a very effective contraceptive and also has been shown to be useful in the control of menorrhagia. Endometrial suppression occurs over a period of 3–4 months when most women will experience irregular bleeding or spotting. By the sixth month, amenorrhea and oligomenorrhea occur in 50% and 25% of women, respectively, and usually persist for the 5-year duration of therapy. A prospective clinical trial that included women with abnormal uterine bleeding and at least one type II submucosal leiomyoma of 5 cm or less demonstrated a 90% reduction in menstrual blood loss at 3 months, 6 months, and 12 months after insertion and a low expulsion rate. These results compare favorably with those seen for the endometrial ablation thermal balloon.

Munro MG. Uterine leiomyomas, current concepts: pathogenesis, impact on reproductive health, and medical, procedural, and surgical management. Obstet Gynecol Clin North Am 2011;38:703–31.

Soysal S, Soysal ME. The efficacy of levonorgestrel-releasing intrauterine device in selected cases of myoma-related menorrhagia: a prospective controlled trial. Gynecol Obstet Invest 2005;59:29–35.

37
Endometriosis

A 30-year-old nulligravid woman reports that she and her partner have been trying to conceive for 18 months without success. She has severe dysmenorrhea and deep dyspareunia. Transvaginal ultrasonography identifies a 6-cm ovarian cyst consistent with an endometrioma. The best next step in management to improve her symptoms and maximize her chances of conception in the next 6 months is

* (A) laparoscopy with excision of the endometrioma
* (B) gonadotropin-releasing hormone agonist therapy for 3 months
* (C) intrauterine insemination
* (D) in vitro fertilization

Women with endometriosis typically have cyclic pelvic pain, dyspareunia, adnexal masses, and infertility. Classical studies suggest that approximately 25–50% of infertile women have endometriosis and that 30–50% of women with endometriosis are infertile. Among women with pelvic pain, the prevalence of endometriosis ranges from 30% to 80%. Although a direct causal mechanism for the role of endometriosis in infertility has not been established, a reasonable body of evidence supports the association between the two disease processes. Mechanisms suggested include altered pelvic anatomy impairing oocyte release or inhibiting ovum capture, altered peritoneal function, altered hormonal and cell-mediated function, abnormal uterotubal transport, poor oocyte quality, impaired implantation, and endocrine and ovulatory abnormalities. Intrauterine insemination or in vitro fertilization alone would not improve this patient's pain symptoms. Although gonadotropin-releasing hormone agonists would improve her pain symptoms, they would impair her fertility.

The American Society for Reproductive Medicine classification system for endometriosis is the most widely accepted staging system for endometriosis. Medical therapy to suppress endometriosis has not been shown to improve fecundity rates and may result only in a delay in the use of more effective treatments to achieve pregnancy. Surgery for stage III or IV endometriosis can be helpful to treat pelvic adhesions that may affect reproductive function.

In stage I–II endometriosis, laparoscopic ablation of endometrial implants has been associated with a small but significant improvement in live-birth rates. In patients with stage III–IV endometriosis, a nonrandomized study demonstrated that the cumulative pregnancy rates in 216 infertile patients monitored up to 2 years after surgical resection were 45–63%. Laparoscopic cystectomy for ovarian endometriomas greater than 4 cm improved fertility compared with cyst drainage and coagulation. The health care provider should make all attempts to preserve functioning ovarian tissue. The possible adverse consequences of loss of ovarian cortex and periovarian adhesion formation must be weighed against the possible improvement in fecundity and pain symptoms. After the first infertility operation, particularly in the setting of diminished ovarian reserve, patients with suspected stage III–IV endometriosis may be better served by using assisted reproductive technologies as the first-line therapy.

Bulun SE. Endometriosis. N Engl J Med 2009;360:268–79.

Chapron C, Vercellini P, Barakat H, Vieira M, Dubuisson JB. Management of ovarian endometriomas. Hum Reprod Update 2002; 8:591–7.

Crosignani PG, Vercellini P, Biffignandi F, Costantini W, Cortesi I, Imparato E. Laparoscopy versus laparotomy in conservative surgical treatment for severe endometriosis. Fertil Steril 1996;66:706–11.

Endometriosis and infertility: a committee opinion. Practice Committee of the American Society for Reproductive Medicine. Fertil Steril 2012; 98:591–8.

Revised American Society for Reproductive Medicine classification of endometriosis: 1996. Fertil Steril 1997;67:817–21.

38
Male factor infertility

A 36-year-old nulligravid woman comes to your office and reports that she has spent 8 months trying to conceive without success. Her male partner is unable to provide a semen specimen because of erectile dysfunction (ED). He reports low energy, low libido, and generalized fatigue; evaluation reveals hyperprolactinemia and a pituitary microadenoma. The treatment most likely to improve his symptoms is

* (A) dopamine agonist
* (B) sildenafil citrate
* (C) testosterone supplement
* (D) clomiphene citrate

Hyperprolactinemia, although a rare condition, often is associated with ED, diminished libido, and orgasmic or ejaculatory dysfunction. The condition should not be neglected in men because many cases result from pituitary tumors. All types of hyperprolactinemia (idiopathic, tumoral, or drug induced) can inhibit most aspects of male sexual behavior. A literature review that encompassed studies that included a total of more than 300 hyperprolactinemic men found sexual dysfunction in 88% of the men, including ED in almost every case. Typically, ED was found to be associated with reduced sexual desire. The nonsexual symptoms of hyperprolactinemia were less frequent: reduced body hair in approximately 40% of the men, gynecomastia in 21%, and galactorrhea in 13%. Systematic determinations of serum prolactin in patients with ED found very low levels of hyperprolactinemia (1–5%). The mechanism suggested for prolactin-induced sexual dysfunction is a decrease in testosterone secretion. Hyperprolactinemia impairs pulsatile luteinizing hormone release, which results in a decrease of serum testosterone secretion.

Dopamine agonist therapy is the first-choice treatment for prolactin-induced sexual dysfunction. In some cases of pituitary adenoma, hypogonadism persists despite return of a normal prolactin level because of definitive interruption of the hypothalamic–pituitary–testicular connection, destruction of the pituitary gonadotrophs by the pituitary tumor, or as a result of pituitary resection. Such patients may require human chorionic gonadotropin and human menopausal gonadotropin injection in addition to dopamine agonist therapy to enhance spermatogenesis.

Direct-to-consumer marketing for testosterone supplementation has increased in recent years and has created a subset of male fertility patients who have unknowingly suffered from iatrogenic hypogonadotropic hypogonadism. Although testosterone supplementation may help hypogonadotropic male patients to improve libido and arousal, spermatogenesis will not improve.

Sildenafil citrate is an effective treatment to improve erectile performance in the male patient with dysfunction. The described patient's symptoms are a result of low testosterone that will not be affected by sildenafil treatment. Clomiphene citrate treatment requires a functioning hypothalamic–pituitary–testicular axis to provide any clinical results. The described patient has hypothalamic–pituitary–testicular axis dysfunction due to hyperprolactinemia. Dopamine agonist is the treatment that is most likely to improve his symptoms.

Buvat J. Hyperprolactinemia and sexual function in men: a short review. Int J Impot Res 2003;15:373–7.

Chao J, Hwang TI. Contemporary management of erectile dysfunction. Urol Sci 2013;24:35–40.

Colao A, Di Sarno A, Guerra E, De Leo M, Mentone A, Lombardi G. Drug insight: Cabergoline and bromocriptine in the treatment of hyperprolactinemia in men and women. Nat Clin Pract Endocrinol Metab 2006;2:200–10.

39

Use of donor sperm for same-sex couples

A 33-year-old woman comes to your office with her 30-year-old female partner to discuss using a known sperm donor for insemination. You meet with the 25-year-old donor to discuss the process of donating his sperm and to obtain a medical and family history from him. He is a healthy man but has two cousins who died in their 20s from sudden cardiac arrest; no further evaluation was performed. You are concerned that this potential donor may be at risk of a genetic condition that would predispose him and his offspring to cardiac arrest. The most appropriate next step would be

* (A) donor rejection by health care provider
* (B) genetic testing of donor
* (C) disclosure of donor's medical history to patient and partner
* (D) genetic counseling for donor
* (E) proceeding with insemination using donor sperm

Artificial insemination is the technique of introducing sperm into the vagina or uterus without sexual intercourse. Clinically, this technique is performed for several indications, eg, in cases of unexplained or male factor infertility, when it usually is used with the male partner's sperm. A therapeutic donor insemination is performed using sperm from a man who is not the patient's partner. A number of circumstances may call for use of therapeutic donor insemination, eg, in men with severe male factor infertility, couples that carry a heritable disease, and, commonly, women who do not have a male partner. The American Society for Reproductive Medicine Ethics Committee guidelines state that single, gay, and lesbian patients should have access to fertility services. Lesbian couples often use therapeutic donor insemination to attempt pregnancy. A 2001 study found that approximately 75% of U.S. fertility clinics provide such services to single and lesbian women.

Many patients select an anonymous donor from a sperm bank, although some elect to have a "known" sperm donor. Regardless of plans to use an anonymous or known donor for donor insemination, recommendations are available on how to select donors in a fashion that is safe for the donor and the recipient. The American Society for Reproductive Medicine's "Guidelines for Gamete and Embryo Donation" provide a summary of the current recommendations, including information from the U.S. Food and Drug Administration about screening donors for communicable diseases. For example, donors must be of legal age (older than 17 years) and ideally younger than 40 years. Donors must undergo a psychologic assessment by a mental health professional to address the emotional, psychologic, and ethical aspects of being a sperm donor.

In cases in which the sperm donor is a friend or a family member of the recipient couple, the donor and recipient often will meet with the mental health professional together, and it is recommended that legal issues (such as parental rights and responsibilities) be addressed before donor insemination because such legal issues may vary from state to state. All donors must undergo an extensive history (personal, medical, sexual, and family history) as well as a physical examination and routine blood tests. Guidelines require that sperm samples from anonymous male donors be quarantined for at least 6 months to allow for repeat sexually transmitted infection testing of the donor before release of the sperm for donor insemination. Female recipients who use a known sperm donor may defer this quarantine in selected situations if they recognize and accept the potential risk of undiagnosed sexually transmitted infections.

The guidelines also recommend that potential donors be screened for common genetic mutations, such as cystic fibrosis, regardless of ethnic background. Other genetic testing should be considered based on the donor's ethnicity and family history. Karyotype testing sometimes is recommended, although it is not a routine requirement.

In cases where the medical or family history of a potential donor raises suspicion for an inheritable disease, further testing may be warranted before the individual is accepted as a donor. It is not necessary to reject a donor solely based on concern for carrier status before appropriate confirmatory testing is completed. In the described scenario, the potential donor's history raises suspicion for an inherited heart disease, such as hypertrophic cardiomyopathy (HCM), given that he had two cousins with early sudden cardiac arrest and death. In one reported case, the sperm from an anonymous sperm donor with undiagnosed HCM resulted in at least 22 offspring. After he was diagnosed with HCM, the children were tested and at least eight were found to be affected. That case and other similar cases raise issues about the proper protocol for genetic testing for potential donors.

Many sperm and egg donors are young men and women in their 20s who have not had the need or inclination to contemplate their own genetic health. Given the complexities associated with a diagnosis of an adult-onset disorder, such as HCM, breast cancer, and Alzheimer disease, donors should be given the opportunity to make an informed choice about proposed genetic testing and their desire to learn the results. Therefore, in the described scenario, it is not appropriate to proceed with genetic testing without first addressing the implications of the testing with the donor. Similarly, although the recipient couple likely has an interest in genetic testing for the selected donor, they are not the ultimate decision makers about whether the donor proceeds with this testing. If he opts not to pursue testing, they may decide not to use him as their donor, given the gap in information about genetic risk for their offspring.

For targeted genetic testing for high-risk individuals, it is recommended that pretest and posttest genetic counseling be part of the testing process to allow individuals to make educated decisions about whether to proceed with testing. In addition, they need information about disease risk if indeed they are carriers of a mutation as well as residual risk if they are not found to be carriers. When a health care provider is evaluating a potential donor, the donor becomes their patient, and the health care provider is then responsible for recommendations about his or her health and testing. The health care provider should thus proceed in a fashion that protects the rights and choices of the donor.

Daar JF, Brzyski RG. Genetic screening of sperm and oocyte donors: ethical and policy implications. JAMA 2009;302:1702–4.

Maron BJ, Lesser JR, Schiller NB, Harris KM, Brown C, Rehm HL. Implications of hypertrophic cardiomyopathy transmitted by sperm donation. JAMA 2009;302:1681–4.

Recommendations for gamete and embryo donation: a committee opinion. Practice Committee of American Society for Reproductive Medicine and the Practice Committee of Society for Assisted Reproductive Technology; Fertil Steril 2013;99:47–62.

40

Phantom human chorionic gonadotropin results

A 28-year-old woman, gravida 4, para 3, underwent methotrexate treatment for an ectopic pregnancy 3 months ago. She has followed up with monthly serum β-hCG quantitative values. These values have consistently been between 127 mIU/mL and 132 mIU/mL. She has been taking oral contraceptives since her most recent visit. The best next step in her management is

    (A) repeat serum β-hCG in 2 days
    (B) chest X-ray
*   (C) parallel urinary β-hCG assay
    (D) pelvic ultrasonography
    (E) repeat methotrexate treatment

False-positive or phantom serum β-hCG test results can lead to misdiagnosis and mismanagement of the patient. It is critical to consider false-positive β-hCG test results when monitoring patients with pregnancy of uncertain location, molar gestation, or gestational trophoblastic neoplasia.

Phantom β-hCG can be caused by the presence of heterophilic antibodies that can result in false elevation of serum β-hCG. Heterophilic antibodies are human antibodies that react against animal-derived antigens used in immunoassay testing. Women who have a history of exposure to animals are at risk of heterophilic antibody development over their lifetime. Such individuals include women laboratory workers who work with animals, women raised on farms, and women who regularly work close to or with animal tissue.

Heterophilic antibodies can cross-react with immunoglobulins from other species. Patients who have heterophilic antibodies may have false-positive serum β-hCG test results. Heterophilic antibodies tend to interfere with two-sided immunometric assays commercially used for β-hCG measurements as well as with other hormonal assays.

To evaluate patients for phantom β-hCG, serum and urine samples should be assessed. Patients with phantom β-hCG usually have no measurable β-hCG in a parallel urine sample. The main limitation in this technique is that patients with low β-hCG levels may have urine levels too low to detect the true presence of β-hCG. Typically, serum β-hCG values less than 25 mIU/mL are considered too low to detect in urine.

Flooding the sample with a mixture of nonspecific animal antibodies or antibodies against heterophilic antibodies is another strategy to assess if β-hCG levels are false. If subsequent testing of the sample does not demonstrate a β-hCG value, then heterophilic antibodies are likely the cause for false β-hCG elevation.

Patients with suspected heterophilic antibodies should have β-hCG levels measured in serial dilutions. The serial dilutions of a true serum β-hCG value should follow a linear β-hCG curve. Serial dilutions that do not follow a linear curve suggest the presence of heterophilic antibodies and, therefore, a false or phantom β-hCG value (see Table 40-1).

The patient is not a candidate to repeat her β-hCG level in 2 days because it is established that she has had consistent β-hCG levels every month. It is inappropriate to order a chest X-ray at this point in her management given that metastatic gestational trophoblastic disease is unlikely. Additionally, it is very rare to have gestational trophoblastic disease after an ectopic pregnancy. Performing ultrasonography of the uterus is only appropriate if the β-hCG value is noted to be real and not a false elevation.

It is inappropriate to administer an additional dose of methotrexate to this patient. She has already been treated medically for an ectopic pregnancy. Other causes for persistent elevations (phantom β-hCG) should be considered.

Avoiding inappropriate clinical decisions based on false-positive human chorionic gonadotropin test results. ACOG Committee Opinion No. 278. The American College of Obstetricians and Gynecologists. Obstet Gynecol 2002;100:1057–9.

Berkowitz RS, Goldstein DP. Current management of gestational trophoblastic diseases. Gynecol Oncol 2009;112:654–62.

Cole LA, Kohorn EI. The need for an hCG assay that appropriately detects trophoblastic disease and other hCG-producing cancers. J Reprod Med 2006;51:793–811.

TABLE 40-1. Test to Assess for Phantom Serum β-hCG

Test		**Explanation**
Urine	Assess β-hCG quantitative or qualitative urinary β-hCG.	Heterophilic antibodies are not in the urine. If urinary β-hCG test is negative and serum testing is positive, heterophilic antibodies are likely present.
Serum	Rerun serum β-hCG with serial dilutions of the serum.	Serial dilutions of serum measuring β-hCG should show a linear curve. Lack of linear β-hCG curve confirms the presence of heterophilic antibodies. Heterophilic antibodies react with reagents in the immunoassay and not β-hCG.
Serum	Laboratory to add heterophilic antibody blocking agents to serum.	If β-hCG is not detected after adding heterophilic antibody blocking agents, heterophilic antibodies can explain false β-hCG elevations.

Avoiding inappropriate clinical decisions based on false-positive human chorionic gonadotropin test results. ACOG Committee Opinion No. 278. American College of Obstetricians and Gynecologists. Obstet Gynecol 2002;100:1057–9.

41
Normal menstrual cycle

A patient with menstrual cycles between 26 days and 30 days comes to your office for evaluation. You tell her that the phase of the menstrual cycle that most influences cycle length is the

* (A) follicular phase
  (B) ovulatory phase
  (C) luteal phase
  (D) secretory phase

The menstrual cycle involves the interaction of different hormones and endocrine glands as well as a responsive uterus. The cycle occurs in three phases: 1) follicular (proliferative), 2) ovulatory, and 3) luteal (secretory). Follicular and luteal refer to ovarian changes and follicle development whereas proliferative refers to endometrial changes. Menstruation is part of the proliferative phase of the menstrual cycle. The menstrual cycle also can be described based on the number of days between the onset of menstrual bleeding in one cycle and the onset of bleeding in the next cycle. Most women have a menstrual cycle length between 25 days and 30 days, with the median duration being 28 days. The variability in length of a menstrual cycle is based on the variable length of the follicular phase; the luteal phase is constant in most women at 14 days in length. Therefore, the luteal phase does not contribute to the variability of the overall menstrual cycle length. However, disorders such as hyperprolactinemia can decrease the length of this phase of the menstrual cycle.

Folliculogenesis takes place during the follicular phase of the menstrual cycle. As follicles grow, a dominant follicle is selected and destined to ovulate. The growth of follicles is dependent on follicle-stimulating hormone stimulation. During this follicle growth, estradiol is produced and is responsible for the proliferation of the endometrial lining of the uterus. Ovulation happens in response to the luteinizing hormone surge, which is released in a positive feedback mechanism from the anterior pituitary due to prolonged exposure to estradiol. Ovulation typically takes place approximately 24 hours after the luteinizing hormone peak. It is also constant and does not contribute to the variability in the menstrual cycle length.

The luteal phase starts after ovulation. During this phase, the remaining granulosa cells that are not released with the oocyte become the corpus luteum, which predominantly secretes progesterone. Peak progesterone production is noted 1 week after ovulation takes place (midluteal). Progesterone converts the proliferative endometrium into a secretory endometrium, allowing for possible embryo implantation. If a pregnancy takes place, human chorionic gonadotropin will maintain the corpus luteum. However, if a pregnancy fails to occur, luteolysis takes place and the corpus luteum is lost. The loss of the corpus luteum and progesterone lead to menstruation due to the instability and sloughing of the endometrium. The length of days that a patient bleeds does not contribute to the total length of the cycle.

Cahill DJ, Wardle PG, Harlow CR, Hull MG. Onset of the preovulatory luteinizing hormone surge: diurnal timing and critical follicular prerequisites. Fertil Steril 1998;70:56–9.

Presser HB. Temporal data relating to the human menstrual cycle. In: Ferin M, Halber F, Richart RM, Van de Wiele R, editors. Biorhythms and human reproduction. New York: John Wiley and Sons; 1974. p. 145–60.

Treloar AE, Boynton RE, Behn BG, Brown BW. Variation of the human menstrual cycle through reproductive life. Int J Fertil 1967;12:77–126.

Vollman RF. The menstrual cycle. Philadelphia: WB Saunders; 1977.

42
Dysmenorrhea

A 16-year-old adolescent woman comes to your office with dysmenorrhea, headache, diarrhea, and back pain during menses that interferes with her ability to attend school. On bimanual examination and ultrasonography, there is no evidence of gynecologic pathology. She used oral contraceptives (OCs) in the past and developed minor depression. She now wishes to avoid using any hormonal preparations. She requests information about other treatment options. The best next step for pain relief from dysmenorrhea in this patient is

    (A) acetaminophen
* (B) nonsteroidal antiinflammatory drugs (NSAIDs)
    (C) presacral neurectomy
    (D) opioids
    (E) uterine nerve ablation

Dysmenorrhea is the occurrence of painful menstrual cramps and other associated symptoms. It is one of the most common gynecologic problems in reproductive-aged women. After ovulation, prostaglandins may be produced in abundance in the late luteal phase, leading to myometrial contractions and pain. Patients also may experience additional symptoms such as headache, nausea, backache, and diarrhea. Such additional symptoms can be attributed to the release of prostaglandin compounds into the systemic circulation. *Primary dysmenorrhea* occurs in the absence of any demonstrable disease. *Secondary dysmenorrhea* refers to menstrual pain resulting from pelvic pathology, such as endometriosis, uterine fibroids, or adenomyosis.

Dysmenorrhea is most common in adolescence. Studies have reported a prevalence of 60–90% among adolescents. Incidence of the disease has been observed to decrease with increasing age. Research has shown that women with dysmenorrhea have higher levels of prostaglandins than asymptomatic females. Endometrial prostaglandin production reaches a peak in the first 48 hours of menstruation, which coincides with the time of greatest menstrual discomfort. Medical therapy for dysmenorrhea includes OCs and NSAIDs, both of which act by suppressing prostaglandin production and reducing the volume of menstrual flow. Although these treatments are very effective, a 20–25% failure rate leads patients to seek additional treatments or surgery.

The secretory endometrium is rich in arachidonic acid, leukotrienes, and prostaglandins. Release of these agents into the systemic circulation leads to uterine smooth-muscle contractions. The result is crampy, spasmodic lower abdominal or back pain and myometrial ischemia. Constriction of the uterine vessels reaches a peak on the first day of menstrual flow.

Dysmenorrhea may be treated with the use of NSAIDs, such as aspirin, propionic acid derivatives (eg, naproxen and ibuprofen), and fenamates (eg, mefenamic acid and diclofenac). It is known that NSAIDs inhibit the cyclo-oxygenase (COX)-1 and COX-2 enzymes. Pain relief comes from their ability to block endometrial prostaglandin production. Clinical trials have shown the effectiveness of NSAIDs to be in the 70–90% range. Long-term use of NSAIDs may increase the risk of serious gastrointestinal adverse effects such as ulceration, bleeding, and perforation. Selective COX-2 inhibitors (eg, celecoxib) have been developed for patients at high risk of serious gastrointestinal tract adverse effects. These agents, however, are expensive and carry an increased risk of serious and potentially fatal cardiovascular thrombotic events, such as myocardial infarction and stroke. Because COX-2 inhibitors interfere with the formation of lipoxins and prostacyclin in blood vessels, vasoconstriction and platelet aggregation can occur, resulting in possible excess clot formation and hypertension. Long-term use can lead to increased rates of myocardial infarction and life-threatening thrombotic events.

Epidemiologic studies have shown that women who take OCs have a lower prevalence of dysmenorrhea. This is thought to be because OCs allow for an atrophic and decidualized endometrium that produces less prostaglandins. By decreasing menstrual flow and prostaglandin-mediated pain, OCs offer a safe and effective choice for the treatment of dysmenorrhea.

Numerous clinical trials have been performed to assess the effectiveness of NSAIDs and OCs for dysmenorrhea. Systematic reviews of randomized controlled trials have shown that NSAIDs are superior to acetaminophen or placebo and are an effective treatment for dysmenorrhea. Insufficient evidence is available to demonstrate the superiority of any individual NSAID for either pain relief or safety. To date, no studies have compared the performance of combination OCs with NSAIDs. Systematic review of cyclic OCs for dysmenorrhea has offered limited evidence for pain improvement for low-dose and medium-dose OC preparations. No evidence has been found of a difference in efficacy between different OC preparations. The data suggest NSAIDs should be the treatment of choice for primary dysmenorrhea. These agents usually are used for short-term therapy just before menses and continued for up to 4–6 days. Oral contraceptives may be a wise choice for any patient who does not tolerate NSAIDs or who concurrently desires contraception.

The best next step for pain relief from dysmenorrhea for the described patient is NSAIDs, given that she has intolerance for the adverse effects of OCs. Acetaminophen and presacral neurectomy would be less effective in her case than NSAIDs. Opioids should be avoided because of their adverse effects, potential for abuse, and possible dependency with long-term use.

Patients with secondary dysmenorrhea should have further investigation of their underlying pathology. Such an investigation will likely guide treatment. Many of the therapies for conditions that cause dysmenorrhea will treat the pain and the underlying pathology. Depot medroxyprogesterone acetate and the levonorgestrel intrauterine device are effective in the treatment of endometriosis and dysmenorrhea. Long-acting gonadotropin-releasing hormone agonists, such as depot leuprolide acetate, are effective for endometriosis and dysmenorrhea.

Women who fail to gain relief from NSAIDs or OCs have a variety of alternatives. Therapies that have been proposed for the relief of dysmenorrhea include exercise, behavioral modification strategies, dietary modification, herbal therapies, vitamins, and application of heat to the affected area. Surgery may be an option when medical treatment has failed.

Uterine nerve ablation and presacral neurectomy have been used for dysmenorrhea. Both procedures are designed to interrupt most cervical sensory pain nerve fibers. Surgical interruption of pelvic nerve pathways also has been studied by systematic review. Insufficient evidence is available to recommend the use of nerve interruption in the management of dysmenorrhea, regardless of the cause.

Archer DF. Menstrual-cycle related symptoms: a review of the rationale for continuous use of oral contraceptives. Contraception 2006; 74:359–66.

Dawood MY. Primary dysmenorrhea: advances in pathogenesis and management [Review]. Obstet Gynecol 2006;74:439–45.

Majoribanks J, Proctor M, Farquhar C, Derks RS. Nonsteroidal antiinflammatory drugs for dysmenorrhoea. Cochrane Database Syst Rev 2010;1:CD001751. DOI: 10.1002/14651858.CD001751.pub2.

Proctor M, Latthe P, Farquhar C, Khan K, Johnson N. Surgical interruption of pelvic nerve pathways for primary and secondary dysmenorrhea. Cochrane Database Syst Rev 2005;4:CD001896.

Speroff L, Fritz MA. Menstrual disorders. In: Speroff L, Fritz MA. Clinical gynecologic endocrinology and infertility. 8th ed. Philadelphia (PA): Lippincott Williams and Wilkins; 2011. p. 579–83.

Wong CL, Farquahar C, Roberts H, Proctor M. Oral contraceptive pill for primary dysmenorrhoea (Review). Cochrane Database Syst Rev 2009;4:CD002120. DOI: 10.1002/14651858.CD002120.pub3.

43
BRCA mutations

A 25-year-old woman who is a known *BRCA1* mutation carrier comes to your office for counseling. You discuss risk-reduction strategies, including surgery, but she tells you that she is not ready to consider mastectomy or salpingo-oophorectomy. In addition to semiannual clinical breast examinations, you advise her that the recommended breast cancer screening strategy for a known *BRCA1* mutation carrier would be annual

* (A) mammography
* (B) breast ultrasonography
* (C) breast magnetic resonance imaging (MRI)
* (D) mammography and breast ultrasonography
* (E) mammography and breast MRI

Hereditary breast and ovarian cancer syndrome is a genetic cancer-susceptibility syndrome. It usually is characterized by multiple family members with breast cancer, ovarian cancer, or both; an individual with breast and ovarian cancer; or an individual with breast cancer at an early age. Under appropriate circumstances, genetic testing can help physicians to identify patients who are at increased risk of breast and ovarian cancer. For at-risk individuals, screening and prevention strategies should be discussed.

The *BRCA1* and *BRCA2* genes are tumor suppressor genes. Mutations in these genes are associated with elevated risks of breast and ovarian cancer. In families with hereditary breast and ovarian cancer syndrome, *BRCA1* and *BRCA2* mutations account for the overwhelming majority of cases. In the general population, however, *BRCA1* and *BRCA2* mutations only account for approximately 10% of cases of ovarian cancer and 3–5% of cases of breast cancer. For ovarian cancer, *BRCA1* mutations carry a 39–46% risk, whereas for breast cancer, *BRCA1* and *BRCA2* mutations carry a risk of approximately 65–74%. In the general population, between 1 in 300 individuals and 1 in 800 individuals are estimated to carry a *BRCA1* or *BRCA2* mutation. In order to help physicians determine which individuals should be offered genetic testing, the Society of Gynecologic Oncologists has established screening guidelines (Box 43-1).

If a patient is found to be a carrier of a mutation, she and her family members should be offered genetic counseling. Mastectomy and salpingo-oophorectomy are the most effective strategies to reduce the risk of breast and ovarian cancer. Risk-reducing salpingo-oophorectomy is recommended for *BRCA1* or *BRCA2* mutation carriers by age 40 years or when childbearing is complete.

Prophylactic salpingo-oophorectomy decreases the risk of ovarian cancer, fallopian tube cancer, and peritoneal cancer by 85–90% and simultaneously decreases the risk of breast cancer by 50%. Similarly, prophylactic mastectomy lowers the risk of breast cancer by more than 90%.

For patients who delay or decline surgery, screening for breast and ovarian cancer is recommended. Screening for ovarian cancer has not been shown to decrease mortality or to improve survival, but, given the risk associated with *BRCA1* and *BRCA2* mutations, screening is recommended. Consensus groups recommend periodic screening with CA 125 and transvaginal ultrasonography starting at approximately age 30 years or 5–10 years earlier than the youngest age of diagnosis of ovarian cancer in an affected relative. Breast cancer surveillance is more reliable than ovarian cancer surveillance. Starting at age 25 years (or earlier based on the youngest age of diagnosis), *BRCA1* and *BRCA2* mutation carriers should undergo semiannual clinical breast examinations in addition to annual mammography and breast MRI. Screening with MRI is more sensitive than mammography. The combination of MRI, mammography, and clinical breast examination has the greatest sensitivity for detecting breast cancer in *BRCA* mutation carriers.

Therefore, the most appropriate breast cancer screening for the described *BRCA1* mutation carrier, in addition to a semiannual clinical breast examination, is a combination of mammography and breast MRI. Mammography alone would be insufficient. Her needs also would not be met by breast MRI alone or a combination of mammography and breast ultrasonography. Breast ultrasonography is not recommended for screening high-risk individuals but is useful for imaging dense masses.

BOX 43-1

Society of Gynecologic Oncologists Guidelines for Genetic Testing

Patients with greater than approximately 20–25% chance of having an inherited predisposition to breast and ovarian cancer and for whom genetic risk assessment is recommended

- Women with a personal history of both breast and ovarian cancer
- Women with ovarian cancer* and a close relative† with breast cancer at 50 years or less or ovarian cancer at any age
- Women with ovarian cancer* at any age who are of Ashkenazi Jewish ancestry
- Women with breast cancer at age 50 years or less and a close relative† with ovarian or male breast cancer at any age.
- Women of Ashkenazi Jewish ancestry and breast cancer at age 40 years or less
- Women with a first- or second-degree relative with a known *BRCA1* or *BRCA2* mutation

Patients with greater than approximately 5–10% chance of having an inherited predisposition to breast and ovarian cancer and for whom genetic risk assessment may be helpful‡

- Women with breast cancer at 40 years or less
- Women with bilateral breast cancer (particularly if the first cancer was at 50 years or less)
- Women with breast cancer at 50 years or less and a close relative† with breast cancer at 50 years or less
- Women of Ashkenazi Jewish ancestry with breast cancer at 50 years or less
- Women with breast or ovarian cancer at any age and two or more close relatives† with breast cancer at any age (particularly if at least one breast cancer was at 50 years or less)
- Unaffected women with a first- or second-degree relative that meets one of the above criteria

*Note that peritoneal and fallopian tube cancers are considered part of the spectrum of hereditary breast–ovarian cancer syndrome.

†A close relative is defined as a first-, second-, or third-degree relative (ie, mother, sister, daughter, aunt, niece, grandmother, granddaughter, first cousin, great grandmother, great aunt).

‡In families with a paucity of female relatives in either lineage, it may also be reasonable to consider genetic risk assessment even in the setting of either an isolated case of breast cancer at age 50 years or more, or an isolated case of ovarian, fallopian tube, or peritoneal cancer at any age.

Lancaster JM, Powell CB, Kauff ND, Cass I, Chen LM, Lu KH, et al. Society of Gynecologic Oncologists Education Committee statement on risk assessment for inherited gynecologic cancer predispositions. Society of Gynecologic Oncologists Education Committee. Gynecol Oncol 2007;107:159–62.

Hereditary breast and ovarian cancer syndrome. ACOG Practice Bulletin No. 103. American College of Obstetricians and Gynecologists. Obstet Gynecol 2009;113:957–66.

Kriege M, Brekelmans CT, Boetes C, Besnard PE, Zonderland HM, Obdeijn IM, et al. Efficacy of MRI and mammography for breast-cancer screening in women with a familial or genetic predisposition. Magnetic Resonance Imaging Screening Study Group. N Engl J Med 2004;351:427–37.

Lancaster JM, Powell CB, Kauff ND, Cass I, Chen LM, Lu KH, et al. Society of Gynecologic Oncologists Education Committee statement on risk assessment for inherited gynecologic cancer predispositions. Society of Gynecologic Oncologists Education Committee. Gynecol Oncol 2007;107:159–62.

44

Contraception for a patient with diabetes mellitus

A 19-year-old woman with a 5-year history of well-controlled type 1 diabetes mellitus has had infrequent vaginal intercourse for 2 years. Although you counsel her about the benefits of long-acting reversible contraception and depot medroxyprogesterone acetate, she rejects those options. Therefore, the best contraceptive method for this patient would be

* (A) combination hormonal contraceptive
* (B) contraceptive sponge
* (C) male condom
* (D) periodic abstinence during ovulation
* (E) vaginal spermicide

Women with type 1 diabetes mellitus who conceive and continue an unplanned pregnancy during periods of poor glycemic control are at increased risk of developing major maternal and fetal complications. Pregnancy should be contemplated when the patient's hemoglobin A_{1c} level is optimal, but studies demonstrate that less than 50% of women with diabetes plan their pregnancies. Moreover, most diabetes specialists feel unqualified to provide effective contraceptive counseling and defer this responsibility to other women's health care providers.

In a study of 107 women with diabetes who answered a research questionnaire survey on their contraceptive knowledge and methods used, 36% of women stated they had not received any contraceptive advice in the past year; 8% of women recalled obtaining information from their diabetes specialist; and the remaining women said they received information from their general practitioner or family planning clinic. Women in this study had a mean age of 31 years with an age range of 16–44 years. They used the following methods for contraception: combination oral contraceptives (OCs) (24%), condoms (16%), progestin-only OCs (7%), intrauterine devices (IUDs) (6%), natural family planning (6%), sterilization (female or male partner) (5%), depot medroxyprogesterone acetate (1%), and progestin implant (1%). Of the women in the study, 34% used no contraceptive method. Although these women were taking insulin and other medications to control their diabetes, the women had concerns about using any additional medication and elected not to use hormonal methods of contraception. Some of this bias may be due to older OC formulations with higher doses of estrogen and progestin. Current low-dose (10–35 micrograms) ethinyl estradiol OCs do not cause metabolic disturbance of a woman's diabetes and are safe for women with well-controlled diabetes. In typical use, approximately 9% of women who use OCs will experience an unintended pregnancy during the first year of use.

In comparison, the failure rates for a number of other methods are all higher with typical use: use of the contraceptive sponge in a nulliparous woman (12%), male condom (18%), periodic abstinence during ovulation (24%), and use of a vaginal spermicide (28%). The failure rate with typical use of other methods often used by women with diabetes included 85% for no method of contraception and 9% for the progestin-only OC, despite the fact that the progestin-only OC only inhibits ovulation in approximately 50% of menstrual cycles. The following methods eliminate or decrease the necessity of patient adherence to the method and all have a lower failure rate than OCs: depot medroxyprogesterone acetate (6%), levonorgestrel IUD (0.2%), copper IUD (0.8%), and progestin implant (0.05%). Because this patient rejects these methods, the most efficacious contraceptive method for her would be a combination hormonal contraceptive.

Shawe J, Smith P, Stephenson J. Use of contraception by women with type 1 or type 2 diabetes mellitus: 'it's funny that nobody really spoke to me about it'. Eur J Contracept Reprod Health Care 2011;16:350–8.

Use of hormonal contraception in women with coexisting medical conditions. ACOG Practice Bulletin. No. 73. American College of Obstetricians and Gynecologists. Obstet Gynecol 2006;107:1453–72.

U.S. Selected Practice Recommendations for Contraceptive Use, 2013: adapted from the World Health Organization selected practice recommendations for contraceptive use, 2nd edition. Division of Reproductive Health, National Center for Chronic Disease Prevention and Health Promotion, Centers for Disease Control and Prevention. MMWR Recomm Rep 2013;62(RR-05):1–60.

45

Congenital adrenal hyperplasia

A 30-year-old woman, gravida 2, para 1, comes to your office for fertility treatment. She has a history of bilateral tubal blockage. She underwent in vitro fertilization (IVF) 3 years ago and gave birth to a severely virilized female infant with 21-hydroxylase deficiency. The child now has learning disabilities. She wishes to avoid pregnancy termination. The best way to prevent virilized female infants in subsequent pregnancies is

* (A) preimplantation genetic diagnosis
* (B) prenatal dexamethasone treatment
* (C) dexamethasone treatment starting at 12 weeks of gestation
* (D) amniocentesis
* (E) chorionic villus sampling

Congenital adrenal hyperplasia (CAH) is caused by a defect in one of the five enzymatic steps in the adrenal cortex necessary in the conversion of cholesterol to cortisol. The most common defect is related to 21-hydroxylase deficiency, an autosomal recessive disorder due to mutations in the *CYP21A2* gene, which accounts for approximately 90–95% of all cases or 1 in 14,500 newborns. The *CYP21A2* gene is part of a complex major histocompatibility locus on chromosome 6p21. To date, more than 170 *CYP21A2* mutations have been identified. Because CAH is autosomal recessive, a patient who gives birth to a child with CAH has a one in four chance of giving birth to a subsequent child with CAH. There is also a two in four chance that the fetus will be unaffected but carry the mutation, and a one in four chance that the fetus will not be a carrier. Given that one half of conceptions are female, the risk of having a female with a genital tract anomaly is one in eight. The 21-hydroxylase enzyme converts 17-hydroxyprogesterone to 11-deoxycortisol in the synthesis of cortisol, in the zona fasciculata of the adrenal cortex; therefore, increased levels of 17-hydroxyprogesterone are used to make the diagnosis. In the zona glomerulosa, 21-hydroxylase converts progesterone to deoxycorticosterone in the synthesis of aldosterone. Deficient aldosterone synthesis is more deleterious to the newborn infant, which causes hyponatremia, hyperkalemia, acidosis, hypovolemia, and, if untreated, death.

Deficient cortisol synthesis leads to increased production of adrenocorticotropin and increased steroidogenesis, which, in the absence of 21-hydroxylation, results in the accumulation of androgenic precursor steroids leading to virilization of the external genitalia in females. The virilization may be sufficiently severe to result in errors in sex assignment in the delivery room, including defects in the urogenital sinus and scrotum, labial fusion, and clitoromegaly. Despite the virilization of the female external genitalia, the ovaries, uterus, and fallopian tubes are normal. Affected males have normal external genital development and may escape initial diagnosis; however, both genders experience varying degrees of electrolyte abnormalities at birth that require immediate attention because of a significant risk of sudden death in the newborn.

Beginning in the 1980s, suraphysiologic corticosteroid treatment was used to decrease the virilization of the external genitalia of affected female fetuses. However, recent clinical observations and animal studies have raised concerns regarding the safety of this prenatal treatment.

The fetal adrenal cortex is composed of a fetal zone and a definitive (adult) zone. The fetal zone primarily secretes dehydroepiandrosterone and dehydroepiandrosterone sulfate, which are converted to estriol by the placenta. Fetal cortisol synthesis is low for much of the gestation, and the placenta prevents most transfer of maternal cortisol to the fetus. Corticosteroids administered to the pregnant woman can cross the placenta. In the placenta, the *HSD11B2* gene encodes the 11β-hydroxysteroid dehydrogenase enzyme, which is responsible for the inactivation of maternal cortisol. However, when not inactivated by placental *HSD11B2*, dexamethasone can have effects on the female fetus at levels 60 times greater than the usual fetal levels.

Prenatal treatment of CAH has different levels of success. When dexamethasone was given throughout the gestation in 49 pregnancies, of the 25 female fetuses who received dexamethasone before week 9 of pregnancy, 11 had normal female genitalia, 11 had minimal virilization (Prader stages 1–2), and 3 were virilized (Prader stage 3). Among 24 female fetuses in whom treatment was begun after week 9, the genitalia averaged a Prader score of 3.0; among untreated females, the score was 3.75. The success rate was approximately 80–85%, thus, even if treatment is initiated in a timely fashion, not all fetuses will escape the presence of genital anomalies.

Prenatal treatment of CAH is associated with other risks to the pregnant woman and her child, including greater pregnancy-associated weight gain, more striae and edema, mild hypertension, and gestational diabetes mellitus. Because of documented teratogenic effects produced by high doses of dexamethasone administered to pregnant animals and human patients (eg, cleft lip and palate), the U.S. Food and Drug Administration classifies dexamethasone as a category C drug. Administered corticosteroids can promote differentiation of the fetal lung and a single course of antenatal corticosteroids can improve pulmonary outcome in preterm infants, but the use of multiple courses is associated with decreased weight, length, and head circumference at birth. Low birth weight has been shown to be associated with increased risks of adulthood hypertension, type 2 diabetes mellitus, and cardiovascular disease. In addition, children who have received short-term prenatal dexamethasone had poor verbal working memory, poor self-perception of scholastic competence, and increased self-rated social anxiety. Some observers have characterized the prenatal treatment of CAH as an unresolved ethical dilemma, eg, exposure to high-dose dexamethasone in seven of eight fetuses (four out of four males and three out of four females) to treat one affected female.

Prenatal treatment to prevent CAH in patients at risk of a virilized child must be started as soon as the woman knows she is pregnant. To start dexamethasone treatment at 12 weeks of gestation will be ineffective. Genetic diagnosis by chorionic villus sampling or by amniocentesis usually is not done until 10–12 weeks or after 14 weeks of gestation, respectively, and will lead to unnecessary exposure of the pregnancy.

Because the described patient will undergo IVF because of bilateral tubal obstruction, the best method to prevent other severely virilized female infants is to perform preimplantation genetic diagnosis. This prenatal genetic testing of the parents will identify the specific mutations carried by the mother and father. In carrying out IVF with preimplantation genetic analysis of all embryos, the only embryos to be transferred will be embryos not affected by the condition, or male or females who will be known to be carriers at birth but will not suffer from genital or metabolic anomalies.

Barker DJ, Eriksson JG, Forsen T, Osmond C. Fetal origins of adult disease: strength of effects and biological basis. Int J Epidemiol 2002; 31:1235–9.

Feldman-Witchel S, Miller WL. Prenatal treatment of congenital adrenal hyperplasia—not standard of care. J Genet Counsel 2012;21: 615–24.

Fernandez-Balsells MM, Muthusamy K, Smushkin G, Lampropulos JF, Elamin MB, Elnour NOA, et al. Prenatal dexamethasone use for the prevention of virilization in pregnancies at risk for classical congenital adrenal hyperplasia because of 21-hydroxylase (*CYP21A2*) deficiency: a systematic review and meta-analyses. Clin Endocrinol 2010;73:436–44.

Speiser PW, Azziz R, Baskin LS, Ghizzoni L, Hensle TW, Merke DP, et al. Congenital adrenal hyperplasia due to steroid 21-hydroxylase deficiency: an Endocrine Society clinical practice guideline. J Clin Endocrinol Metab 2010;95:4133–60.

46

Macroadenoma

A 23-year-old woman visits your clinic with a 6-month history of headaches, amenorrhea, and bilateral galactorrhea. Laboratory results show a prolactin level of 325 ng/mL and a thyroid-stimulating hormone (TSH) level of 1.25 mU/L. Cranial magnetic resonance imaging (MRI) shows a 13-mm enclosed pituitary adenoma. Formal visual field testing is normal. Additional hormone testing shows normal levels of insulin-like growth factor 1, free thyroxin, and morning and afternoon cortisol. By contrast, follicle-stimulating hormone and luteinizing hormone levels are abnormally low. The most appropriate management for this patient is

    (A) observation and repeat MRI in 3 months
* (B) dopamine agonist
    (C) somatostatin analog
    (D) evaluation by neurosurgeon for transsphenoidal surgery

Macroadenomas are benign pituitary tumors greater than 10 mm in diameter. The morbidity from these tumors includes direct oversecretion of active anterior pituitary hormones, impairment of hormone secretion by the normal pituitary tissue, and mass effect on the surrounding central nervous system structures. Elevated circulating levels of all anterior pituitary hormones have been reported with macroadenomas. Prolactin and growth hormone elevations are the most commonly seen increases, and many anterior pituitary tumors will cosecrete both hormones. With prolactin-secreting macroadenomas, the prolactin level correlates with tumor size and invasiveness. Clinically active adrenocorticotropic hormone, TSH, and gonadotropin tumors are much less common. Tumors that do not produce active hormones have been referred to as nonsecreting, but most of these tumors secrete glycoprotein chains, including free alpha subunit, follicle-stimulating hormone beta subunit, or luteinizing hormone beta subunit. Free alpha subunit may be assayed and used as a tumor marker if present. Compression on the normal pituitary, interference with blood supply, and feedback from secreted hormones may impair normal pituitary function. Mass effect on the optic chiasm may lead to peripheral vision loss. Bitemporal vision loss is the most common vision loss but the degree is typically asymmetrical. Blurred vision is the most common symptom, but it is not unusual for patients not to realize that they have vision loss. Macroadenomas also may invade the cavernous sinus (a collection of thin-walled veins lateral to the sella turcica through which runs the internal carotid artery and cranial nerves III, IV, V, and VI). Symptoms from involvement of the cavernous sinus are unusual but decrease the chance of complete surgical resection.

Pituitary apoplexy is the most serious complication of a macroadenoma, although it is uncommon (approximately 2% of cases). It involves bleeding into the tumor with sudden expansion. Predisposing factors include hypertension, sudden hypotension, closed head injury, anticoagulation therapy, previous irradiation, treatment with a dopamine agonist, and dynamic pituitary function testing. Presenting symptoms include severe headache with nausea and vomiting, visual changes, lethargy, and possible altered consciousness. Unrecognized panhypopituitarism with loss of adrenal function may result in mortality. Because of the risk of pituitary apoplexy, patients with a macroadenoma should wear a medical alert device until treatment has resulted in microadenoma or tumor removal.

Central nervous system symptoms, primarily headache or abnormal anterior pituitary hormone findings, typically lead to the diagnosis of macroadenoma using MRI. The finding of a macroadenoma requires complete evaluation of the anterior pituitary function. Baseline hormones should include prolactin, insulin-like growth factor 1 to evaluate growth hormone function, morning and afternoon cortisol, adrenocorticotropic hormone to evaluate the cortisol axis, TSH level, free thyroxin level for thyroid function, and free alpha subunit as a tumor marker. Abnormally low levels may require dynamic testing for an absolute diagnosis of deficiency. Additionally, abnormally high levels may require suppression testing for verification. All patients with a macroadenoma should have formal visual field testing by ophthalmology.

The treatment of macroadenomas depends on the type of hormone secreted and the degree of mass effect. Although each case has to be individualized, prolactin-secreting macroadenomas (Fig. 46-1) are most commonly treated medically, whereas tumors that secrete other hormones or that are nonsecreting are treated surgically. Prolactin-secreting macroadenomas may be treated initially with a dopamine agonist, such as bromocriptine or cabergoline. Cabergoline may be preferable because of

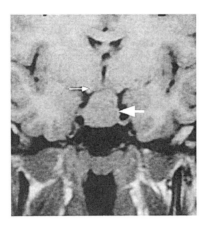
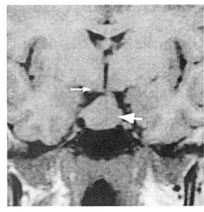

FIG. 46-1. Patient with amenorrhea/galactorrhea due to a prolactin-secreting macroadenoma. Despite compression of the optic chiasm (small arrow), visual field testing was normal. Follow-up magnetic resonance imaging 3 months after initiation of dopamine agonist therapy shows tumor shrinkage away from the chiasm. (Price TM, Bates GW. Adolescent amenorrhea. In: Carpenter SEK, Rock JA, editors. Pediatric and adolescent gynecology. 2nd edition. Philadelphia: Lippincott, Williams & Wilkins; 2000. p. 203.)

its greater potency and generally less pronounced adverse effects of nausea and orthostatic hypotension. Response to a dopamine agonist tends to be rapid. Approximately three quarters of women will return to normal prolactin levels within 3 months. A decrease in prolactin typically indicates a decrease in tumor size, but the degree of decrease does not correlate with the magnitude of tumor shrinkage. A repeat MRI examination is reasonable as early as 3 months after initiation of therapy but no later than 6 months. Either an increase in tumor size or no change in tumor size by 6 months suggests the need for surgery. Many adenomas express somatostatin receptors. Macroadenomas that cosecrete prolactin and growth hormone may have a better response to a dopamine agonist and a somatostatin analog. Discontinuation of treatment with a dopamine agonist also may result in rebound growth of a macroadenoma, so patient education is crucial. For the described patient, the most appropriate management is treatment with a dopamine agonist.

In macroadenomas that are not secreting prolactin, transsphenoidal resection is usually the treatment of choice. The success of complete resection correlates with the expertise of the neurosurgeon and with whether the tumor remains within the sella turcica. Published success rates vary from 6% to 74% for initial complete removal. Expansion of the tumor into the cavernous sinus commonly precludes complete resection. Complications from transsphenoidal resection include vision loss (1.5%), infection (0.5%), vascular injury or stroke (0.6%), cerebrospinal fluid rhinorrhea (3.3%), and oculomotor nerve palsy (0.6%). Persistent hyperprolactinemia after the resection of prolactin-secreting macroadenomas is seen in most cases, requiring continued medical therapy.

Radiation therapy usually is reserved for tumors that cannot be controlled medically or surgically. Some degree of hypopituitarism is commonly seen after radiation therapy.

Patients with a macroadenoma should avoid pregnancy until the tumor has regressed in size to less than 10 mm in diameter. With prolactin-secreting macroadenomas, approximately 39% will increase in size with pregnancy. Thus, continuation of dopamine agonist therapy is reasonable in the event of pregnancy. A change in symptoms warrants reevaluation with MRI and pituitary function testing.

Ikeda H, Watanabe K, Tominaga T, Yoshimoto T. Transsphenoidal microsurgical results of female patients with prolactinomas. Clin Neurol Neurosurg 2013;115:1621–5.

Klibanski A, Zervas NT. Diagnosis and management of hormone-secreting pituitary adenomas. N Engl J Med 1991;324:822–31.

Nawar RN, AbdelMannan D, Selman WR, Arafah BM. Pituitary tumor apoplexy: a review. J Intensive Care Med 2008;23:75–90.

47

Fertility options after tubal ligation

A 28-year-old woman, gravida 3, para 3, is seeking to conceive after tubal sterilization. She underwent laparoscopic tubal occlusion with the use of the titanium clip lined with silicone rubber 3 years ago. She requests tubal reversal in order to conceive. The most important factor in the success of tubal reanastomosis is

    (A) type of sterilization procedure
    (B) location of tubal occlusion
*   (C) patient's age at the time of reanastomosis
    (D) length of tube after reanastomosis

Tubal sterilization is a commonly used method of contraception. After tubal sterilization, patients sometimes desire restoration of their fertility. The most common reason that a patient requests fertility restoration is a change in her marital status. Some women also may encounter regret after tubal sterilization. Women who desire fertility restoration after tubal ligation have the option of in vitro fertilization or microsurgical tubal reanastomosis in order to conceive. After the introduction of microsurgical techniques, the successful outcomes of tubal reanastomosis procedures have improved. Pregnancy rate after a tubal reanastomosis is influenced by many factors, including the patient's age, her prior fertility, the type of sterilization procedure performed, and tubal length after repair of the tubes as well as her partner's semen parameters. These factors should all be considered in the preoperative evaluation of a woman seeking tubal reanastomosis.

Cumulative pregnancy rates 12 months after tubal reanastomosis range between 57% and 84%. However, patients older than 36 years have been shown to have a lower pregnancy rate and to require a longer interval to achieve pregnancy. Nonetheless, women between ages 40 years and 45 years have a cumulative pregnancy rate of approximately 45–50%. The data show that tubal reanastomosis remains a viable option for women after age 40 years. In assessing what is the most important influence on the success of tubal reanastomosis, the patient's age appears to play the most major role. Type of sterilization (eg, tubal ligation versus titanium clip), location of the tubal occlusion (isthmic versus isthmic–ampullary), and length of the tube affect pregnancy rates after tubal reanastomosis. However, their role is not as significant as the patient's age at the time of tubal reanastomosis.

Gordts S, Campo R, Puttemans P, Gordts S. Clinical factors determining pregnancy outcome after microsurgical tubal reanastomosis. Fertil Steril 2009;92:1198–202.

Sreshthaputra O, Sreshthaputra RA, Vutyavanich T. Factors affecting pregnancy rates after microsurgical reversal of tubal sterilization. J Reconstr Microsurg 2013;29:189–94.

Yoon TK, Sung HR, Kang HG, Cha SH, Lee CN, Cha KY. Laparoscopic tubal anastomosis: fertility outcome in 202 cases. Fertil Steril 1999;72:1121–6.

48
Androgen insensitivity syndrome

A 17-year-old adolescent comes to your office with primary amenorrhea. She has Tanner stage 4 breast development and a blind-ending vagina with no palpable uterus. Her pubic hair is Tanner stage 1 and she has no axillary hair. The most likely diagnosis is

* (A) androgen insensitivity syndrome
* (B) müllerian dysgenesis
* (C) Perrault syndrome
* (D) Swyer syndrome

Androgen insensitivity syndrome (AIS) occurs in patients with a normal male karyotype (46,XY) and is caused by an inactivating mutation in the gene that encodes the intracellular androgen receptor. Anecdotal reports suggest that AIS has been recognized for a number of centuries. Joan of Arc and Queen Elizabeth I of England are thought to have had this syndrome.

Affected individuals have testes that make testosterone and antimüllerian hormone, but the mutation makes the androgen receptor insensitive to androgen action. As a result, male internal and external genitalia fail to undergo normal embryologic development. At the time of puberty, estrogen, derived from peripheral conversion of high circulating testosterone levels, causes feminization because there is no androgen action to oppose it. In the complete form of AIS, the phenotype is female. Incomplete androgen resistance is associated with a variety of clinical presentations that usually involve sexual ambiguity at birth. The biologic action of androgen is mediated by a single-gene–encoded intracellular androgen receptor on the long arm of the X chromosome, Xq. Mutations in this gene result in varying degrees of androgen receptor dysfunction; however, the phenotypes often show poor concordance with the genotype.

The incidence of the syndrome may be as high as 1 in 20,000 individuals. In one series of primary amenorrhea cases, androgen insensitivity was the third most common cause of primary amenorrhea after gonadal dysgenesis and müllerian dysgenesis. Diagnosis at the time of menarche is usually straightforward. The classic presentation involves a phenotypic female who presents with primary amenorrhea, normal breasts, little or no pubic or axillary hair, an absent uterus, a 46,XY karyotype, and serum testosterone levels in the normal male range. The internal genitalia are characterized by the near absence of all structures except testes. The testes may be located in the abdomen, inguinal canal, or labia majora and may present as a hernia. The lack of androgen action results in scant or absent pubic and axillary hair. Most of these patients are slightly taller than their peers and have a short vaginal pouch, which usually can be lengthened either by the use of an intravaginal dilator or by intercourse. Some cases might require surgical creation of a neovagina.

Key issues in this disorder include decisions on sexual assignment, the timing of gonadectomy, and genetic and psychologic counseling. Because the disorder is caused by an X-linked mutation, the patient may have female siblings with the same disorder. In terms of sexual assignment, opinions diverge in regard to when and how a woman with androgen insensitivity should be told about her diagnosis. Most experts recommend that the decision to have such a conversation be individualized based on the age and maturity of the patient. It should be emphasized that in complete AIS, the assignment is female. Psychologic counseling should be recommended.

Gonads carry a risk of gonadoblastoma or invasive cancer. This risk appears to increase with age. More recent data has suggested that the overall risk of malignancy may be lower than was previously believed. Because of the low incidence of tumor development before puberty, gonadectomy should be planned after sexual maturation and breast development are complete. After surgical removal, estrogen therapy is recommended to maintain secondary sexual characteristics.

The differential diagnosis of AIS is relatively uncomplicated. The typical patient with AIS has breasts, normal-to-above-average height, absent pubic and axillary hair, a blind-ending vagina, and possibly an inguinal hernia. The condition that is most often confused with AIS is müllerian dysgenesis (Table 48-1). Patients with müllerian dysgenesis usually exhibit normal axillary and pubic hair, a short vagina and absent uterus, normal female serum testosterone levels, and a normal female karyotype (46,XX).

The described patient does not have Perrault syndrome (46,XX; gonadal dysgenesis and sensorineural deafness) or Swyer syndrome (46,XY; gonadal dysgenesis). In Perrault syndrome and Swyer syndrome, the gonads fail

TABLE 48-1. Comparison of Characteristics of Androgen Insensitivity Syndrome and Müllerian Dysgenesis

	Androgen Insensitivity Syndrome	Müllerian Dysgenesis
Karyotype	46,XY	46,XX
Breasts	++	++
Uterus	−	−
Pubic hair	−	+
Axillary hair	−	+
Testosterone levels	Male	Female

+, present; −, absent.

to develop and the patients do not undergo puberty or develop breasts because of the lack of estrogen action. In addition, patients with either Perrault syndrome or Swyer syndrome would be expected to have a uterus because of the lack of production of antimüllerian hormone.

Cools M, Drop SL, Wolffenbuttel KP, Oosterhuis JW, Looijenga LH. Germ cell tumors in the intersex gonad: old paths, new directions, moving frontiers. Endocr Rev 2006;27:468–84.

Hughes IA, Deeb A. Androgen resistance. Best Pract Res Clin Endocrinol Metab 2006;20:577–98.

Sultan C, Lumbroso S, Paris F, Jeandel C, Terouanne B, Belon C, et al. Disorders of androgen action. Semin Reprod Med 2002;20:217–28.

Wisniewski AB, Migeon CJ, Meyer-Bahlburg HF, Gearhart JP, Berkovitz GD, Brown TR, et al. Complete androgen insensitivity syndrome: long-term medical, surgical, and psychosexual outcome. J Clin Endocrinol Metab 2000;85:2664–9.

49

Depot medroxyprogesterone acetate and unscheduled vaginal bleeding

A 22-year-old nulligravid woman began receiving depot medroxyprogesterone acetate (DMPA) injections 2 months ago. She reports daily painless spotting. An office speculum examination reveals a normal ectropion. Bimanual examination is consistent with a small nontender uterus without an adnexal mass. The characteristic that is most likely to be revealed through transvaginal ultrasonography is

* (A) endometrial polyp
* (B) 15-mm endometrial stripe
* (C) adenomyosis
* (D) myoma
* * (E) 4-mm endometrial stripe

Erratic, "unscheduled" vaginal bleeding continues to be the most common adverse effect associated with widespread use of long-acting, progestin-only methods of contraception. As a consequence, it is also the main reason for premature discontinuation of these methods. Progestin contraceptives have the advantage of offering long-term contraception, which improves efficacy and is highly acceptable to many women. During the first months of use, episodes of unpredictable bleeding and spotting lasting 7 days or longer are common. At 1 year, approximately 50% of users experience amenorrhea, increasing to approximately 75% with long-term use.

Research has increased knowledge about the multiple mechanisms that contribute to the appearance of superficial, thin-walled fragile vessels within the endometrium of many women with troublesome bleeding. Progestin exposure affects all cellular elements of the endometrium, and it is likely that unscheduled bleeding arises because of changes in the tightly regulated system of vascular, stromal, and epithelial interaction seen during the normal menstrual cycle. The exact mechanisms are not fully understood, but superficial vascular fragility is a key feature. In addition, changes in endometrial steroid response, structural integrity, tissue perfusion, and local angiogenesis all contribute to unscheduled bleeding.

Long-acting progestin exposure alters endometrial vascular development leading to irregularly distributed, superficially enlarged, fragile capillaries and venules embedded in a collapsed stromal extracellular matrix. On transvaginal ultrasonographic evaluation, the endometrial stripe would appear thin and homogeneous, typically less than 5 mm. It is unlikely that a patient who is taking DMPA would develop an endometrial polyp, adenomyosis, a myoma, or a thickened endometrium.

Mifepristone and other progesterone antagonists or receptor modulators may be effective treatments for abnormal bleeding with progestin-only methods. In new users of DMPA, mifepristone taken once every 2 weeks was found to reduce unscheduled bleeding. Some studies have shown that estrogen supplementation may terminate a current bleeding episode but does not improve long-term bleeding patterns.

D'Arcangues C. Management of vaginal bleeding irregularities induced by progestin-only contraceptives. Hum Reprod 2000;15(suppl 3):24–9.

Hickey M, Fraser IS. Iatrogenic unscheduled (breakthrough) endometrial bleeding. Rev Endocr Metab Disord 2012;13:301–8.

Jain JK, Nicosia AF, Nucatola DL, Lu JJ, Kuo J, Felix JC. Mifepristone for the prevention of breakthrough bleeding in new starters of depo-medroxyprogesterone acetate. Steroids 2003;68:1115–9.

50

Alternative therapies for menopause

A 53-year-old woman underwent a mastectomy for invasive breast cancer without metastasis 2 years ago. She comes to your office to report that she is experiencing severe, frequent hot flushes. She reports that she has more than 14 hot flushes per week and asks about non-estrogen-containing therapy. The most efficacious therapy with the fewest adverse effects for reduction in the number of hot flushes is

* (A) venlafaxine
  (B) gabapentin
  (C) fluoxetine
  (D) clonidine

Vasomotor symptoms are reported to be more severe, distressing, and of greater duration in breast cancer survivors than in control women. This may be because of the patient's age at diagnosis (frequently older than age 50 years), abrupt discontinuation of existing use of hormone therapy, induction of premature menopause secondary to therapy (chemotherapy with or without oophorectomy), or predisposition to estrogen deficiency due to therapy with tamoxifen citrate or aromatase inhibitors. To prevent hormonal stimulation of potential residual breast cancer cells, non-estrogen-containing therapy often is prescribed to alleviate the number of hot flushes that occur in breast cancer survivors. Various antidepressant medications from the selective serotonin reuptake inhibitor (SSRI) and selective noradrenergic reuptake inhibitor classes have been found to be effective for the treatment of vasomotor symptoms. These medications include the SSRI fluoxetine and the selective noradrenergic reuptake inhibitor venlafaxine. Gabapentin, a gamma-aminobutyric acid analog, also has been found to be effective against hot flushes. Clonidine is a centrally acting α_2 adrenergic agonist used to treat hypertension.

In a multicenter, randomized, crossover trial of venlafaxine versus gabapentin for the treatment of hot flushes in breast cancer survivors, both were found to be effective in reducing vasomotor symptoms (66% reduction), but patients significantly preferred venlafaxine (68%) over gabapentin (32%), ($P=.01$). The study was an open-label, 4-week treatment group-sequential trial of venlafaxine. A total of 56 women used either venlafaxine or gabapentin initially and then were switched to the other medication. Participants had baseline levels of at least 14 bothersome hot flushes per week in the month before study enrollment. During the study, women kept a diary to record the incidence, duration, and severity of their hot flushes. While the study was in progress, patients were not allowed to use any other therapy for the alleviation of hot flushes. Although patients significantly preferred venlafaxine to gabapentin, both medications were associated with adverse effects. Treatment with venlafaxine was associated with greater nausea ($P=.02$), appetite loss ($P=.003$), and constipation ($P=.05$) and fewer negative mood changes ($P=.01$) than gabapentin. Compared with venlafaxine, gabapentin was associated with more dizziness ($P=.005$) and increased appetite ($P<.001$).

In two studies to compare the use of clonidine with placebo in breast cancer patients who took tamoxifen, neither study achieved statistical significance for the reduction of hot flush frequency or severity. The clonidine group reported an increase in sleep problems compared with the placebo group. In breast cancer survivors, the SSRI fluoxetine has been effective for vasomotor symptoms, but the use of some SSRI and selective noradrenergic reuptake inhibitor medications may inhibit the cytochrome P450 2D6 isoenzymes (CYP2DG) that are important in converting tamoxifen to its active metabolite, endoxifen. The SSRIs fluoxetine, paroxetine, and bupropion have been shown to adversely affect the circulating endoxifen concentration level. The selective noradrenergic reuptake inhibitor venlafaxine appears to have less effect on endoxifen levels and appears to be safer than the SSRI medications.

The described patient should be counseled that the leading patient-preferred non-estrogen-containing therapy for the reduction in the number of hot flushes is venlafaxine. In 2013, low-dose paroxetine was approved by the U.S. Food and Drug Administration for the treatment of hot flushes. To date, few comparative studies of the efficacy of low-dose paroxetine for the treatment of hot flushes have been reported.

Bordeleau L, Pritchard KI, Loprinzi CL, Ennis M, Jugovic O, Warr D, et al. Multicenter, randomized, cross-over clinical trial of venlafaxine versus gabapentin for the management of hot flashes in breast cancer survivors. J Clin Oncol 2010;28:5147–52.

Hall E, Frey BN, Soares CN. Non-hormonal treatment strategies for vasomotor symptoms: a critical review. Drugs 2011;71:287–304.

Kelly CM, Juurlink DN, Gomes T, Duong-Hua M, Pritchard KI, Austin PC, et al. Selective serotonin reuptake inhibitors and breast cancer mortality in women receiving tamoxifen: a population based cohort study. BMJ 2010;340:c693.

Morrow PK, Mattair DN, Hortobagyi GN. Hot flashes: a review of pathophysiology and treatment modalities. Oncologist 2011;16: 1658–64.

Portman DJ, Kaunitz AM, Kazempour K, Mekonnen H, Bhaskar S, Lippman J. Effects of low-dose paroxetine 7.5 mg on weight and sexual function during treatment of vasomotor symptoms associated with menopause. Menopause 2014 Feb 17. [Epub ahead of print]

Simon JA, Portman DJ, Kaunitz AM, Mekonnen H, Kazempour K, Bhaskar S, et al. Low-dose paroxetine 7.5 mg for menopausal vasomotor symptoms: two randomized controlled trials. Menopause 2013;20:1027–35.

Tan O, Pinto A, Carr BR. Hormonal and non-hormonal management of vasomotor symptoms: a narrated review. J Endocrinol Diabetes Obes 2013;1(2):1009–24.

51

Obesity and pregnancy

A 27-year-old, gravida 2, para 0, comes to your office to discuss her recent pregnancy losses. Within the past year, she has had two first-trimester pregnancy losses (under 10 weeks) while taking clomiphene citrate. She has a body mass index (BMI) of 36 (calculated as weight in kilograms divided by height in meters squared) and has irregular menstrual cycles. Her thyroid-stimulating hormone and prolactin levels are normal, as is transvaginal ultrasonography. She shows no evidence of diabetes mellitus on a 2-hour glucose tolerance test. She has no clinical signs of hyperandrogenemia. In addition to initiating a recurrent pregnancy loss evaluation, the best next step is to

* (A) initiate a nutritional and weight-loss program
* (B) use luteal vaginal progesterone support
* (C) obtain a thrombophilia panel
* (D) start metformin hydrochloride with her next menses
* (E) prescribe low-dose aspirin during her next pregnancy

Increasing data indicate that obesity may increase the risk of miscarriage. A greater risk of miscarriage occurs in obese women after natural conception, ovulation induction, and assisted reproduction. The mechanism for early miscarriage in the obese patient has not been fully elucidated. Obesity in pregnancy is a risk for the woman and her fetus. Maternal complications include mortality, gestational diabetes mellitus, preeclampsia, hypertension, cesarean delivery, thromboembolism, postpartum hemorrhage, and anesthesia complications. Fetal complications include miscarriage, stillbirth, congenital malformations, and fetal growth abnormalities. These complications may be preventable with preconception lifestyle modifications. Many reproductive centers around the world are implementing BMI guidelines that postpone care until a desired BMI is met. A team approach (nutritionist, nurse, physician) is needed to help support and motivate these patients to lose weight and become healthier. Often the goal of family building can motivate an obese patient to adopt a healthier lifestyle. With weight loss, menstrual irregularities often revert to regular ovulatory cycles. In addition to initiating a recurrent pregnancy loss evaluation, this patient should join a nutritional weight-loss program.

The corpus luteum provides progesterone, which induces a secretory maturation of the endometrium in preparation for implantation. A normal luteal phase is 14 days, whereas a short luteal phase is less than 10 days. Abnormal thyroid-stimulating hormone and prolactin levels can be associated with a short luteal phase. In this patient with oligo-ovulation, ovulation induction with clomiphene will correct a luteal phase deficiency by increasing follicular phase follicle-stimulating hormone levels, which in turn causes multiple ovulation. Increased pituitary gonadotropin release induces follicular development and produces sufficient progesterone levels from the corpus luteum for adequate endometrial maturation.

Ordering a thrombophilia panel is not indicated for the described patient. She had two miscarriages before 10 weeks of gestation, which is before placental function is present. The vast majority of first-trimester losses are due to chromosomal anomalies. A thrombophilia panel is indicated when losses occur after 10 weeks, when thrombosis can occur in the vasculature of the placenta.

Metformin hydrochloride is an insulin sensitizer used in the treatment of impaired glucose tolerance and type 2 diabetes mellitus. Obese women with or without polycystic ovary syndrome (PCOS) are often insulin resistant, resulting in an increased risk of developing impaired glucose tolerance and type 2 diabetes mellitus. When PCOS is suspected (menstrual irregularities, hyperandrogenism, polycystic-appearing ovaries), a 2-hour oral glucose tolerance test is recommended. Although a hemoglobin A_{1c} test may be a potential substitute for the 2-hour glucose test, to date, the efficacy of the hemoglobin A_{1c} test in patients with PCOS has not been completely proven.

Metformin is indicated if impaired glucose tolerance or type 2 diabetes mellitus is diagnosed. Metformin can decrease the progression of impaired glucose tolerance to diabetes mellitus. This patient has irregular menstrual cycles but is not clinically hyperandrogenic and does not have polycystic-appearing ovaries. The next step in her management would not be to initiate metformin unless she has impaired glucose tolerance or type 2 diabetes mellitus.

Pregnancy is a hypercoagulable state. Aspirin, an anticoagulant, has been used for women with recurrent pregnancy loss to try to increase the subsequent live-birth rate in women with inherited thrombophilias, antiphospholipid syndrome, and unexplained recurrent pregnancy loss. There are no data to suggest that empiric aspirin therapy decreases pregnancy loss associated with obesity. Therefore, empiric aspirin use in this patient is not indicated.

de Jong PG, Goddijn M, Middeldorp S. Testing for inherited thrombophilia in recurrent miscarriage. Semin Reprod Med 2011;29:540–7.

Kaandorp S, Di Nisio M, Goddijn M, Middeldorp S. Aspirin or anticoagulants for treating recurrent miscarriage in women without antiphospholipid syndrome. Cochrane Database of Systematic Reviews 2009, Issue 1. Art. No.: CD004734. DOI: 10.1002/14651858.CD004734.pub3.

Obesity in pregnancy. Committee Opinion No. 549. American College of Obstetricians and Gynecologists. Obstet Gynecol 2013; 121:213–7.

52

Bulimia nervosa and binge-eating disorder

A 24-year-old woman comes to your office for her annual well-woman examination. When reviewing her medical history, she tells you that she has experienced several months of bulimia nervosa for which she has not sought treatment. You discuss nutritional counseling. She has a body mass index of 22 (calculated as weight in kilograms divided by height in meters squared) and her physical examination is normal. The first-line treatment that you recommend is

    (A) supportive psychotherapy
    (B) behavioral therapy
    (C) interpersonal psychotherapy
* (D) cognitive–behavioral therapy

Bulimia nervosa is characterized by recurrent episodes of eating unusually large amounts of food accompanied by feelings of loss of control and recurrent inappropriate compensatory behaviors in order to prevent weight gain. With bulimia nervosa, binge eating and inappropriate compensatory behaviors occur on average at least once a week for 3 months. The disturbance does not occur exclusively during episodes of anorexia nervosa. Compensatory behaviors to prevent weight gain include self-induced vomiting; misuse of laxatives, diuretics, and enemas; fasting; or excessive exercise. The individual's self-esteem is typically strongly influenced by body shape and weight.

By contrast, with binge-eating disorder, the binge eating does not occur during the course of bulimia nervosa or anorexia nervosa and is not associated with the recurrent use of inappropriate compensatory behavior as observed with bulimia nervosa. Binge-eating disorder may occur in patients with normal weight but often occurs in conjunction with obesity.

Bulimia nervosa and binge-eating disorder are associated with a number of medical complications. Patients often hide their eating disorder from their physician and may present with other nonspecific complaints, such as fatigue, menstrual irregularity, abdominal pain and bloating, or constipation. Physical examination findings that may suggest underlying bulimia nervosa include tachycardia, parotid gland swelling, hypotension, dry skin, and dental enamel erosion. Associated medical complications may involve multiple organ systems, specifically gastrointestinal, cardiovascular, renal, musculoskeletal, and endocrine disorders. Severe complications, such as electrolyte abnormalities, esophageal rupture, or cardiac arrhythmias, are rare but can be life threatening.

Several approaches have been proposed in regard to the long-term outpatient treatment of eating disorders. Nutritional counseling and rehabilitation are important but are unlikely to be effective alone. The goal of nutritional rehabilitation is to restore a consistent eating pattern that typically consists of three meals and two snacks a day. A Cochrane review including more than 100 trials supports the efficacy of cognitive–behavioral therapy as a first-line therapy in the treatment of individuals with bulimia nervosa. The therapy focuses on modifications of the specific behaviors and ways of thinking that contribute to the patient's eating disorder. The cognitive aspect of therapy focuses on modifying the dysfunctional thoughts and attitudes associated with bulimia nervosa. The behavioral component strives to improve the patient's behavioral responses to their dysfunctional thoughts. Typically, approximately 20 individual sessions over 4–5 months are necessary.

Cognitive–behavioral therapy has demonstrated low relapse rates and appears to be superior to other forms of psychotherapy. Remission rates with cognitive–behavioral therapy are significantly greater than with interpersonal psychotherapy, behavioral therapy, hypnosis plus behavioral therapy, or supportive psychotherapy. One third to one half of patients who undergo cognitive–behavioral therapy will make a complete and lasting recovery. The remaining patients exhibit responses that range from greatly improved to no improvement. For patients with bulimia nervosa who are overweight or obese, cognitive–behavioral therapy is likely to help reduce binge eating and purging but may not result in weight loss. Patients with personality disorders, particularly borderline personality disorder, may be less responsive to cognitive–behavioral therapy or to therapy in general.

When cognitive–behavioral therapy is not sufficient, antidepressant drugs, particularly selective serotonin reuptake inhibitors, are effective in treating bulimia nervosa. Higher doses than those used to treat major depression are recommended. These medications typically result in a dramatic reduction in the frequency of binge eating and purging as well as an improvement in mood. Bupropion hydrochloride is contraindicated in the treatment of bulimia nervosa because there is an increased seizure risk in patients who are actively purging.

American Psychiatric Association. Diagnostic and statistical manual of mental disorders: DSM-5. 5th ed. Arlington (VA): American Psychiatric Association; 2013.

Hay PPJ, Bacaltchuk J, Stefano S, Kashyap P. Psychological treatments for bulimia nervosa and binging. Cochrane Database of Systematic Reviews 2009, Issue 4. Art. No.: CD000562. DOI: 10.1002/14651858.CD000562.pub3.

Mehler PS. Clinical practice. Bulimia nervosa. N Engl J Med 2003; 349:875–81.

Treasure J, Claudino AM, Zucker N. Eating disorders. Lancet 2010; 375:583–93.

53
Postpartum thyroiditis

A 24-year-old woman, gravida 1, para 1, comes to your clinic for a follow-up visit at 6 months postpartum. She has nothing to report, but her thyroid is noted to be symmetrically enlarged without nodules. Her pulse is 90 beats per minute, her thyroid-stimulating hormone (TSH) level is low, and her free thyroxine level is 2.4 ng/dL. The best treatment option is

    (A) a beta blocker
    (B) propylthiouracil
    (C) radioactive iodine
    (D) methimazole
*   (E) observation

Thyroid disease is a common clinical endocrine disorder in women of reproductive age. Hypothyroidism and hyperthyroidism can occur during pregnancy and are associated with maternal and fetal morbidity and mortality if undiagnosed or untreated.

Postpartum thyroiditis is defined as a transient autoimmune thyroid disorder that may occur in 5–10% of all women during the first year after childbirth. It occurs as thyrotoxicosis, hypothyroidism, or thyrotoxicosis followed by hypothyroidism in the first year postpartum in women who demonstrated no clinical evidence of thyroid disease before pregnancy. It appears that an autoimmune-induced release of preformed hormone occurs from the thyroid. Postpartum thyroiditis almost exclusively affects thyroid antibody-positive women. Among women who are euthyroid in the first trimester but who test positive for thyroid autoantibodies, approximately 50% will develop postpartum thyroiditis. Consequently, women who are known to be positive for thyroid peroxidase antibodies should have their TSH level measured at 6 months postpartum or sooner if clinically indicated. Postpartum thyroiditis has an approximately 70% risk of recurrence and also can occur after a pregnancy loss.

Postpartum thyroiditis may be related to a change in the immune system in postpartum women. In general, the syndrome initially presents as transient hyperthyroidism within 13 weeks after delivery followed by transient hypothyroidism. This can be associated with thyroid enlargement. If the thyroid is biopsied, a lymphocytic infiltration may be observed. Some of the symptoms that may occur after delivery are vague and nonspecific, such as depression and memory impairment. Such symptoms can make it difficult to differentiate postpartum thyroiditis from postpartum depression or the postpartum blues. Primary risk factors other than thyroid antibodies include a history of thyroid dysfunction or a family history of thyroid disease or autoimmune disease. Up to 25% of women with type 1 diabetes mellitus develop postpartum thyroiditis. Similarly, the prevalence of postpartum thyroiditis also is increased in women with Graves disease in remission and in women with chronic viral hepatitis. Consequently, the TSH level should be checked at 3 months and 6 months postpartum in patients with any of these conditions.

Transient hyperthyroidism results from excessive release of thyroid hormone from the glandular disruption caused by the antibodies. Symptoms may include a painless but enlarged goiter, palpitations, or fatigue. Approximately 30% of patients with postpartum thyroiditis are asymptomatic during the thyrotoxic phase. This phase usually occurs between 1 month and 6 months postpartum and generally resolves after 1–2 months. The hypothyroid phase usually occurs between 3 months and 8 months postpartum and typically lasts 4–6 months.

The best treatment option for the described patient is observation. Antithyroid drugs would not be indicated or required. Patients that develop significant hypothyroid symptoms during the hypothyroid transient period may benefit from low-dose levothyroxine. Asymptomatic women with postpartum thyroiditis who have an elevated TSH level that is less than 10 mIU/L and who are not immediately planning another pregnancy do not necessarily require intervention but should be remonitored in 4–8 weeks. If the patient's TSH remains elevated, treatment should be initiated with levothyroxine. Because the disorder may last up to 1 year, levothyroxine can be continued and then tapered off. Patients may redevelop hypothyroidism later in life and yearly assessment of the TSH level is recommended. Beta blockers, propylthiouracil, radioactive iodine, and methimazole would be of use only if the patient exhibits signs and symptoms of overt hyperthyroidism.

De Groot L, Abalovich M, Alexander EK, Amino N, Barbour L, Cobin RH, et al. Management of thyroid dysfunction during pregnancy and postpartum: an Endocrine Society clinical practice guideline. J Clin Endocrinol Metab 2012;97:2543–65.

Lucas A, Pizarro E, Granada ML, Salinas I, Foz M, Sanmarti A. Postpartum thyroiditis: epidemiology and clinical evolution in a nonselected population. Thyroid 2000;10:71–7.

Stagnaro-Green A, Pearce E. Thyroid disorders in pregnancy. Nat Rev Endocrinol 2012;8:650–8.

Thyroid disease in pregnancy. ACOG Practice Bulletin No. 37. American College of Obstetricians and Gynecologists. Obstet Gynecol 2002;100:387–96.

54

Anorexia and nutrition

A 28-year-old nulligravid woman with oligo-ovulation comes to your office with a 2-year history of primary infertility. Her body mass index is 19 (calculated as weight in kilograms divided by height in meters squared). Hysterosalpingography, thyroid-stimulating hormone level, and partner semen analysis are normal. She is a compulsive exerciser. Her physical examination reveals dry skin with normal blood pressure. The most appropriate next step in the management of this patient is

(A) clomiphene citrate and insemination
(B) a trial of antidepressant medication
* (C) to administer a screening questionnaire for eating disorders
(D) to prescribe prenatal vitamins

Eating disorders affect approximately 5–7% of women of childbearing age. The three categories of eating disorders are 1) anorexia nervosa, 2) bulimia nervosa, and 3) eating disorders not otherwise specified. The vast majority of patients with eating disorders do not disclose their disorder to their physician. One study noted that 58% of infertility patients with menstrual irregularities had clinical indicators for eating disorders, but none of these patients disclosed them to their health care provider.

Anorexic women have a distorted body image and refuse to maintain a normal body weight for their height. They have a fear of gaining weight even though they are underweight and are often amenorrheic. Anorexic patients binge and purge or restrict their food intake. Commonly, anorexic patients use compulsive exercise to lose weight or to maintain body weight. The common thread among all eating disorders is the continual drive to be thin. Treatment of anorexia hinges on a team that includes a nutritionist, a mental health professional, and a clinical coordinator. Weight gain with a decrease in exercise is the ultimate goal in this patient. The goal can be accomplished with the use of a team approach.

Eating disorders often manifest as secondary amenorrhea. This may be the clinician's first clue to a possible eating problem. The described woman is oligomenorrheic and is thin with a body mass index of 19. In addition, she is a self-proclaimed intense exerciser. Because her cycles are irregular, she is having difficulty timing intercourse and becoming pregnant. The SCOFF questionnaire is a simple set of five yes or no questions (Box 54-1). A score of 2 or more indicates a likely case of anorexia nervosa or bulimia nervosa. This patient has initiated an evaluation, and the goal of motherhood is often a motivating force for patients to initiate treatment of these disorders. Once she is in remission, her cycles may normalize as her hypothalamic–pituitary–ovarian axis is released from inhibition. She can use timed intercourse or an ovulation predictor kit to help her become pregnant. Attempting pregnancy before remission increases the risk of maternal malnutrition and

---

BOX 54-1

The SCOFF Questions*

- Do you make yourself **S**ick because you feel uncomfortably full?
- Do you worry you have lost **C**ontrol over how much you eat?
- Have you recently lost **O**ver 14 lb (6.4 kg) in a 3-month period?
- Do you believe yourself to be **F**at when others say you are too thin?
- Would you say that **F**ood dominates your life?

*One point for every "yes"; a score of 2 or more indicates a likely case of anorexia nervosa or bulimia.

Morgan JF, Reid F, Lacey JH. The SCOFF questionnaire: assessment of a new screening tool for eating disorders. BMJ 1999;319:1467–8. Copyright 1999 BMJ Publishing Group Ltd., reproduced with permission.

giving birth to a newborn who is small for gestational age.

Clomiphene citrate should not be used as a first-line agent in an anovulatory patient with an eating disorder. Efforts should be made to correct the patient's eating disorder before ovulation is initiated. Without weight gain, clomiphene is only occasionally effective. Anorexic patients often have hypothalamic amenorrhea and should be treated accordingly.

A trial of antidepressant medication, such as a selective serotonin reuptake inhibitor, is reserved for the patient who, in addition to anorexia, has a psychologic comorbidity. The anorexic patient should be screened for depression and anxiety. If depression or anxiety is detected, the problem should be treated in conjunction with the eating disorder. The risks and benefits of psychiatric medications in pregnancy need to be discussed with the patient because there may be fetal risks. Selective serotonin reuptake inhibitors are part of the management for patients with bulimia nervosa but not for patients with anorexia.

The patient with oligo-ovulation needs to be screened for thyroid disease. Checking a thyroid-stimulating hormone level in this patient is part of her evaluation for her eating disorder and oligo-ovulation. Thyroid disease needs to be corrected before attempting pregnancy because hypothyroidism and hyperthyroidism can affect maternal and fetal health.

All women of childbearing age should take preconceptional folic acid. In addition to a balanced diet, women need extra iron and folic acid during pregnancy. To reduce the risk of neural tube defects, a minimum of 0.4 mg of folic acid should be taken daily before pregnancy and then through the first 12 weeks of gestation.

Andersen AE, Ryan GL. Eating disorders in the obstetric and gynecologic patient population. Obstet Gynecol 2009;114:1353–67.

Fritz MA, Speroff L. Amenorrhea. In: Fritz MA, Speroff L, editors. Clinical gynecologic and infertility. 8th ed. Philadelphia (PA): Lippincott Williams & Wilkins; 2011. p. 486–9.

Hoffman ER, Zerwas SC, Bulik CM. Reproductive issues in anorexia nervosa. Expert Rev Obstet Gynecol 2011;6:403–14.

55

Contraception for an older patient who smokes

A 37-year-old patient continues to smoke cigarettes despite your advice to quit. You counsel her that the contraceptive method that presents the highest risk of morbidity and mortality for her is

    (A) progestin-only oral contraceptives (OCs)
    (B) depot medroxyprogesterone acetate injections
    (C) progestin implant
    (D) copper-containing intrauterine device (IUD)
*   (E) combination hormonal contraceptives

Thromboembolic events are classified into two types: 1) venous thromboembolism and 2) arterial thrombosis. Venous thromboembolism is characterized by a low-flow circulatory system, a high circulating fibrinogen level, and a low platelet count. Clinically, venous thromboembolism manifests as either a deep vein thrombosis or a pulmonary embolism. Arterial thrombosis is characterized by a high-flow circulatory system, a low circulating fibrinogen level, and a high platelet count. Clinically, arterial thrombosis manifests as either a myocardial infarction (MI) or a stroke.

Because of their pharmacologic levels of estrogen, estrogen-containing contraceptive methods (pill, patch, or ring) increase the production of clotting factors, which in turn increases the risk of thrombotic events. The route of estrogen administration does not affect this increased risk of thrombosis. When compared with pregnancy, the use of an estrogen-containing contraceptive method is less likely to lead to a nonfatal venous thromboembolism event.

Women who have risk factors for hypercoagulability or who are 35 years or older and smoke cigarettes are at increased risk of thromboembolic events and should not use estrogen-containing contraceptive methods. The type of progestin also can influence venous thromboembolism risk. In a recent case–control study of 186 newly diagnosed idiopathic cases compared with 681 controls, the incidence rates for venous thromboembolism were 30.8 (95% confidence interval [CI], 25.6–36.8) per 100,000 woman-years among users of OCs containing drospirenone and 12.5 (95% CI, 9.61–15.9) per 100,000 woman-years among users of OCs containing levonorgestrel. Use of OCs increases risk of venous thromboembolism over baseline; the age-adjusted incidence ratio for a venous thromboembolism event with a drospirenone-containing OC compared with a levonorgestrel-containing OC is 2.8 (95% CI, 2.1–3.8). Prior studies involving the contraceptive patch also have shown a twofold to threefold increased risk of venous thromboembolism compared with users of levonorgestrel-containing contraceptives.

In OC users who smoke, the increased thrombotic risk exists because of a nicotine-mediated decrease in circulating prostacyclin and an increase in thromboxane levels, leading to increased platelet aggregation. In OC users, procoagulatory increases in fibrinogen and fibrinopeptide A levels are balanced by a compensatory increase in antithrombin III levels. However, in an OC user who smokes cigarettes, there is no corresponding increase in antithrombin III levels, which predisposes this patient to an increased hypercoagulable state.

When examining the risk of arterial thrombosis, women who are older than 35 years and smoke cigarettes should not use combination OCs. In a 15-year Danish historical cohort study, the absolute risk of thrombotic stroke and MI was increased by a factor of 0.9–1.7 with 20 micrograms of ethinyl estradiol OCs and by a factor of 1.3–2.3 with 30–40 micrograms of ethinyl estradiol OCs, with relatively small differences in risk according to the type of associated progestin used. In contrast, none of the women who used progestin-only contraceptive methods (progestin-only OCs, progestin-containing IUD, or progestin implant) had a significant increased risk of a thrombotic stroke or MI.

Women older than 35 years who smoke should be counseled to use a non-estrogen-containing contraceptive because of the increased risk of a thromboembolic event with the use of estrogen-containing contraceptive methods. Acceptable contraceptive methods include the progestin-only OC, the depot medroxyprogesterone acetate injection, the progestin implant, and either the progestin-containing IUD or the copper-containing IUD.

Jick SS, Hernandez RK. Risk of non-fatal venous thromboembolism in women using oral contraceptives containing drospirenone compared with women using oral contraceptives containing levonorgestrel: case-control study using United States claims data. BMJ 2011;342:d2151.

Lidegaard O, Lokkegaard E, Jensen A, Skovlund CW, Keiding N. Thrombotic stroke and myocardial infarction with hormonal contraception. N Engl J Med 2012;336:2257–66.

56

Leiomyomas and heavy menstrual bleeding

A 30-year-old nulligravid woman is trying to conceive. She has a history of heavy menstrual bleeding. Pelvic ultrasonography and endometrial biopsy fail to reveal any anatomic abnormality. The medical treatment most likely to improve her heavy menstrual bleeding is

   (A) depot gonadotropin-releasing hormone (GnRH) analog
   (B) mefenamic acid
   (C) norethindrone
 * (D) tranexamic acid
   (E) medroxyprogesterone acetate

The U.S. Food and Drug Administration recently approved an oral formulation of tranexamic acid for clinical use in the United States. The pharmacologic mechanism of tranexamic acid involves reversibly blocking lysine-binding sites on plasminogen, which in turn stops the interaction of plasmin and fibrin polymer. This prevents fibrin degradation and leads to stabilization of clots and reduced bleeding. Tranexamic acid has been used extensively to reduce blood loss and the need for blood transfusion during and after surgical procedures, such as coronary artery bypass, scoliosis surgery, and knee arthroplasty. Tranexamic acid has been used for the treatment of abnormal uterine bleeding (AUB) outside of the United States for decades. Tranexamic acid is superior to placebo, diclofenac, mefenamic acid, ethamsylate, and luteal phase norethisterone for reduction of blood loss in idiopathic AUB and reduction of blood loss associated with intrauterine devices. A recent study indicates that tranexamic acid appears to reduce heavy bleeding, including bleeding caused by leiomyomas.

Use of a GnRH analog would place the patient in a state of amenorrhea and improve her AUB but would not be advisable given the adverse effects and cost. Therapy with a GnRH analog also would be a temporary measure; it cannot be given for extended periods of time without estrogen add-back therapy.

Misoprostol is a synthetic prostaglandin E_1 analog that is used for the prevention of nonsteroidal antiinflammatory drug (NSAID)-induced gastric ulcers and to help with the induction of labor. Misoprostol is not used for the treatment of AUB.

Norethindrone is a synthetic progestin used in combination oral contraceptives and for the induction of menses. Although it can be used to control anovulatory amenorrhea, for the prevention of endometrial hyperplasia, and for cycle control, it is not as effective as tranexamic acid for the management of AUB.

Mefenamic acid is a NSAID that is used to treat pain, including dysmenorrhea. It decreases uterine contractions by an unknown mechanism. It has been shown that NSAIDs are effective in reducing blood loss compared with placebo in cases of heavy menstrual bleeding. A recent meta-analysis, however, has shown such drugs to be less effective than tranexamic acid. Medroxyprogesterone acetate is also less effective than tranexamic acid.

Lethaby A, Duckitt K, Farquhar C. Non-steroidal anti-inflammatory drugs for heavy menstrual bleeding. Cochrane Database of Systematic Reviews 2013, Issue 1. Art. No.: CD000400. DOI: 10.1002/14651858.CD000400.pub3.

Lukes AS, Moore KA, Muse KN, Gersten JK, Hecht BR, Edlund M, et al. Tranexamic acid treatment for heavy menstrual bleeding: a randomized controlled trial. Obstet Gynecol 2010;116:865–75.

Munro MG. Uterine leiomyomas, current concepts: pathogenesis, impact on reproductive health, and medical, procedural, and surgical management. Obstet Gynecol Clin North Am 2011;38:703–31.

Naoulou B, Tsai MC. Efficacy of tranexamic acid in the treatment of idiopathic and non-functional heavy menstrual bleeding: a systematic review. Acta Obstet Gynecol Scand 2012;91:529–37.

57
Absolute risk versus relative risk

A 23-year-old healthy nonsmoking nulligravid woman has been taking combined oral contraceptives (OCs) for the past 2 years with no problems. However, she became concerned after her friend informed her that individuals who use OCs have a risk of venous thromboembolism twice as high as those who do not use OCs. She has no personal or family history of venous thromboembolism. You inform her that her risk of venous thromboembolism is higher while she is taking OCs compared with women who do not use OCs, although her actual risk while she is taking OCs is only 0.0015% per year. This concept is described as

(A) negative predictive value
(B) positive predictive value
(C) sensitivity versus specificity
(D) type 1 error
* (E) relative risk versus absolute risk

Women who use estrogen-containing OCs have increased risks of venous thromboembolism compared with nonusers. Although the risk of venous thromboembolism is lower with currently available OCs, which contain a much lower concentration of estrogen, the risk is not eliminated. Certain factors increase the risk. For example, obese women, older women, and women who smoke have a significantly increased risk of venous thromboembolism if they use OCs. Progestin-only pills appear not to increase the risk of venous thromboembolism.

It is important for health care providers to have meaningful conversations with patients to review the risks and benefits of OC use. Absolute risk is a number or rate, whereas relative risk (also known as risk ratio) is a ratio. Relative risk describes the risk compared with a standard. Commonly, cohort studies will present data demonstrating the risk of disease in exposed individuals versus unexposed individuals. The risk for unexposed persons is generally set as equal to one and the experimental group risk is expressed as a fraction below or above that value. Absolute risk or incidence is the number of cases of disease per population size per unit of time. In this case, the relative risk of venous thromboembolism in young women who use low-dose OCs is 2.0 compared with the risk in young women who do not use OCs. However, the absolute risk of venous thromboembolism for nonusers and users is low: 5–10 per 10,000 women per year versus 10–20 per 10,000 women per year, respectively. Because venous thromboembolism is a comparatively rare event in young women, discussions with such patients about OC use should include discussion of the absolute and relative risks of venous thromboembolism while taking OCs to permit them to make educated decisions (Table 57-1).

The terms positive predictive value (PPV) and negative predictive value (NPV) are used when evaluating how well a certain test identifies patients who do and do not have a disease. Specifically, PPV refers to the likelihood that a person who has a positive test actually has the disease in question and NPV refers to the likelihood that a person who has a negative test actually does not have the disease. The predictive values (negative and positive) depend on the prevalence of the disease in the population in question. However, PPV and NPV are terms reserved

TABLE 57-1. Relative Risk and Actual Incidence of Venous Thromboembolism

Population	Relative Risk	Incidence (per 10,000 Women per Year)
Young women, general population	1	5–10
Pregnant women	12	60–120
High-dose OCs	6–10	30–100
Low-dose OCs	2	10–20
Factor V Leiden mutation carrier	6–8	30–80
Factor V Leiden mutation carrier and OCs	10–15	50–100
Leiden mutation, homozygous	80	400–800

Abbreviation: OCs, oral contraceptives.

Speroff L, Fritz MA. Oral contraception. In: Speroff L, Fritz MA, editors. Clinical gynecologic endocrinology and infertility. 8th ed. Philadelphia (PA): Lippincott Williams and Wilkins; 2011. p. 981.

for evaluation of a test's utility and are not applicable in the described case.

Similarly, the terms sensitivity and specificity refer to measures of the performance of a diagnostic test. Sensitivity is the number of patients with a positive test who have a disease divided by all the patients who have the disease. Specificity is the number of patients who have a negative test and who do not have the disease divided by the total number of patients without the disease. The interdependence of sensitivity and specificity often are shown using a receiver operating characteristic curve. The concepts of sensitivity and specificity are related to test properties and are not related to risk of disease from a specific exposure.

Type 1 error is a term used when making assumptions about the validity of data. There is a presumed truth about a situation (such as the true relation between OC use and venous thromboembolism), and authors of studies attempt to gather data that will approximate this truth. When testing a hypothesis, type 1 error (also called alpha error) refers to the probability of incorrectly concluding that a statistically significant difference exists between the groups, and type 2 error refers to the probability of incorrectly assuming that there was no difference between groups in a study. These terms are reserved for considerations about the validity of hypothesis testing and conclusions when interpreting study results.

Heit JA, O'Fallon WM, Petterson TM, Lohse CM, Silverstein MD, Mohr DN, et al. Relative impact of risk factors for deep vein thrombosis and pulmonary embolism: a population-based study. Arch Intern Med 2002;162: 1245–8.

Lidegaard O, Nielsen LH, Skovlund CW, Skjeldestad FE, Lokkegaard E. Risk of venous thromboembolism from use of oral contraceptives containing different progestogens and oestrogen doses: Danish cohort study, 2001–9. BMJ 2011;343:d6423.

Speroff L, Fritz MA. Oral contraception. In: Speroff L, Fritz MA, editors. Clinical gynecologic endocrinology and infertility. 8th ed. Philadelphia (PA): Lippincott Williams and Wilkins; 2011. p. 980–1.

White RH. The epidemiology of venous thromboembolism. Circulation 2003;107:I-4–I-8.

58

Depot medroxyprogesterone acetate and bone loss

A 36-year-old nulligravid woman with a body mass index of 26.5 (calculated as weight in kilograms divided by height in meters squared) comes to your office for her annual well-woman examination. She is sexually active and has been adherent to her depot medroxyprogesterone acetate (DMPA) regimen for the past 2 years. She smokes one pack of cigarettes a day. She is concerned about reports that point to the danger of low bone mineral density (BMD) in women who take DMPA. You counsel her that the best strategy to reduce her risk of low BMD is

* (A) smoking cessation
(B) weight loss
(C) discontinuation of DMPA
(D) calcium supplementation
(E) oral bisphosphonates

Cigarette smoking is a risk factor for BMD loss, but the mechanisms are not completely understood. Less efficient calcium absorption in smokers compared with nonsmokers has been suggested as a possible cause, as have local and systemic toxic effects on bone collagen synthesis and alterations in metabolism of adrenal, cortical, and gonadal hormones. Longitudinal studies indicate that rates of bone loss are approximately 1.5–2 times greater among current smokers compared with individuals who do not smoke. Rates of bone loss, particularly in postmenopausal smokers, are higher when observed over the same period compared with nonsmokers. For the described patient, smoking cessation at her age will limit recidivism and help counteract the increased risk of BMD changes typically seen after menopause.

In recent years, observations of reduced BMD in current DMPA users have led to concerns that DMPA-induced bone loss might lead to osteopenia and increase the long-term risk of fractures, particularly in young women who have not yet attained their peak bone mass and among perimenopausal women who may be starting to lose bone mass. The effect of DMPA on BMD is

59

Ovulation induction in a patient with polycystic ovary syndrome

A 29-year-old nulliparous woman has a history of irregular menstruation every 70–90 days. Her body mass index is 29 (calculated as weight in kilograms divided by height in meters squared). She is hirsute in appearance and has an ovarian volume of 12 cm^3. Her male partner has a normal semen analysis. The ovulation induction medication most likely to result in high-order multiple gestation is

    (A) metformin hydrochloride
* (B) gonadotropin
    (C) clomiphene citrate
    (D) letrozole

Polycystic ovary syndrome (PCOS) is the most common ovulation disorder; it affects approximately 5–10% of women of reproductive age. At a meeting in Rotterdam, Netherlands in 2003, international experts on PCOS reached a consensus regarding diagnosis of the syndrome. The meeting was endorsed by the European Society for Human Reproduction and Embryology and the American Society for Reproductive Medicine, and its proceedings were published. In 2007, in Thessaloniki, Greece, an American Society for Reproductive Medicine and European Society for Human Reproduction and Embryology-Sponsored PCOS Consensus Workshop met to agree on how to manage infertility in patients with PCOS and incorporate concepts from a recent PCOS randomized treatment protocol organized by the Reproductive Medicine Network, which found no value in prescribing metformin alone. No additional benefit was found when metformin was combined with clomiphene compared with clomiphene alone.

Clomiphene remains the treatment of choice for induction of ovulation in anovulatory women with PCOS. It is inexpensive, is taken orally, is patient friendly, has few adverse effects, and requires little ovarian-response monitoring. Abundant clinical data are available in regard to its safety. The mechanism of action is not entirely known, but it is thought to involve the blockade of the negative feedback mechanism in the pituitary gland and hypothalamus that results in increased secretion of follicle-stimulating hormone. The main factors that predict outcome of treatment are obesity, hyperandrogenemia, and age. Ovarian volume and menstrual status are additional factors that help to predict responsiveness to clomiphene.

The starting dose of clomiphene generally should be 50 mg/d for 5 days. Confirmation of ovulation should be obtained, and if the patient is ovulatory, treatment of up to 6 months is appropriate. If the patient is nonovulatory, the dose may be increased up to a maximum of 150 mg/d for 5 days because there is no clear evidence of efficacy at higher doses. Approximately 75–80% of patients with PCOS will ovulate after clomiphene, with a conception rate of up to 22% per cycle.

Insulin-sensitizing agents currently are being used to treat diabetes mellitus and include metformin, a biguanide, and thiazolidinediones, such as pioglitazone and rosiglitazone. There is considerable interest in the use of metformin for ovulation induction given earlier reports that indicated an increased rate of ovulation in patients taking metformin, but this did not translate to significantly higher pregnancy rates. Two randomized controlled trials have indicated that metformin does not increase live-birth rates higher than those observed with clomiphene alone in either obese or normal-weight women with PCOS. The Reproductive Medicine Network trial demonstrated a selective disadvantage to metformin compared with clomiphene and no apparent advantage to adding metformin to clomiphene, except perhaps in women who have a body mass index greater than 35 and in those with clomiphene resistance. In addition, there are insufficient data to document any advantage to the use of thiazolidinediones over metformin.

The ovulation induction medication most likely to result in high-order multiple gestation is gonadotropin. Gonadotropin therapy initiates and maintains follicle growth by increasing follicle-stimulating hormone above a threshold dose for sufficient duration to generate a limited number of developing follicles. Patients with PCOS are prone to excessive multiple-follicle development, which increases multiple births. Therefore, gonadotropin therapy is discouraged as a first-line agent for use in patients with PCOS.

Letrozole is a promising medication to induce ovulation. Recent data have been reassuring with regard to its safety and ability to induce ovulation at a similar or higher rate to clomiphene. However, this medication is not approved in the United States for noncancer use and it is also more expensive than clomiphene.

Bates GW Jr, Propst AM. Polycystic ovarian syndrome management options. Obstet Gynecol Clin North Am 2012;39:495–506.

Consensus on infertility treatment related to polycystic ovary syndrome. Thessaloniki ESHRE/ASRM-Sponsored PCOS Consensus Workshop Group. Fertil Steril 2008;89:505–22.

Legro RS, Barnhart HX, Schlaff WD, Carr BR, Diamond MP, Carson SA, et al. Clomiphene, metformin, or both for infertility in the polycystic ovary syndrome. Cooperative Multicenter Reproductive Medicine Network. N Engl J Med 2007;356:551–66.

60

Congenital adrenal hyperplasia

A 24-year-old nulligravid woman comes to your office for prenatal counseling. Her medical history is significant for congenital adrenal hyperplasia (CAH). The most common form of congenital adrenal hyperplasia is deficiency in

    (A) 17α-hydroxylase
*   (B) 21-hydroxylase
    (C) 11β-hydroxylase
    (D) 3β-hydroxysteroid dehydrogenase

The virilizing congenital adrenal hyperplasias are a group of autosomal recessive disorders characterized by mutations in steroidogenic enzyme genes. The virilizing forms of CAH are due to decreased activity of 21-hydroxylase (*P450c21*), 11β-hydroxylase (*P450c11B1*), or 3β-hydroxysteroid dehydrogenase type 2. The common features of these disorders are impaired cortisol production, loss of cortisol-negative feedback inhibition, increased pituitary adrenocorticotropic hormone secretion, accumulation of steroid intermediates proximal to the deficient enzyme, and increased adrenal androgen secretion. The most common form of CAH that accounts for approximately 90–95% of cases is 21-hydroxylase deficiency. The reported incidence of classic 21-hydroxylase deficiency ranges from 1 in 9,000 to 1 in 15,000 with a variation between ethnic and racial backgrounds. The incidence of nonclassic CAH is approximately 1 in 1,000 but also varies among populations.

In female fetuses with CAH, the adrenal gland produces supraphysiologic doses of androgens during the crucial time of sexual differentiation (9–15 weeks of gestation), which can induce virilization of the external genitalia. In classic CAH, the degree of genital virilization may range from mild clitoral enlargement alone to, in rare cases, a penile urethra. However, the development of the ovaries, uterus, and fallopian tubes remains normal because their development is unaffected by the degree of androgens present.

Congenital adrenal hyperplasias are autosomal recessive inherited disorders. Therefore, if a pregnant patient is a carrier of CAH, her offspring could be at risk of classic 21-hydroxylase deficiency. The patient's partner should be offered screening for carrier status for 21-hydroxylase deficiency. If her partner is not a carrier, it is unlikely that the offspring will be affected. Recurrence risk for future pregnancies is 25% if both parents are carriers. For a woman with CAH, the risk of having a child with CAH depends on the probability that the father is a carrier. Carrier status is usually 1 in 60. Therefore, the risk of a CAH fetus being born to a heterozygous mother is 1:120 and the risk of a CAH female fetus is 1:240.

Prenatal dexamethasone may be administered to the pregnant patient who is a carrier at or before 9 weeks of gestation because it can prevent the genital virilization of female infants with classic CAH. A postulated potential benefit is reduced androgenization of the fetal female brain secondary to exposure to large levels of androgens. However, this treatment remains controversial.

A small percentage of patients with CAH may be hypertensive secondary to mineralocorticoid deficiency. These patients typically have a deficiency in 11β-hydroxylase enzyme. Loss of this enzyme causes a decrease in cortisol production, a compensatory increase in adrenocorticotropic hormone secretion, and increased production of androstenedione, 11-deoxycortisol, 11-deoxycorticosterone, and dehydroepiandrosterone. This condition only represents 5–8%, or 1 in 100,000, of all individuals with CAH.

The 3β-hydroxysteroid dehydrogenase is involved in the synthesis of corticosteroids, mineralocorticoids, testosterone, and estradiol. This enzyme deficit is considered extremely rare. Typically these patients would have a markedly elevated dehydroepiandrosterone level and low testosterone level. These patients would warrant further screening for 3β-hydroxysteroid dehydrogenase.

Patients who fail to achieve puberty, who are hypertensive, or who have primary amenorrhea should be evaluated for 17α-hydroxylase deficiency. Laboratory tests typically demonstrate hypokalemia, low levels of androgens (testosterone and dehydroepiandrosterone), corticosteroid, and high levels of adrenocorticotropic hormone and progesterone. A finding of 17α-hydroxylase deficiency is unlikely in the described patient because of the rarity of the condition.

Merce Fernandez-Balsells M, Muthusamy K, Smushkin G, Lampropulos JF, Elamin MB, Abu Elnour NO, et al. Prenatal dexamethasone use for the prevention of virilization in pregnancies at risk for classical congenital adrenal hyperplasia because of 21-hydroxylase (CYP21A2) deficiency: a systematic review and meta-analyses. Clin Endocrinol (Oxf) 2010;73:436–44.

Miller WL, Auchus RJ. The molecular biology, biochemistry, and physiology of human steroidogenesis and its disorders [published erratum appears in Endocr Rev 2011;32:579]. Endocr Rev 2011;32:81–151.

Nimkarn S, New MI. Congenital adrenal hyperplasia due to 21-hydroxylase deficiency: A paradigm for prenatal diagnosis and treatment. Ann N Y Acad Sci 2010;1192:5–11.

Speiser PW, Azziz R, Baskin LS, Ghizzoni L, Hensle TW, Merke DP, et al. Congenital adrenal hyperplasia due to steroid 21-hydroxylase deficiency: an Endocrine Society clinical practice guideline. Endocrine Society [published erratum appears in J Clin Endocrinol Metab 2010;95:5137]. J Clin Endocrinol Metab 2010;95: 4133–60.

61

Androgen disorders

A 5-year-old girl is brought to your clinic by her mother. She is short for her age and walks with a slight limp. Her mother asks you about her significant skin discoloration. On examination, she has Tanner stage 1 breasts and hyperpigmented lesions on her face, chest, and left arm (Fig. 61-1; see color plate). Imaging shows that her ovaries and brain are normal. Follicle-stimulating hormone and luteinizing hormone levels are low. The best medical treatment for her is

    (A) gonadotropin-releasing hormone (GnRH) agonist
    (B) GnRH antagonist
*   (C) aromatase inhibitor
    (D) medroxyprogesterone acetate

McCune–Albright syndrome is a rare endocrinologic disorder characterized by a triad of cutaneous pigmentations (café au lait spots which have irregular borders and stop at the midline), polyostotic fibrous dysplasia, and multiple endocrine abnormalities. At least two of these three features must be present in a patient in order to make the diagnosis. Fibrous dysplasia can involve a single or multiple skeletal sites. Patients often exhibit a limp with or without pain and, occasionally, a pathologic fracture. Scoliosis is common and may be progressive.

Precocious puberty is classified as gonadotropin dependent (central or true) or gonadotropin independent (peripheral or pseudo). Central and peripheral precocious puberty are differentiated by the level of activity of the hypothalamic–pituitary–ovarian axis. The hypothalamic–pituitary–ovarian axis is active in physiologic puberty as well as central precocious puberty. Peripheral precocious puberty is independent of gonadotropin secretion from the pituitary and there is no activation of the hypothalamic–pituitary–ovarian axis. The source of the sex steroids is either exogenous (accidental ingestion) or endogenous (McCune–Albright syndrome or sex steroid-producing tumor).

The most common endocrinopathy is precocious puberty due to overfunction of the ovary. Other endocrine abnormalities include excessive function of the thyroid gland, adrenal gland, and pituitary gland (over production of growth hormone or prolactin). The presence of higher levels of sex steroids in peripheral precocious puberty may cause the activation of the hypothalamic–pituitary–ovarian axis and the patient may develop central precocious puberty.

Evaluation of a patient who presents with premature development of secondary sexual characteristics starts with an examination that pays close attention to the Tanner stages of the breasts and pubic hair as well as evidence of McCune–Albright syndrome, often characterized by hyperpigmented skin lesions. Laboratory evaluation consists of testing for gonadotropin levels (follicle-stimulating hormone and luteinizing hormone), sex steroid level (estradiol), and thyroid hormone levels. Pelvic ultrasonography may be ordered to evaluate for ovarian masses or cysts. In addition to laboratory testing, X-ray of the dominant hand should be done to evaluate for bone age.

The treatment goal is to preserve the height potential of the child as well as to minimize secondary sexual characteristics to avoid psychologic effects when the child compares herself with her peers. In cases of central precocious puberty, the goal is to suppress the excessive gonadotropin production. This can be achieved through the use of GnRH analogs such as leuprolide acetate. Leuprolide will cause suppression of the gonadotropin production from the pituitary, which will in turn reduce sex steroid production from the ovaries. Because peripheral precocious puberty is independent of gonadotropin production, leuprolide therapy will not be effective in halting the progression of pubertal development. Therefore, the underlying cause of sexual precocity should be addressed first. Aromatase inhibitors have been effectively used to treat sexual precocity in McCune–Albright syndrome since the 1980s. These medications inhibit aromatase activity by competitively binding to the cytochrome P450 subunit of the aromatase enzyme, decreasing estrogen biosynthesis. Other aromatase inhibitors, such as testolactone, a first-generation aromatase inhibitor, have been shown to be somewhat effective, but high doses are required to achieve the desirable effect. The newer third-generation aromatase inhibitors, anastrozole and letrozole, have the advantage of once-daily dosing.

In the described case, because the cause of precocious puberty is secondary to McCune–Albright syndrome, which causes precocious puberty independent of the hypothalamus and pituitary, use of GnRH agonists or antagonists would not be useful. The use of medroxyprogesterone acetate in McCune–Albright syndrome does not prevent closure of the epiphyses and patients may remain short.

Alves C, Silva SF. Partial benefit of anastrozole in the long-term treatment of precocious puberty in McCune-Albright syndrome. J Pediatr Endocrinol Metab 2012;25:323–5.

Berberoglu M. Precocious puberty and normal variant puberty: definition, etiology, diagnosis and current management. J Clin Res Pediatr Endocrinol 2009;1:164–74.

Bercaw-Pratt JL, Moorjani TP, Santos XM, Karaviti L, Dietrich JE. Diagnosis and management of precocious puberty in atypical presentations of McCune–Albright syndrome: a case series review. J Pediatr Adolesc Gynecol 2012;25:e9–e13.

Brito VN, Latronico AC, Arnhold IJ, Mendonça BB. Update on the etiology, diagnosis and therapeutic management of sexual precocity (Erratum in Arq Bras Endocrinol Metabol 2008;52:576). Arq Bras Endocrinol Metabol 2008;52:18–31.

Haddad N, Eugster E. An update on the treatment of precocious puberty in McCune-Albright syndrome and testotoxicosis. J Pediatr Endocrinol Metab. 2007;20:653–61.

Mieszczak J, Lowe ES, Plourde P, Eugster EA. The aromatase inhibitor anastrozole is ineffective in the treatment of precocious puberty in girls with McCune-Albright syndrome. J Clin Endocrinol Metab 2008;93:2751–4.

Wit JM, Hero M, Nunez SB. Aromatase inhibitors in pediatrics. Nat Rev Endocrinol 2011;8:135–47.

62

Premature adrenarche

A 6-year-old white girl comes to your office with her mother. The girl has a 6-month history of pubic hair growth. Physical examination shows a body mass index of 27 (calculated as weight in kilograms divided by height in meters squared). She has Tanner stage 1 breast development and Tanner stage 3 pubic hair development. Bone age is slightly advanced. Blood tests reveal normal levels of androstenedione, testosterone, 17-hydroxyprogesterone, and thyroid-stimulating hormone. Her dehydroepiandrosterone sulfate (DHEAS) level is slightly elevated for her age. The disease that this patient will most likely develop as an adult is

* (A) polycystic ovary syndrome (PCOS)
  (B) adult-onset congenital adrenal hyperplasia (CAH)
  (C) Cushing disease
  (D) androgen-producing adrenal adenoma
  (E) ovarian tumor

Premature adrenarche has historically been defined as pubic or axillary hair development in girls before age 8 years and refers to early activation of the hypothalamic–pituitary–adrenal axis with increased production of the adrenal C19 steroids, DHEAS, and androstenedione (Fig. 62-1). Normal adrenarche occurs around age 6–8 years with increased production of dehydroepiandrosterone (DHEA) and DHEAS from the adrenal reticularis. The DHEA is metabolized to androstenedione and, subsequently, to testosterone in peripheral tissues and in the gonads. Although production of DHEA is under the control of the adrenocorticotropic hormone (ACTH), adrenarche is not associated with an increase in ACTH. Thus, the increase in DHEA is due to changes in steroidogenic enzyme activities, including 17α-hydroxylase (converts pregnenolone to 17-hydroxypregnenolone) and 17,20 lyase (converts 17-hydroxypregnenolone to DHEA). Because 3β-hydroxysteroid dehydrogenase activity remains low, this decreases the conversion of Δ5 compounds (pregnenolone, 17-hydroxypregnenolone, DHEA) to the respective Δ4 compounds (progesterone, hydroxyprogesterone, androstenedione). Physical manifestations of adrenarche include development of pubic and axillary hair and increased apocrine sweat gland activity resulting in body odor.

In the general population, an increase in adrenal C19 steroids is seen in girls aged 6–8 years, but racial variations in axillary and pubic hair development exist. (Mean ages for pubic hair growth: African Americans, 8.8 years; whites, 10.5 years. Mean ages for axillary hair growth: African Americans, 10 years; whites, 11.8 years.) Normal sexual hair development is dependent on androgen production by the adrenal glands and the gonads. Decreased androgen production by either organ, such as with primary hypoadrenalism or with primary ovarian failure (Turner syndrome), results in decreased sexual hair

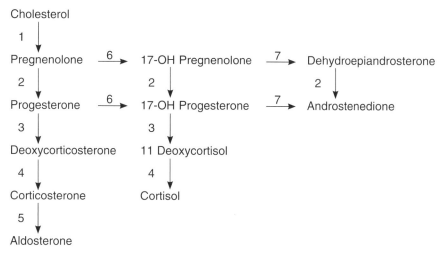

FIG. 62-1. Adrenal steroid biosynthesis in the adrenal cortex.

but not the total absence of it. Traditionally, activation of the hypothalamic–pituitary–adrenal axis (adrenarche) and the hypothalamic–pituitary–gonadal axis (gonadarche) have been considered independent because either can occur without the other in pathologic conditions. Yet physiologically they are interrelated. Peripheral conversion of DHEA to androgens and estrogens affects gonadotropin-releasing hormone pulsatility. Children with early adrenarche have slightly early pubertal onset.

The term premature adrenarche is often incorrectly used interchangeably with premature pubarche. Premature adrenarche consists of abnormally elevated DHEA and DHEAS levels in addition to physical findings of body odor, oily skin, acne, or sexual hair development. The differential diagnoses for premature adrenarche include idiopathic premature adrenarche, congenital adrenal hyperplasia due to enzyme defects (21-hydroxylase deficiency, 11β-hydroxylase deficiency, and 3β-hydroxysteroid dehydrogenase deficiency), Cushing disease, virilizing adrenal or ovarian tumors, and exogenous exposure to androgens. Central precocious puberty may rarely initially present with pubarche. By far the most cases are idiopathic premature adrenarche.

Evaluation for premature adrenarche includes a history, physical examination, review of growth records, and laboratory testing for DHEAS, follicle-stimulating hormone, luteinizing hormone, testosterone, thyroid-stimulating hormone, and morning 17-hydroxyprogesterone levels. Typically, DHEAS levels are elevated compared with chronologic age but are normal for the pubertal stage. An ACTH stimulation test for CAH should be performed in cases with screening 17-hydroxyprogesterone levels greater than 200 ng/dL. In patients with hypertension, additional testing should include blood deoxycortisol, deoxycorticosterone levels to evaluate for 11β-hydroxylase deficiency, and 24-hour urinary free cortisol to evaluate for Cushing disease. Adrenal and ovarian tumors are rare and most often will be accompanied by signs of virilization. Pelvic ultrasonography is indicated in patients with testosterone levels greater than 200 ng/dL. Adrenal imaging may be necessary in patients with abnormal cortisol levels and in individuals with greatly elevated DHEAS levels (greater than 700 micrograms/dL). Height and bone age are advanced for chronologic age but normal for pubertal age, and predicted adult height is normal. In contrast to central precocious puberty, follicle-stimulating hormone and luteinizing hormone levels are consistent with age and show a prepubertal response to gonadotropin-releasing hormone stimulation.

Children with premature adrenarche are more likely to develop adult hyperandrogenemia with clinical criteria consistent with PCOS. Prepubertal children with premature adrenarche, compared with controls, may show biochemical characteristics of PCOS, including insulin resistance, dyslipidemia, increased inflammatory markers, and abnormal adipose distribution. These findings are more prominent with associated obesity. Other conditions (eg, CAH, Cushing disease, ovarian tumors, and adrenal tumors) may cause premature adrenarche but are not more likely in adults with a history of childhood premature adrenarche.

A girl with idiopathic premature adrenarche may be at higher risk as an adult of PCOS along with other metabolic derangements. Additional testing for impaired glucose tolerance and lipid abnormalities may be considered in children with premature adrenarche and obesity or a history of low birth weight with subsequent catch-up growth.

Barker DJ, Winter PD, Osmond C, Margetts B, Simmonds SJ. Weight in infancy and death from ischaemic heart disease. Lancet 1989;2:577–80.

Ibanez L, Dimartino-Nardi J, Potau N, Saenger P. Premature adrenarche--normal variant or forerunner of adult disease? Endocr Rev 2000;21:671–96.

Idkowiak J, Lavery GG, Dhir V, Barrett TG, Stewart PM, Krone N, et al. Premature adrenarche: novel lessons from early onset androgen excess. Eur J Endocrinol 2011;165:189–207.

63
Uterine fibroid embolization

A 40-year-old woman with symptomatic leiomyomas is considering her treatment options. She is interested in learning more about uterine fibroid embolization. You counsel her that the most common complication that patients experience after uterine fibroid embolization is

(A) infection
(B) hemorrhage
(C) chronic malodorous vaginal discharge
* (D) postembolization syndrome
(E) vaginal passage of a myoma

Uterine fibroid embolization, also known as uterine artery embolization, is a radiologic technique to decrease the size and symptoms of uterine myomas. Polyvinyl alcohol particles or microspheres are injected via a transcutaneous femoral artery approach into both uterine arteries. Supplemental metal coils also may be used to augment vascular occlusion. The goal is to produce ischemia, devascularization, and involution of myomas without damaging normal uterine tissue.

Postembolization syndrome is the most common complication experienced by patients who undergo uterine fibroid embolization. It is characterized by pelvic pain, cramping, nausea, vomiting, fever, fatigue, myalgias, malaise, and leukocytosis. The symptoms occur in the first 48 hours after the procedure and will improve gradually over the next week. Because of the intensity of postembolization syndrome, most patients are hospitalized overnight for pain control, and they are administered antipyretics and antiemetics.

Other complications of uterine fibroid embolization include infection, hemorrhage, adhesions, chronic malodorous vaginal discharge, vaginal passage of a myoma, and ovarian insufficiency. Infection is possible with uterine fibroid embolization or surgical myomectomy, but this complication is uncommon with both procedures. Hemorrhage rarely occurs with uterine fibroid embolization and is more of a concern with surgical myomectomy. Pelvic and intrauterine adhesions commonly occur after uterine fibroid embolization and surgical myomectomy and studies suggest incidence rates as high as 59% for pelvic adhesions. Chronic malodorous vaginal discharge is unique to uterine fibroid embolization and would not be expected with surgical myomectomy. Approximately 4–7% of patients experience chronic malodorous vaginal discharge after uterine fibroid embolization, presumably due to fluid extrusion from infarcted myomas into the endometrial cavity. The discharge will ultimately resolve spontaneously in most cases, but some patients will require hysteroscopic removal of the necrotic tissue.

Vaginal passage of a myoma occurs in up to 10% of patients after uterine fibroid embolization. This complication also appears to be unique to uterine fibroid embolization and would not be expected with surgical myomectomy.

The safety of uterine fibroid embolization in women who desire future childbearing has not been established. In patients who wish to preserve their fertility, some unique factors need to be taken into consideration. Transient or permanent ovarian failure has been described in up to 15% of patients after uterine fibroid embolization. The etiology is unclear, but embolic material may travel to the ovarian blood supply. Women older than 45 years may be particularly susceptible to ovarian insufficiency after undergoing the procedure. In one investigation, reproductive-aged women younger than 45 years who underwent uterine fibroid embolization showed no change in basal follicle-stimulating hormone testing, whereas women older than 45 years exhibited an elevation in follicle-stimulating hormone levels after the procedure.

Although uterine fibroid embolization is less likely to adversely affect ovarian reserve in reproductive-aged women younger than 45 years, this possibility should be discussed when counseling women who are considering the procedure. Intrauterine adhesions are another concern after either uterine fibroid embolization or myomectomy (abdominal or hysteroscopic). Ischemic endometrium after uterine fibroid embolization may predispose these patients to intrauterine adhesion formation and future placentation abnormalities. Patient selection is critical in determining the appropriateness of uterine fibroid embolization. Surgical myomectomy remains first-line therapy for women with symptomatic myomas who desire fertility preservation. In addition, women with submucosal myomas or subserosal myomas with small stalks are not ideal candidates.

A number of studies have examined the long- and short-term outcomes of uterine fibroid embolization. Overall, 85–95% of women report improvement in abnormal bleeding and 60–96% report improvement in bulk-related symptoms. In terms of long-term outcomes, 75% of patients report improved or normal uterine bleeding 5 years or more after uterine fibroid embolization, but 20% of patients report that they have undergone a second procedure to manage leiomyoma-related symptoms. In appropriately selected candidates, uterine fibroid embolization is a safe and effective option for the management of uterine myomas.

Alternatives to hysterectomy in the management of leiomyomas. ACOG Practice Bulletin No. 96. American College of Obstetricians and Gynecologists. Obstet Gynecol 2008;112:387–400.

Edwards RD, Moss JG, Lumsden MA, Wu O, Murray LS, Twaddle S, et al. Uterine-artery embolization versus surgery for symptomatic uterine fibroids. Committee of the Randomized Trial of Embolization versus Surgical Treatment for Fibroids. N Engl J Med 2007;356:360–70.

Toor SS, Jaberi A, Macdonald DB, McInnes MD, Schweitzer ME, Rasuli P. Complication rates and effectiveness of uterine artery embolization in the treatment of symptomatic leiomyomas: a systematic review and meta-analysis. AJR Am J Roentgenol 2012;199:1153–63.

van der Kooij SM, Hehenkamp WJ, Volkers NA, Birnie E, Ankum WM, Reekers JA. Uterine artery embolization vs hysterectomy in the treatment of symptomatic uterine fibroids: 5-year outcome from the randomized EMMY trial. Am J Obstet Gynecol 2010;203:105.e1–13.

64

Ectopic pregnancy

A 22-year-old woman, gravida 1, para 0, comes to your office 8 weeks after her last menstrual period. Ultrasonography demonstrates a 4-cm left-sided adnexal mass. There is a discernible fetal heart rate, and she has a β-hCG level of 15,000 mIU/mL. The most appropriate next step is

    (A) methotrexate injection into adnexal mass
    (B) intramuscular injection of methotrexate
* (C) surgery
    (D) repeat serum β-hCG level in 2 days

Early pregnancy failure is the most common complication of pregnancy, and approximately 1–2% of all pregnancies will be ectopic. It is a leading cause of maternal morbidity and mortality. An expedient diagnosis of an ectopic pregnancy may be lifesaving in many circumstances. Approximately 25% of recognized pregnancies end in miscarriage.

Human chorionic gonadotropin is the most widely studied test of early pregnancy. As a single value it is nondiagnostic; however, serial serum β-hCG levels are helpful in identifying patients who require closer surveillance for early pregnancy failure or ectopic pregnancy. The expected β-hCG rise in 48 hours for a viable intrauterine pregnancy is at least 53%, although a recent study suggests that the optimal accuracy for correctly identifying an intrauterine pregnancy may require a more conservative threshold of a 35% rise in 48 hours. Patients who demonstrate a slow increase in β-hCG require vigilant care until an ectopic pregnancy can be excluded.

The β-hCG discriminatory zone for a singleton intrauterine pregnancy is between 1,500 mIU/L and 2,500 mIU/L. The absence of an intrauterine gestational sac of approximately 5 mm in size in the presence of a β-hCG value of 2,000 mIU/L may be an ectopic pregnancy, a multiple pregnancy not yet discernible by ultrasonography, or a failed pregnancy.

Once a concern for ectopic pregnancy is established, the patient must be monitored carefully. Most ectopic pregnancies occur within the fallopian tube. However, there have been reported cases within the ovary, cervix, uterine corneal region, and intra-abdominal sites. Patients are at high risk of intra-abdominal bleeding and hemorrhage. This puts the patient at ultimate risk of death if hemorrhage remains untreated.

The described patient has a clear diagnosis of ectopic pregnancy. Therefore, the patient should not have serial serum β-hCG levels because the location of the ectopic pregnancy is clear.

Methotrexate is an antimetabolite that impairs DNA replication and cell proliferation acting through inhibition of the dihydrofolate reductase enzyme. It acts on trophoblastic and normal cells. Methotrexate injected either into the gestational sac or intramuscularly is contraindicated in this patient because she has a 4-cm gestational sac and embryonic cardiac motion. Other contraindications for methotrexate include a gestational sac greater than 3.5 cm and visible cardiac activity on ultrasonography. See Box 64-1 for criteria for the use of intramuscular methotrexate in tubal pregnancies.

> **BOX 64-1**
>
> **Criteria for the Use of Intramuscular Methotrexate in Tubal Pregnancies**
>
> *Absolute indications*
> - Hemodynamically stable without active bleeding or signs of hemoperitoneum
> - Nonlaparoscopic diagnosis
> - Patient desires future fertility
> - General anesthesia poses a significant risk
> - Patient is able to return for follow-up care
> - Patient has no contraindications to methotrexate
>
> *Relative indications*
> - Unruptured mass 3.5 cm or less at its greatest dimension
> - No fetal cardiac motion detected
> - Patients whose β-hCG level does not exceed a predetermined value (6,000–15,000 mIU/mL)
>
> *Absolute contraindications*
> - Breastfeeding
> - Overt or laboratory evidence of immunodeficiency
> - Alcoholism, alcoholic liver disease, or other chronic liver disease
> - Preexisting blood dyscrasias (eg, bone marrow, hypoplasia, leukopenia, thrombocytopenia, or significant anemia)
> - Known sensitivity to methotrexate
> - Active pulmonary disease
> - Peptic ulcer disease
> - Hepatic, renal, or hematologic dysfunction
>
> *Relative contraindications*
> - Gestational sac larger than 3.5 cm
> - Embryonic cardiac motion
>
> Medical management of ectopic pregnancy. ACOG Practice Bulletin No. 94. American College of Obstetricians and Gynecologists. Obstet Gynecol 2008;111: 1479–85.

The described patient would be best served by surgical management of her ectopic pregnancy. Compared with surgery, medical therapy is less likely to succeed. It may lead to ectopic pregnancy rupture and a life-threatening clinical scenario that places the patient's life at risk and compromises future fertility.

Barnhart K, Sammel MD, Chung K, Zhou L, Hummel AC, Guo W. Decline of serum human chorionic gonadotropin and spontaneous complete abortion: defining the normal curve. Obstet Gynecol 2004;104:975–81.

Medical management of ectopic pregnancy. ACOG Practice Bulletin No. 94. American College of Obstetricians and Gynecologists. Obstet Gynecol 2008;111:1479–85.

Morse CB, Sammel MD, Shaunik A, Allen-Taylor L, Oberfoell NL, Takacs P, et al. Performance of human chorionic gonadotropin curves in women at risk for ectopic pregnancy: exceptions to the rules. Fertil Steril 2012;97:101–106e2.

65

Condom failure and morning-after contraception

A 19-year-old woman calls you to report that she engaged in intercourse 4 days earlier and the condom she used broke. She has heard that there are pills to take to prevent pregnancy. You counsel her that, in this situation, the most appropriate oral morning-after contraception is

* (A) single-dose ulipristal acetate
* (B) high-dose combined oral contraceptives (OCs)
* (C) single-dose regimen of levonorgestrel
* (D) two-dose regimen of levonorgestrel
* (E) no treatment because of the timeframe

Emergency contraception, also called the morning-after pill or postcoital contraception, is used to prevent pregnancy after unprotected or insufficiently protected sexual intercourse. Most commonly, the need for emergency contraception arises when OC use is inconsistent, condom failure occurs, or no contraceptive has been used.

In 1974, Yuzpe and colleagues described the first emergency contraception method, which comprised high-dose combination OCs. In 1998, the U.S. Food and Drug Administration approved the first product specifically intended for emergency contraception. This method consisted of four tablets of ethinyl estradiol 50 micrograms and norgestrel 0.5 mg, with two tablets taken initially and two more tablets taken 12 hours later. Other methods of emergency contraception that have since been approved include progestin-only OCs, antiprogestin OCs, and the copper-containing intrauterine device (IUD).

The current levonorgestrel-only method for emergency contraception consists of a 1.5 mg total dose of levonorgestrel. The single-dose protocol involves taking two 0.75 mg levonorgestrel pills at the same time. The two-dose regimen requires the patient to take one pill immediately and another 12 hours later. Both protocols are equally efficacious, but the single-dose regimen may improve compliance. Both levonorgestrel-only regimens appear to be more effective than combination estrogen–progestin regimens and also are associated with less nausea and vomiting. Because of these advantages, levonorgestrel-only regimens are preferred over combination estrogen–progestin options. Levonorgestrel regimens are highly efficacious up to 72 hours after intercourse (although moderate efficacy may persist up to 120 hours after intercourse). Even in the first 72 hours, the timing of treatment is important. When levonorgestrel regimens are used in the first 24 hours, 95% of pregnancies are prevented. When treatment is delayed 25–48 hours after intercourse, 85% of pregnancies are prevented. At 49–62 hours after intercourse, only 58% of pregnancies are prevented. Efficacy also may decline with obesity.

Single-dose ulipristal acetate is efficacious as emergency contraception up to 120 hours after unprotected intercourse. Unlike levonorgestrel regimens, efficacy of ulipristal acetate does not appear to decrease over time (up to 120 hours after intercourse). For the patient described, ulipristal acetate is the most appropriate oral option given the time interval since she and her partner experienced condom failure. A copper-containing IUD is also very effective as emergency contraception. Published trials report IUD insertions for emergency contraception up to 5 days after unprotected intercourse with pregnancy rates ranging from 0% to 0.2%.

Hormonal forms of emergency contraception work by inhibiting or delaying ovulation. Emergency contraception is not effective once a pregnancy has been established and should not be confused with medical abortion. If hormonal emergency contraception is inadvertently taken after a pregnancy is already established, there is no increased teratogenic risk to the developing embryo. Emergency contraception, particularly levonorgestrel-only emergency contraception, is considered safe and should be available to all women, including women with contraindications to conventional combined OCs. The Centers for Disease Control and Prevention's "Medical Eligibility Criteria for Contraceptive Use" lists no conditions in which the risks of emergency contraception outweigh the benefits.

The American College of Obstetricians and Gynecologists recommends that emergency contraception be taken as soon as possible after an episode of unprotected or inadequately protected intercourse. Although efficacy is highest with early use, emergency contraception should be made available to patients who request it up to 5 days after unprotected intercourse. Additionally, neither a physical examination nor a pregnancy test is required before providing emergency contraception, although counseling regarding long-term contraceptive methods is recommended. In particular, the copper-containing IUD may be appropriate for some women who desire either emergency contraception or long-acting contraception.

Emergency contraception. Practice Bulletin No. 112. American College of Obstetricians and Gynecologists; Obstet Gynecol 2010;115:1100–9.

Fine P, Mathe H, Ginde S, Cullins V, Morfesis J, Gainer E. Ulipristal acetate taken 48-120 hours after intercourse for emergency contraception. Obstet Gynecol 2010;115:257–63.

Glasier AF, Cameron ST, Fine PM, Logan SJ, Casale W, Van Horn J, et al. Ulipristal acetate versus levonorgestrel for emergency contraception: a randomised non-inferiority trial and meta-analysis. Lancet 2010;375:555–62.

Raine TR, Harper CC, Rocca CH, Fischer R, Padian N, Klausner JD, et al. Direct access to emergency contraception through pharmacies and effect on unintended pregnancy and STIs: a randomized controlled trial. JAMA 2005;293:54–62.

66

Contraception for a patient with systemic lupus erythematosus

A 32-year-old obese woman, gravida 1, para 1, with systemic lupus erythematosus (SLE) requests reversible contraception. Her disease has been complicated by lupus nephritis, vasculitis, and antiphospholipid antibodies. You counsel her that the reversible contraceptive with the greatest risk for her is

* (A) copper intrauterine device (IUD)
* (B) levonorgestrel IUD
* (C) depot medroxyprogesterone acetate
* (D) combination hormonal contraceptives

For women with coexisting medical conditions, reliable contraception is especially important. No option is completely risk free, so health care providers must weigh the risks of contraception against the risks of an unplanned pregnancy. In SLE, avoiding disease flares or disease progression is particularly important. In many cases of SLE, the coexistence of antiphospholipid antibodies increases a woman's baseline risk of thromboembolism. Complications of SLE, such as systemic vascular disease, nephritis, or uncontrolled hypertension, also can make the contraceptive choices for these patients more difficult.

Twelve months of combination hormonal contraceptive use by women with mild SLE but without antiphospholipid antibodies has been shown to carry a risk no different than use of a progestin-only oral contraceptive or the copper IUD in terms of disease progression, flares, thromboses, or infections. Experts agree that women with SLE and antiphospholipid antibodies carry a baseline increased risk of thromboembolism and, thus, are not good candidates for combination hormonal contraceptives. Additionally, combination hormonal contraceptive use and obesity represent independent risk factors for venous thromboembolism. Theoretical concerns have been raised regarding the risk of infection with IUD insertion in immunosuppressed patients with autoimmune disease. Limited evidence exists to address this issue, but IUD insertion does not appear to be associated with worsening disease activity or infection in women with SLE.

For the described obese patient with antiphospholipid antibodies, combination hormonal contraceptives carry the greatest risk of the available options. In nonobese women with uncomplicated SLE, however, combination hormonal contraceptives are a reasonable choice based on data from two different clinical trials. For patients with complicated SLE, progestin-only contraceptives (oral contraceptives, injections, implants, and hormonal IUDs) or nonhormonal contraceptives (copper IUD or permanent surgical options) are preferred over combination hormonal contraceptives. Approximately 25% of pregnant women with SLE ultimately terminate their pregnancies, which emphasizes the importance of effective birth control for this patient population.

Culwell KR, Curtis KM, del Carmen Cravioto M. Safety of contraceptive method use among women with systemic lupus erythematosus: a systematic review. Obstet Gynecol 2009;114:341–53.

Petri M, Kim MY, Kalunian KC, Grossman J, Hahn BH, Sammaritano LR, et al. Combined oral contraceptives in women with systemic lupus erythematosus. OC-SELENA Trial. N Engl J Med 2005;353:2550–8.

Sanchez-Guerrero J, Uribe AG, Jimenez-Santana L, Mestanza-Peralta M, Lara-Reyes P, Seuc AH, et al. A trial of contraceptive methods in women with systemic lupus erythematosus. N Engl J Med 2005;353:2539–49.

Use of hormonal contraception in women with coexisting medical conditions. ACOG Practice Bulletin. No. 73. American College of Obstetricians and Gynecologists. Obstet Gynecol 2006;107:1453–72.

67

Contraception for a patient with diabetes mellitus

A 36-year-old primigravid woman with type 1 diabetes mellitus is currently at 35 weeks of gestation. She is considering different contraceptive methods postpartum. In the absence of appropriate counseling, the contraceptive method that this patient is most likely to choose is

    (A) progestin-only oral contraceptives
    (B) intrauterine device (IUD)
    (C) subdermal progestin implant
    (D) progestin injection every 3 months
*   (E) tubal sterilization

Women with diabetes mellitus have many misperceptions about appropriate contraceptive choices. The misperceptions vary depending on the patient's age. In a study of contraceptive methods practiced by women in a large managed care organization where contraceptive prescriptions and services were covered benefits, 11.6% of women ages 35–44 years with diabetes chose tubal sterilization compared with 6.9% of women with no chronic condition (Table 67-1). Women ages 35–44 years with diabetes were also more likely to choose highly effective methods (sterilization, hysterectomy, IUD, and subdermal progestin implant) versus women with no chronic condition at a rate of 24.8% versus 15.6%. In comparison, 18.7% of women ages 35–44 years with no chronic condition were more likely to use moderately effective contraceptive methods, ie, combination oral contraceptives, patch, or ring, versus 9% of women with diabetes. Some of the use of highly effective over moderately effective contraceptive methods may be because of health care provider counseling in this age group, but even among the women who chose the highly effective methods, the 19% use of irreversible methods (tubal sterilization, hysterectomy) in the diabetes group exceeds the 8% use in the women with no chronic condition. Similarly, within the group of women with diabetes aged 35–44 years, the 19% use of irreversible methods is greater than the 5.8% use of reversible methods (IUD or subdermal implant). For the group of women aged 35–44 years with no chronic disease, the 7.6% use of reversible, highly effective methods is just slightly below the 8% use of irreversible, highly effective methods. In summary, patients aged 35–44 years with diabetes tend to use more irreversible, highly effective methods, such as tubal sterilization and hysterectomy, than reversible,

TABLE 67-1. Proportion of Nonpregnant Women Aged 35–44 Years Who Received Contraceptive Services by Disease History (%)

Contraceptive Method	Women With Diabetes Mellitus (n=5,788)	Women With No Chronic Condition (n=48,570)
Highly effective	24.8	15.6
Irreversible	19.0	8.0
Sterilization*	11.6	6.9
Hysterectomy	7.4	1.1
Reversible	5.8	7.6
Intrauterine device	5.8	7.6
Subdermal implant	0.03	0.02
Moderately effective		
Combination oral contraceptive, patch, or ring	9.0	18.7

*Tubal sterilization including transcervical sterilization.

Modified from Schwarz EB, Postlethwaite D, Hung YY, Lantzman E, Armstrong MA, Horberg MA. Provision of contraceptive services to women with diabetes mellitus. J Gen Intern Med 2012;27: 196–201.

highly effective methods, such as the IUD and subdermal implant.

Some of these preferences expressed by patients aged 35–44 years with diabetes may have their origins in prior counseling sessions with health care providers when the patient was much younger or be based upon the patient's perception of her diabetes. In the United States, approximately 50% of all pregnancies are unintended. When an unintended pregnancy occurs, a patient with diabetes is more likely than a patient without diabetes to be in less-than-optimal glycemic control. As a consequence, the birth defect rate among women with diabetes is estimated to be 9%, 2–3 times greater than the risk for the general population. These factors could be responsible for a health care provider bias to counsel women to use more effective contraceptive methods and may account for the observed differences in the contraceptive choices of patients aged 35–44 years with diabetes. Additional data suggest that the increased prevalence of tubal sterilization and hysterectomy procedures in patients with diabetes over highly effective reversible contraceptive methods might reflect a knowledge gap among health care providers who are unfamiliar with these less invasive, less expensive, long-acting reversible contraceptive methods. Short-term utilization of long-acting reversible contraceptive methods may avoid patient regret over an irreversible hysterectomy or tubal sterilization. Even the use of moderately effective methods, such as low-dose combination oral contraceptives, may be preferable to an irreversible method, and their use is safe among women with well-controlled diabetes. Further data from the aforementioned research study in which women with diabetes were monitored in a large managed care organization revealed that 61.3% of the women did not receive contraceptive counseling or practice a contraceptive method compared with 39.7% of the women without a chronic condition. Clinicians appear to provide less preconception counseling and contraceptive therapy to women with diabetes than women without a chronic medical condition.

Similarly, adolescents with type 1 diabetes have further misconceptions and have access to less reliable information about contraceptive methods. In a recent study, 43% of adolescents with type 1 diabetes felt that all birth control methods were less effective in women with diabetes versus women without diabetes and only 69% felt comfortable about asking a health care professional about birth control. Clearly, an increased education effort is needed to inform adolescent women with diabetes about the necessity, efficacy, and safety of current reversible forms of contraception. This educational effort needs to be continued for the group of women in the 35–44-year age range, where reversible forms of highly effective contraceptive methods need to be emphasized as suitable options to the more prevalent current use of irreversible methods, such as tubal sterilization and hysterectomy. This education effort needs to be further extended to all health care providers who counsel women with diabetes about their contraceptive options.

Schwarz EB, Postlethwaite D, Hung YY, Lantzman E, Armstrong MA, Horberg MA. Provision of contraceptive services to women with diabetes mellitus. J Gen Intern Med 2012;27:196–201.

Schwarz EB, Sobota M, Charron-Prochownik D. Perceived access to contraception among adolescents with diabetes: barriers to preventing pregnancy complications. Diabetes Educ 2010;36:489–94.

68
Maternal virilization in hyperreactio luteinalis

A 33-year-old woman with worsening third-trimester hirsutism and increasing clitoromegaly is found to have enlarged ovaries at cesarean delivery (Fig. 68-1; see color plate). The most appropriate management is

      (A) bilateral oophorectomy
      (B) unilateral oophorectomy with frozen section
*   (C) 3-month postpartum pelvic ultrasonography
      (D) postoperative testing for CA 125 and carcinoembryonic antigen
      (E) intraoperative fine-needle aspiration with cytology

Virilization of the mother during pregnancy, although uncommon, is most often due to benign ovarian conditions and rarely due to neoplasia, placental enzyme disorder, or iatrogenic causes. Normally, circulating levels of the androgen testosterone and the androgen precursor androstenedione increase in pregnancy because of human chorionic gonadotropin (hCG) stimulation of ovarian thecal cells. In contrast, circulating dehydroepiandrosterone and dehydroepiandrosterone sulfate (DHEAS) levels decrease during pregnancy despite increased production by the adrenal gland. This is because of the primary conversion of dehydroepiandrosterone and DHEAS to estrogen in the placenta.

Two principal mechanisms prevent androgen manifestations during pregnancy: 1) an increase in sex hormone-binding globulin (SHBG) and 2) high aromatase activity in the placenta. Levels of SHBG increase dramatically during pregnancy because estrogen stimulates the liver; SHBG binds a high proportion of dihydrotestosterone (DHT) and testosterone, thus decreasing the availability of free androgens. Meanwhile, the high aromatase content of the placenta converts maternal androstenedione to estrone and testosterone to estradiol. It should be noted that DHT is not a substrate for aromatase. Thus, virilization of the pregnant woman occurs when the level of androgens exceeds the capability of these mechanisms or there is a defect in placental aromatization.

Ovarian luteoma and hyperreactio luteinalis are benign conditions of the ovary that are unique to pregnancy and lead to hirsutism and virilization. Luteoma typically presents as a solid ovarian mass and is bilateral 30–50% of the time. Luteomas vary in size, but typically are 6–10 cm. Approximately one third of luteomas will produce adequate androgens to result in maternal virilization. In addition to hormonal manifestations, the solid mass may increase the risk of ovarian torsion. Cut surface of the mass appears yellow because of the high cholesterol content.

Hyperreactio luteinalis is defined as the presence of multiple theca-lutein cysts bilaterally due to hCG stimulation. The condition is more common with multiple gestations, molar pregnancy, and choriocarcinoma but may be seen in apparently normal pregnancies. In approximately 70% of cases, the presentation occurs in the third trimester or after delivery in cases of persistently high hCG. Approximately 30% of women experience virilization. The condition resembles ovarian hyperstimulation syndrome observed in women with multiple theca-lutein cysts, which are caused by drugs taken to induce ovulation. Occasionally, hyperreactio luteinalis will mimic ovarian hyperstimulation syndrome, with the loss of ovarian vascular integrity leading to ascites and pleural effusion. As long as there is not a condition leading to persistence of hCG after delivery, hyperreactio luteinalis resolves postpartum. As with any mass effect, an increased risk exists of ovarian torsion, cyst rupture, and bleeding.

Although not typical, women with significant polycystic ovary syndrome-associated hyperandrogenism may experience worsening of hirsutism during pregnancy. This rarely causes virilization. Symptoms improve after pregnancy but tend to have a high recurrence rate in subsequent pregnancies.

Ovarian tumors that may coincidentally develop during pregnancy include Sertoli–Leydig cell, granulosa–theca cell, thecoma, cystadenoma, Brenner, cystadenocarcinoma, and Krukenberg tumors. These entities cause hyperandrogenism either by direct production of androgens or by inducing a surrounding theca hyperplasia responsible for androgen production. Maternal virilization is seen most of the time. Adenocarcinoma of the adrenal gland and Cushing disease are rare causes of maternal virilization.

Placental aromatase deficiency is an autosomal recessive mutation of the *CYP19* gene in the fetus. The inability of the placenta to convert androstenedione to estrone and testosterone to estradiol results in maternal virilization. A female fetus also will be androgenized with ambiguous genitalia. The diagnosis is suspected when low levels of estriol are observed in the maternal circulation.

Rarely, maternal virilization is caused by exposure to synthetic androgen. Androgenization of a female fetus in disorders leading to maternal virilization is dependent on the level and type of androgen present. In women who exhibit virilization from a luteoma, approximately 60–70% will have some degree of masculinization of a female infant. In contrast, masculinization of a female infant is essentially never seen with hyperreactio luteinalis or pregnancy exacerbation of polycystic ovary syndrome. In the case of fetal aromatase deficiency, female infants will always show some degree of masculinization.

Evaluation of maternal virilization includes a detailed history in regard to all drugs used and onset of symptoms along with laboratory testing, including ovarian ultrasonography and laboratory testing for testosterone, DHEAS, DHT, hCG, and estriol levels. The finding of elevated hCG and bilateral multiple ovarian cysts is consistent with hyperreactio luteinalis and requires postpartum follow-up with ultrasonography and laboratory evaluation to show normalization of hCG and androgen levels. There is no indication for oophorectomy, biopsy, or fine-needle aspiration. With the appearance of multiple ovarian cysts and no other lesions in the pelvis, CA 125 and carcinoembryonic antigen tumor markers are not indicated.

A solid mass in the ovary with maternal virilization is a more difficult problem. Luteomas and ovarian tumors associated with maternal virilization are very rare. The decision to remove a solid mass must take into account the gestational age and other findings that may indicate malignancy, such as ascites, weight loss, elevated tumor markers, or findings suggestive of metastatic disease. Luteomas regress rapidly with delivery, whereas other ovarian tumors will not be substantially affected.

Kanova N, Bicikova M. Hyperandrogenic states in pregnancy. Physiol Res 2011;60:243–52.

Morishima A, Grumbach MM, Simpson ER, Fisher C, Qin K. Aromatase deficiency in male and female siblings caused by a novel mutation and the physiological role of estrogens. J Clin Endocrinol Metab 1995;80:3689–98.

Sir-Petermann T, Maliqueo M, Angel B, Lara HE, Perez-Bravo F, Recabarren SE. Maternal serum androgens in pregnant women with polycystic ovarian syndrome: possible implications in prenatal androgenization. Hum Reprod 2002;17:2573–9.

69

Recurrent pregnancy loss

A 28-year-old woman, gravida 2, comes to your office for evaluation after three first-trimester pregnancy losses. She and her husband are in good health. Her thyroid-stimulating hormone and prolactin levels, pelvic ultrasonography, and hysterosalpingography are all normal. To evaluate the couple's recurrent pregnancy loss, you should order

    (A) antithrombin III analysis
    (B) endometrial biopsy
*   (C) karyotype analysis
    (D) fragile X analysis

Recurrent pregnancy loss is defined as two or more failed clinical pregnancies. Evaluation of pregnancy loss should be initiated after two first-trimester pregnancy losses. The most common contributors to recurrent pregnancy loss are maternal uterine malformations, maternal antiphospholipid syndrome, and cytogenetic abnormalities in either partner.

Cytogenetic abnormalities account for approximately 5% of recurrent pregnancy loss etiologies. Peripheral blood for karyotyping should be collected after two or three losses, depending on family history. Reciprocal and robertsonian translocations can be identified in couples with recurrent pregnancy loss. Translocations are an exchange of material between nonhomologous chromosomes. If there is no loss of genetic material, the translocation is considered balanced; however, if genetic material is lost, it is considered unbalanced. Robertsonian translocations are similar to reciprocal translocations but are specific to the exchange of material between two acrocentric chromosomes (chromosomes 13, 14, 15,

21, and 22) with subsequent loss of their short arms. Translocations are more common in women. Therefore, karyotyping is more likely to detect a translocation in the female partner if a chromosomal abnormality exists. However, this should not be a reason for excluding the male partner from evaluation. Treatment for couples with parental cytogenetic abnormalities should start with genetic counseling. Couples also may opt for preimplantation genetic diagnosis with transfer of unaffected embryos.

Up to 20% of recurrent pregnancy loss patients will test positive for antiphospholipid antibodies. The most widely accepted tests for this condition include tests for lupus anticoagulant, anticardiolipin antibodies, and anti-β_2-glycoprotein. Abnormalities detected with the use of any of these tests require repeat testing in 12 weeks for confirmation. Prophylactic therapies for patients with a finding of antiphospholipid antibodies include daily baby aspirin and heparin or low-molecular weight heparin. Although congenital müllerian anomalies usually affect pregnancies in the second trimester, they also may have a role in first-trimester losses. It is therefore wise to evaluate the endometrial cavity, often starting with pelvic ultrasonography. Saline-infusion ultrasonography, hysterosalpingography, and hysteroscopy can be used to further evaluate the endometrial cavity. Treatment typically involves repair of the anomaly or use of a gestational carrier if repair is not possible. In patients with no personal or close family history of thrombotic events, screening for thrombophilia (factor V Leiden mutations and protein C and S deficiency) is not warranted.

An endometrial biopsy to define luteal phase deficiency is not warranted because the diagnosis of luteal phase deficiency is difficult and its treatment remains controversial. Use of progesterone in unexplained recurrent pregnancy loss has not been shown to improve pregnancy outcome. Fragile X testing will not be helpful in providing information regarding this couple's etiology for recurrent pregnancy loss. Fragile X carrier testing is more beneficial in the evaluation of the patient with premature ovarian failure or with diminished ovarian reserve because premutations may predispose the female partner to ovarian deficiency. Antithrombin III would not be helpful; patients with antithrombin III deficiency usually have thrombotic events before pregnancy loss, which is not the case for the described patient.

Antiphospholipid syndrome. Practice Bulletin No. 132. American College of Obstetricians and Gynecologists. Obstet Gynecol 2012; 120:1514–21.

Evaluation and treatment of recurrent pregnancy loss: a committee opinion. Practice Committee of the American Society for Reproductive Medicine. Fertil Steril 2012;98:1103–11.

Hirshfeld-Cytron J, Sugiura-Ogasawara M, Stephenson MD. Management of recurrent pregnancy loss associated with a parental carrier of a reciprocal translocation: a systematic review. Semin Reprod Med 2011;29:470–81.

Miyakis S, Lockshin MD, Atsumi T, Branch DW, Brey RL, Cervera R, et al. International consensus statement on an update of the classification criteria for definite antiphospholipid syndrome (APS). J Thromb Haemost 2006;4:295–306.

Rai R, Regan L. Recurrent miscarriage. Lancet 2006;368:601–11.

Stephenson MD, Awartani KA, Robinson WP. Cytogenetic analysis of miscarriages from couples with recurrent miscarriage: a case-control study. Hum Reprod 2002;17:446–51.

70

Sheehan syndrome

A 29-year-old woman, gravida 2, para 2, comes to your office with secondary amenorrhea. She is otherwise healthy. She stopped having menstrual cycles after her last delivery 2 years ago, which was complicated by severe postpartum hemorrhage. Magnetic resonance imaging of the patient's brain is as shown (Fig. 70-1). The test that would best identify the most life-threatening manifestation of this condition is a test for the level of

(A) growth hormone
* (B) cortisol
(C) prolactin
(D) free thyroxine (T_4)
(E) follicle-stimulating hormone (FSH)

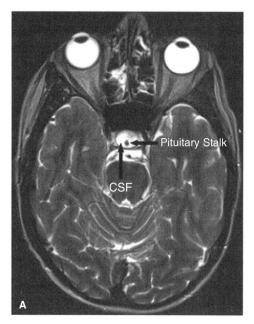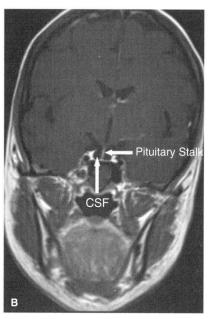

FIG. 70-1

Postpartum infarction of the pituitary, also known as Sheehan syndrome, usually occurs after a significant postpartum hemorrhage involving hypotension and transfusion of multiple units of blood. For severe cases, patients may have immediate severe hypopituitarism with lethargy, anorexia, and weight loss. In mild-to-moderate cases, the initial symptom may be absent lactation due to diminished prolactin, although many patients have a delayed presentation with secondary amenorrhea. Some patients may consult their physician weeks, months, or years after the precipitating event. In developed countries, the incidence of Sheehan syndrome is decreasing because of improvements in obstetric care. However, for patients from countries without access to high-quality medical care, postpartum hemorrhage and subsequent Sheehan syndrome is a common cause of hypopituitarism.

Patients with Sheehan syndrome can have abnormal levels of some or all hormones that originate from the anterior and posterior pituitary. In addition, magnetic resonance imaging of the brain can show a partially or completely empty sella as seen in Figure 70-1. Clinical manifestations of Sheehan syndrome can be quite heterogeneous. Although the most common hormonal alterations include changes in levels of growth hormone, prolactin, and gonadotropins, many patients also have defects in adrenocorticotropic hormone (ACTH) and thyroid-stimulating hormone (TSH) secretion. Growth hormone deficiency in adults tends to have signs and symptoms associated with body composition changes, including a decrease in lean body mass and an increase in fat mass. Growth hormone deficiency in adults is not life threatening and replacement of growth hormone is controversial outside of the pediatric population. Prolactin levels may be chronically low but have no clinical significance in a premenopausal woman with secondary amenorrhea, such as the described patient.

Deficiency in ACTH secretion is a serious condition associated with Sheehan syndrome. Chronic absence of ACTH can cause adrenal atrophy and inability to mount an adequate physiologic response to a crisis such as a serious infection or other acute medical problem. Because adrenal insufficiency can result in a life-threatening adrenal crisis, making it the most serious condition for which to screen, testing for adrenal insufficiency with early-morning serum cortisol levels is essential. When testing for adrenal insufficiency, it is important to wait at least 6 weeks after the pituitary insult because adrenal atrophy can take time to develop and premature testing can yield falsely normal results. After the diagnosis of adrenal insufficiency is confirmed, replacement of corticosteroids is essential as the first step in treatment of Sheehan syndrome. Replacement of thyroid hormone in cases of untreated adrenal insufficiency can cause an even more profound drop in cortisol levels. This may be due to increased metabolic activity that cannot be handled by the low levels of cortisol.

Thyroid hormone levels can be abnormal in patients with Sheehan syndrome because relative deficiencies in TSH can cause secondary hypothyroidism. Serum levels of free T_4 can be low to normal, and TSH levels may be low or normal, demonstrating inability of the pituitary to compensate for the low levels of thyroid hormone. Therefore, practitioners must always check for TSH and free T_4 when secondary hypothyroidism is suspected. Common symptoms of hypothyroidism include fatigue, cold intolerance, weight gain, and constipation. Most cases of hypothyroidism are not life threatening.

Gonadotropin abnormalities are noted frequently in women with pituitary insufficiency. Such abnormalities often present with secondary amenorrhea. Testing for FSH, luteinizing hormone (LH), and estradiol often can help determine if amenorrhea is due to a central cause (low FSH, LH, and estradiol levels) or an ovarian cause (high FSH and LH levels and low estradiol levels). If fertility is desired, medical therapy with exogenous gonadotropins can be helpful. Otherwise, low levels of gonadotropins and, subsequently, estrogen, can precipitate bone loss and osteoporosis, necessitating supplemental estrogen therapy.

Sheehan HL. The recognition of chronic hypopituitarism resulting from postpartum pituitary necrosis. Am J Obstet Gynecol 1971;111:852–4.

Speroff L, Fritz MA. Amenorrhea. In: Speroff L, Fritz MA, editors. Clinical gynecologic endocrinology and infertility. 8th ed. Philadelphia (PA): Lippincott Williams and Wilkins; 2011. p. 483–4.

Tessnow AH, Wilson JD. The changing face of Sheehan's syndrome. Am J Med Sci 2010;340:402–6.

71
Preimplantation genetic screening

A 41-year-old woman, gravida 2, aborta 2, with a history of recurrent first-trimester pregnancy loss would like to undergo in vitro fertilization (IVF) to screen embryos for chromosomal abnormalities. The technique associated with IVF most likely to identify a genetically normal ongoing pregnancy is

    (A) polar body biopsies
    (B) sperm DNA fragmentation
    (C) blastomere biopsy
*   (D) trophectoderm biopsy

Preimplantation genetic diagnosis and screening were established more than 2 decades ago in order to select genetically normal embryos from among others created through IVF to avoid inherited diseases and give the highest potential to achieve pregnancies. Couples with a family history of genetic disorders or structural chromosomal abnormalities had few options before the implementation of these technologies.

Various methodologies have been implemented to screen embryos for either aneuploidy or single-gene disorders to help assess "normal" embryos for embryo transfer in an IVF cycle. These methodologies include polar body biopsy, blastomere biopsy from an embryo at the multicellular stage (day 2 or 3), blastocyst trophectoderm biopsy (day 5), and various DNA analysis techniques which include but are not limited to fluorescence in situ hybridization (FISH) analysis, comparative genomic hybridization analysis, and microarray technologies.

The initial preimplantation genetic diagnosis (PGD)–preimplantation genetic screening standard technique for detecting chromosomes was FISH analysis. It used specific probes for the chromosomes most commonly involved in aneuploidy. The presence or absence of a normal pair of chromosomes could be identified, thereby allowing the selection of only chromosomally normal embryos for transfer. Choosing embryos with normal chromosomes should increase implantation rates and live-birth rates and reduce miscarriage rates. Prospective analysis of this technology did not justify its use and in many instances suggested a decrease in overall pregnancy rates.

Until recently, cleavage-stage embryo biopsy was the prevailing method of embryo assessment for PGD–preimplantation genetic screening. However, the well-documented chromosome mosaicism that exists in early human embryos suggests that a biopsied cell may not be truly representative of the rest of the embryo. Mosaicism could complicate the accuracy of diagnosis, thus contributing to the failure of preimplantation genetic screening to show a benefit and transfer of abnormal embryos believed to be free of disease, or else free of structural chromosomal abnormalities. This technology hinges on the assessment of a single cell for analysis. Some studies have suggested that aneuploidy screening by FISH analysis may be inaccurate up to 60% of the time. In contrast, blastocyst trophectoderm biopsy is the analysis of hundreds of cells. Therefore, it is more representative of the embryo itself. The resulting larger DNA yield may improve the rates of diagnosis when compared with other methods. A number of laboratory strategies are available for analysis of trophectoderm-biopsied cells, such as microarray, polymerase chain reaction, isothermal rolling-circle-based amplification, and complete genomic hybridization. To date, there have been no randomized controlled trials to assess their efficacy in preimplantation genetic screening for advanced maternal age or the advantages of one technique over another in the analysis of trophectoderm-biopsied cells. However, promising data suggests that identifying euploid embryos may improve implantation and the likelihood of live birth.

To date, it has not been demonstrated that aneuploidy screening for women of advanced maternal age with IVF increases their likelihood of pregnancy. In fact, preimplantation genetic screening data for day-3 embryo biopsy demonstrates that the preimplantation genetic screening process may do more harm than good. Additionally, polar body biopsy on either day 1 or day 3 is considered inferior to blastomere biopsy. The trend in preimplantation genetic screening and PGD is toward blastocyst trophectoderm biopsy, given the plethora of cells available for analysis. However, embryos often need to be frozen before transfer to allow time for the analysis of the embryo itself. Data are limited as to whether preimplantation genetic screening for advanced maternal age is worth the effort. Studies have shown excellent pregnancy rates with 24-chromosome screening of trophectoderm-biopsied embryos. This demonstrates that if preimplantation genetic screening is performed in this patient population, day-5 embryo biopsy is the best way to assess the embryo.

Patients are encouraged to work with their obstetrician–gynecologist to assess the need for aneuploidy screening once they become pregnant. Obstetrician–gynecologists are encouraged to offer fetal screening once patients conceive. Options for screening include but are not limited to chromosome-selective sequencing of maternal plasma cell-free DNA, chorionic villi sampling, and amniocentesis.

DNA fragmentation testing is a measure of sperm oxidative DNA damage within a semen specimen. It is theorized that subfertile men carry an increased likelihood of having single- or double-stranded DNA breaks packaged within the head of the sperm. Additionally, oxidative damage can decrease sperm motility, leading to premature acrosome reaction, lipid peroxidation, apoptosis, and reduction in mitochondria membrane potential. These effects on sperm are thought to contribute to subfertility in male patients.

Al-Asmar N, Peinado V, Vera M, Remohi J, Pellicer A, Simon C, et al. Chromosomal abnormalities in embryos from couples with a previous aneuploid miscarriage. Fertil Steril 2012;98:145–50.

Al-Asmar N, Peinado V, Vera M, Remohi J, Pellicer A, Simon C, et al. Oxidative stress in an assisted reproductive techniques setting. Fertil Steril 2006;86:503–12.

Cohen-Bacrie P, Belloc S, Menezo YJ, Clement P, Hamidi J, Benkhalifa M. Correlation between DNA damage and sperm parameters: a prospective study of 1,633 patients. Fertil Steril 2009;91:1801–5.

Hardarson T, Hanson C, Lundin K, Hillensjo T, Nilsson L, Stevic J, et al. Preimplantation genetic screening in women of advanced maternal age caused a decrease in clinical pregnancy rate: a randomized controlled trial. Hum Reprod 2008;23:2806–12.

Harton GL, Magli MC, Lundin K, Montag M, Lemmen J, Harper JC. ESHRE PGD Consortium/Embryology Special Interest Group--best practice guidelines for polar body and embryo biopsy for preimplantation genetic diagnosis/screening (PGD/PGS). European Society for Human Reproduction and Embryology (ESHRE) PGD Consortium/Embryology Special Interest Group. Hum Reprod 2011;26:41–6.

Scott RT Jr, Ferry K, Su J, Tao X, Scott K, Treff NR. Comprehensive chromosome screening is highly predictive of the reproductive potential of human embryos: a prospective, blinded, nonselection study. Fertil Steril 2012;97:870–5.

72

Low-dose estrogen therapy

A 62-year-old woman underwent natural menopause at age 52 years. She tells you she has constant vaginal dryness with irritation. She tried combination hormone therapy in the past but was intolerant to the progestin because of constipation. Her physical examination shows vaginal atrophy. She does not wish to wear a device in her vagina, such as an estradiol ring. The best treatment option for her is

    (A) oral conjugated equine estrogens daily
    (B) vaginal lubricant daily
    (C) transdermal estradiol patch twice weekly
*   (D) vaginal estradiol tablets twice weekly

The Women's Health Initiative trial and other randomized clinical trials have dramatically changed the approach to postmenopausal hormone therapy (HT). Before these trials, observational studies suggested a benefit to postmenopausal HT in the prevention of coronary artery disease and osteoporosis. These benefits were thought to outweigh the known risks of breast cancer and venous thrombotic events. Instead of heart protection, the trials found an increased risk of myocardial infarction, stroke, and dementia, and they confirmed the risks of breast cancer and venous thrombotic events. Although the parameters of such studies, such as the age of subjects and the type of HT used, continue to be debated, the results led to a dramatic shift in the philosophy of HT. Currently, the prevailing attitude toward HT is to use the lowest effective dose for the shortest period of time to relieve menopausal symptoms, including hot flushes and vaginal atrophy.

Hormone therapy consists of estrogen replacement therapy with the addition of progesterone or a progestin to prevent endometrial hyperplasia. This is the only proven benefit of progesterone–progestin. As a consequence, HT is used in women with a uterus, whereas estrogen replacement therapy is used in women with a previous hysterectomy. Many of the risks of estrogen replacement therapy are dose dependent, including breast cancer and venous thrombotic events. For this reason, greater attention has been paid to the use of low-dose estrogen to minimize risks. There is no well-established definition of low-dose estrogen, but common choices include 0.3–0.45 mg conjugated equine estrogen, 0.5 mg oral micronized estradiol, and 0.05 mg transdermal estradiol. Low-dose vaginal estrogen has been obtained by either reducing the amount of standard preparations or by using specially designed products. Two products designed to give low-dose vaginal estrogen include coated estradiol tablets and an estradiol-impregnated vaginal ring that releases estradiol for 3 months.

The U.S. Food and Drug Administration-approved indications for estrogen therapy in postmenopausal women include treatment of vasomotor symptoms (hot flushes), vaginal symptoms from atrophy, and prevention of postmenopausal osteoporosis. This last indication now conflicts with the recommendation of short-term therapy for symptoms.

Treatment of vasomotor symptoms requires systemic therapy. The dose of estrogen correlates with the relief of hot flushes. All of the aforementioned low-dose medications decrease hot flushes when compared with placebo. The addition of progesterone or a progestin to low-dose estrogen replacement therapy does not detract from the effect on hot flushes. High-dose oral progesterone or synthetic progestins also decrease hot flushes compared with placebo, but not nearly to the extent that estrogen does.

The dose of estrogen also correlates with increased bone density. The aforementioned low-dose preparations have been shown to increase bone mineral density (BMD) at the hip or spine compared with placebo. The addition of progesterone or a progestin to estrogen does not detract from the increase in BMD. Alone, progesterone, medroxyprogesterone acetate, and megestrol have no significant impact on BMD. In contrast, the progestin norethindrone acetate does offer bone protection, possibly via aromatization to small amounts of ethinyl estradiol. Vaginal administration of estrogen has not been shown to increase BMD.

Systemic and vaginal therapies alleviate the symptoms of vaginal atrophy, including vaginal dryness, burning, and pain during intercourse.

Common risks of estrogen replacement therapy include endometrial hyperplasia and cancer, venous thrombotic events, and gallbladder disease. The risk of endometrial hyperplasia is estrogen-dose dependent and may be seen with low-dose systemic administration and standard dose vaginal administration. For this reason, these preparations require the use of a continuous or sequential progesterone or progestin for endometrial protection. The dose of progesterone–progestin is dependent on whether it is given continuously or sequentially and on the dose of estrogen being used. The only estrogen preparations that do not require progesterone or a progestin are estradiol tablets and the estradiol-impregnated vaginal ring. Estradiol tablets are used daily for 2 weeks, then twice weekly. The inert synthetic estradiol-impregnated vaginal ring is changed after 3 months. Circulating estradiol levels for the products are in the ranges of 5–11 pg/mL and 7–8 pg/mL, respectively. These levels are within the range of a normal postmenopausal female. Studies for up to a year have shown no evidence of endometrial hyperplasia. Given that this patient does not want a vaginal ring and has an adverse reaction to progesterone–progestin, vaginal estradiol tablets twice weekly would be the best regimen for her.

The risk of venous thrombotic events and gallbladder disease is dependent upon the dose and the route of estrogen administration. Higher odds of both are seen with oral versus nonoral therapy. The initial pass (first-pass effect) of estrogen through the liver causes a greater increase in coagulation factors, whereas the same first pass inhibits bile acid synthesis, resulting in higher biliary cholesterol. Low-dose vaginal estrogen preparations are not known to change the risk of venous thrombotic events or gallbladder disease. The use of a daily vaginal lubricant would not correct this patient's atrophic changes.

Furness S, Roberts H, Marjoribanks J, Lethaby A. Hormone therapy in postmenopausal women and risk of endometrial hyperplasia. Cochrane Database of Systematic Reviews 2012, Issue 8. Art. No.: CD000402. DOI: 10.1002/14651858.CD000402.pub4.

Peeyananjarassri K, Baber R. Effects of low-dose hormone therapy on menopausal symptoms, bone mineral density, endometrium, and the cardiovascular system: a review of randomized clinical trials. Climacteric 2005;8:13–23.

73

Polycystic ovary syndrome and estrogen breakthrough bleeding

A 37-year-old hirsute woman, gravida 2, para 2, has a body mass index of 37 (calculated as weight in kilograms divided by height in meters squared). She comes to your office with irregular, heavy menses that recently required her to be seen in the emergency department, where she received a transfusion because of anemia resulting from her menses. Thyroid-stimulating hormone and prolactin levels are normal. Transvaginal ultrasonography reveals bilateral polycystic-appearing ovaries and a 15-mm endometrium. The finding that an endometrial biopsy is most likely to reveal is

    (A) secretory endometrium
    (B) chronic endometritis
    (C) endometrial hyperplasia with atypia
* (D) disordered proliferative endometrium
    (E) scant endometrium inadequate for evaluation

Polycystic ovary syndrome (PCOS) is the most common reproductive endocrine disorder in women, with a prevalence of approximately 6–15%, depending on the criteria used to diagnose it. The clinical expression of PCOS varies but commonly includes oligoanovulation or anovulation, hyperandrogenism (either clinical or biochemical), and the presence of polycystic ovaries. The condition is heterogeneous, and the etiology of the syndrome remains obscure.

Biopsy of the endometrium should be considered in patients older than 35 years with unopposed estrogen production and amenorrhea that exceeds 12 months. Anovulatory bleeding from unopposed estrogen created in a stagnant follicular phase can cause uncontrolled proliferation of the endometrium. Thus, the finding in the described patient would not be scant endometrium inadequate for evaluation. This endocrine abnormality does not predispose a patient to infection, such as chronic endometritis, and if biopsied, disordered proliferative endometrium often is identified. Although progression to endometrial hyperplasia with atypia may occur, disordered proliferative endometrium is the more common finding. Secretory endometrium is seen after ovulation and exposure of the endometrium to progesterone. This would not been seen in the anovulatory patient.

Although menstrual cycle abnormalities are common during the reproductive years, women with PCOS may ovulate spontaneously. In this minority of PCOS patients, secretory endometrium would be observed. Amenorrheic women with PCOS usually have the most severe hyperandrogenism and high antral follicle counts compared with women who have oligomenorrhea or regular menstrual cycles. In addition, women with PCOS are more likely to have upper body fat distribution compared with weight-matched controls. Greater abdominal or visceral adiposity is associated with greater insulin resistance, which could exacerbate the reproductive and metabolic abnormalities. It is known that obesity is associated with PCOS, but its causal role in this condition has yet to be determined.

Disordered proliferative endometrium is the most common finding, but some data indicate that women with PCOS have an approximately 2.7-fold increased risk of endometrial cancer, which can be preceded by hyperplasia with atypia. Most cases of endometrial cancer are well differentiated and have a good prognosis. Scant endometrium would be more common in a hypoestrogenic state such as menopause or hypothalamic amenorrhea.

Consensus on women's health aspects of polycystic ovary syndrome (PCOS). Amsterdam ESHRE/ASRM-Sponsored 3rd PCOS Consensus Workshop Group. Hum Reprod 2012;27:14–24.

Giudice LC. Endometrium in PCOS: Implantation and predisposition to endocrine CA. Best Pract Res Clin Endocrinol Metab 2006;20:235–44.

Holm NS, Glintborg D, Andersen MS, Schledermann D, Ravn P. The prevalence of endometrial hyperplasia and endometrial cancer in women with polycystic ovary syndrome or hyperandrogenism. Acta Obstet Gynecol Scand 2012;91:1173–6.

74
Hyperthyroidism

A 32-year-old multiparous patient comes to your office for her annual gynecologic examination. She has been experiencing palpitations and slight tremors in her hands. On physical examination, you find that she has a symmetrically enlarged, smooth, and painless thyroid gland. She has a thyroid-stimulating hormone (TSH) level of 0.05 mIU/mL and free thyroxin (T_4) level of 8 ng/mL, results that confirm an overactive thyroid gland. The next step in her evaluation is

* (A) thyroid ultrasonography
* (B) thyrotropin receptor antibody level
* (C) radioiodine uptake
* (D) total triiodothyronine (T_3) to total T_4 ratio

In the United States, the prevalence of hyperthyroidism is approximately 1.2%, with 0.5% being overt and 0.7% subclinical. The most common causes of hyperthyroidism are Graves disease, toxic multinodular goiter, and toxic adenoma. Serum TSH level measurement has the highest sensitivity and specificity of any single blood test used in the evaluation of suspected hyperthyroidism and should be used as an initial screening test. However, when hyperthyroidism is strongly suspected, diagnostic accuracy has been shown to improve when serum TSH and free T_4 levels are assessed at the time of the initial evaluation.

In a patient with a symmetrically enlarged thyroid, recent onset of ophthalmopathy, and moderate-to-severe hyperthyroidism, the diagnosis of Graves disease is likely. A radioiodine uptake test is indicated when the diagnosis is in question (except when contraindicated during pregnancy or when breastfeeding). The test distinguishes the causes of thyrotoxicosis by detecting elevated or normal radioiodine uptake from near-absent uptake (Box 74-1). This test is equally useful whether the thyroid has nodules or is smooth.

Ultrasonography does not generally contribute to the differential diagnosis of thyrotoxicosis. However, ultrasonography may be useful when the radioiodine uptake test is contraindicated or unavailable. In a patient with recent iodine exposure, increased color Doppler flow may be helpful in confirming a diagnosis of thyroid hyperactivity.

An alternative way to diagnose Graves disease is measurement of the thyrotropin receptor antibody level. This approach is used when a thyroid scan or radioiodine uptake test is contraindicated or not available. Use of this test alone will not eliminate the possibility of a different underlying cause of hyperthyroidism.

The total T_3 to total T_4 ratio can be useful in assessing the etiology of thyrotoxicosis when scintigraphy is contraindicated. Because relatively more T_3 is synthesized than T_4 in a hyperactive gland, the ratio (ng/microgram) is usually greater than 20 in Graves disease and toxic nodular goiter and less than 20 in painless or postpartum thyroiditis.

BOX 74-1

Causes of Thyrotoxicosis

Thyrotoxicosis associated with a normal or elevated radioiodine uptake over the neck
* Graves disease
* Toxic adenoma or toxic multinodular goiter
* Trophoblastic disease
* TSH-producing pituitary adenomas
* Resistance to thyroid hormone (T_3 receptor mutation)

Thyrotoxicosis associated with a near-absent radioiodine uptake over the neck
* Painless (silent) thyroiditis
* Amiodarone-induced thyroiditis
* Subacute (granulomatous, de Quervain) thyroiditis
* Iatrogenic thyrotoxicosis
* Factitious ingestion of thyroid hormone
* Struma ovarii
* Acute thyroiditis
* Extensive metastases from follicular thyroid cancer

Abbreviations: TSH, thyroid-stimulating hormone; T_3, triiodothyronine.

Bahn (Chair) RS, Burch HB, Cooper DS, Garber JR, Greenlee MC, Klein I, et al. Hyperthyroidism and other causes of thyrotoxicosis: management guidelines of the American Thyroid Association and American Association of Clinical Endocrinologists. American Thyroid Association and the American Association of Clinical Endocrinologists [published errata appear in Thyroid 2011;21:1169. Thyroid 2012;22:1195]. Thyroid 2011;21:593–646.

75

Fertility preservation techniques

A 21-year-old nulligravid woman comes to your office after she has received a diagnosis of cancer. She is single and interested in future fertility. She would like to explore her options ahead of undergoing gonadotoxic chemotherapy. The fertility preservation technique that you recommend as most appropriate for her is

      (A) ovarian tissue cryopreservation for autotransplantation
*   (B) oocyte cryopreservation
      (C) embryo cryopreservation
      (D) gonadotropin-releasing hormone (GnRH) agonist treatment

Use of fertility preservation has increased in recent years, especially with improvement of techniques such as oocyte cryopreservation. Fertility preservation offers hope for a male or female patient who has received the diagnosis of cancer, given that cancer treatments may be gonadotoxic, limit future fertility, or in some cases render the patient sterile. The level of loss of ovarian function is dependent on the specific chemotherapeutic agent, length of use, and dose. Chemotherapy agents that may cause loss of ovarian function include alkylating agents, such as cyclophosphamide. Patients should be advised that online chemotherapy fertility risk calculators are available.

Cryopreservation now plays an integral role in infertility services. Men can have sperm cryopreserved before anticipated chemotherapy, radiation, or surgery. The relatively lower cytoplasmic content found in sperm compared with other cells in the body potentially provides an advantage that allows for successful cryopreservation. Sperm cryopreservation has permitted the use of cryopreserved sperm decades after freezing. Cryopreservation also has been used successfully to store for future use excess embryos produced during in vitro fertilization cycles ahead of chemotherapy and radiation therapy. Thawed embryos have provided pregnancy success rates similar to fresh embryo transfer cycles.

Oocyte cryopreservation is no longer considered experimental, but it is more technically difficult than sperm or embryo cryopreservation. The large cytoplasmic content of oocytes increases the risk of water crystal damage upon freezing and thawing. Also, the high sensitivity of oocyte microtubules to temperature changes leads to poor viability or damage after cryopreservation. Early studies of oocyte cryopreservation yielded disappointing results in which the survival of an oocyte after thawing was reported to be as low as 25% and as high as 58% with poor pregnancy rates.

Advancements in freezing technology have led to an improvement in oocyte survival. The two main techniques used in oocyte cryopreservation are slow cooling and vitrification. In the slow-cooling technique, freezing takes place at a rate of 0.3–2°C/min (32.5–35.6°F/min). When the cooling is slow, water is removed from the cells by the action of the cryoprotectant, which enables avoidance of damaging intracellular ice crystallization. The oocytes are then stored in liquid nitrogen at –196°C (–320.8°F). To avoid water recrystallization during thawing, the oocytes are thawed quickly. Vitrification, a procedure of ultrarapid cooling of samples, has gained popularity in recent years. Vitrification is the solidification of a solution in a glassy state, which avoids damaging water crystal formation in the oocyte and the surrounding solution. Vitrification also minimizes the exposure time of the oocyte to the toxic cryoprotectant because freezing is done very rapidly. Some researchers have reported an oocyte survival rate as high as 99% and a pregnancy rate of 32.5%. With the advancement of technology, improved oocyte survival rates and improved pregnancy rates have been achieved.

Because of the relative newness of oocyte cryopreservation technology, the long-term outcomes of individuals conceived from cryopreserved oocytes are not known.

A small study has shown that vitrification of oocytes after ovarian stimulation does not appear to be associated with adverse pregnancy outcomes. The major limitation of oocyte cryopreservation is the lack of long-term studies to evaluate offspring born by means of oocyte cryopreservation technology.

Although the survival rate for cryopreserved oocytes is low, it has improved via vitrification. Pregnancy rates per oocyte also have improved via vitrification. Genetic abnormalities in offspring from cryopreserved oocytes have not been shown. However, the cost of oocyte cryopreservation may prove prohibitive for many patients.

Tissue cryopreservation with later autotransplantation does hold promise, but improved freezing techniques are needed before offering this as a treatment option. This process remains experimental and carries the risk of reintroducing cancer cells via the autotransplanted tissue.

Embryo cryopreservation is probably the best way to preserve future fertility but requires a partner or the use of donor sperm. This may not be an option for women without a partner who decline to use a donor specimen for oocyte fertilization. In addition, there are many ethical and legal implications to consider for embryos that will not be used. Oocyte cryopreservation has the advantage of minimizing these issues when considering disposition of stored oocytes.

One reason for the use of GnRH agonist treatment is that less oocyte damage is seen in prepubertal girls who are exposed to gonadotoxic chemotherapeutics compared with postpubertal women. However, the use of GnRH agonists during gonadotoxic treatment has been shown to provide limited benefits in terms of fertility preservation. Indeed, in many cases, the use of GnRH agonists has not been shown to preserve fertility in women who are undergoing gonadotoxic treatment.

Ayensu-Coker L, Bauman D, Lindheim SR, Breech L. Fertility preservation in pediatric, adolescent and young adult female cancer patients. Pediatr Endocrinol Rev 2012;10:174–87.

Cobo A, Garcia-Velasco JA, Domingo J, Remohi J, Pellicer A. Is vitrification of oocytes useful for fertility preservation for age-related fertility decline and in cancer patients? Fertil Steril 2013;99:1485–95.

Elgindy EA, El-Haieg DO, Khorshid OM, Ismail EI, Abdelgawad M, Sallam HN, et al. Gonadatrophin suppression to prevent chemotherapy-induced ovarian damage: a randomized controlled trial. Obstet Gynecol 2013;121:78–86.

Kim SS, Donnez J, Barri P, Pellicer A, Patrizio P, Rosenwaks Z, et al. Recommendations for fertility preservation in patients with lymphoma, leukemia, and breast cancer. ISFP Practice Committee [published erratum appears in J Assist Reprod Genet 2012;29:1155]. J Assist Reprod Genet 2012;29:465–8.

Loren AW, Mangu PB, Beck LN, Brennan L, Magdalinski AJ, Partridge AH, et al. Fertility preservation for patients with cancer: American Society of Clinical Oncology clinical practice guideline update. American Society of Clinical Oncology. J Clin Oncol 2013;31:2500–10.

Mature oocyte cryopreservation: a guideline. Practice Committees of the American Society for Reproductive Medicine and the Society for Assisted Reproductive Technology. Fertil Steril 2013;99:37–43.

76
Müllerian anomaly

A 14-year-old nulligravid girl comes to your office with cyclic pelvic pain that increases with menses. Her medical history is notable for placement of a Harrington rod for treatment of scoliosis. You find a palpable pelvic mass on bimanual rectal examination and suspect a müllerian anomaly. The best next step in the evaluation of this patient is

    (A) two-dimensional ultrasonography
* (B) three-dimensional ultrasonography
    (C) hysterosalpingography
    (D) sonohysterography

Although the true prevalence of müllerian anomalies is unknown because of the high percentage of asymptomatic patients, estimates of prevalence have ranged from 7% in the general population to 8–10% in women with recurrent pregnancy loss. Accurate diagnosis is essential to minimize complications, such as endometriosis and infertility, as well as to prevent unnecessary surgery. Precise classification of a uterine anomaly is of clinical importance because the type of intervention depends on its distinction. Furthermore, accurate classification of uterine anomalies has prognostic importance with respect to obstetric and gynecologic complications.

Historically, conventional two-dimensional transvaginal ultrasonography was considered a good screening tool for the detection of uterine anomalies with high sensitivity (approximately 90–92%). However, the technique is operator dependent and does not reconstruct the coronal plane. Traditionally, patients have been screened with hysterosalpingography, but imaging of the uterine cavity alone without fundal contour is inadequate in many clinical scenarios and in guiding surgical intervention (uterine septum versus bicornuate uterus). This preoperative uncertainty led to laparoscopy with concurrent hysteroscopy as the reference standard for evaluation of müllerian anomalies. In young patients, it is best to avoid invasive procedures such as hysterosalpingography and sonohysterography. Brief fluoroscopy is needed for hysterosalpingography and is contraindicated in patients with iodine contrast allergy.

Three-dimensional transvaginal ultrasonography enables the clinician to assess uterine morphologic characteristics completely, thus alleviating the need for invasive tests. It allows for accurate depiction of the endometrium and serosal superior area of the uterus in the midcoronal plane and a precise measurement of the distance between the midfundus and a line connecting the two internal tubal ostia. With coronal reconstruction, it provides complete information about the nature and extent of uterine masses and congenital anomalies. This information, in addition to lower overall procedure-related risks and patient discomfort, makes it the best diagnostic modality for the described patient.

Renal anomalies are commonly seen in association with müllerian anomalies, such as renal agenesis, ectopic pelvic kidneys, horseshoe kidneys, malrotated kidneys, a duplicated renal pelvis, and unilateral medullary sponge kidney. The renal tract in such patients can be evaluated by means of ultrasonography or magnetic resonance imaging. Although magnetic resonance imaging usually would be considered the best approach to evaluate uterine anomalies, it is contradicted in this patient because of the placement of a Harrington rod.

Acien P, Acien M, Sanchez-Ferrer M. Complex malformations of the female genital tract. New types and revision of classification. Hum Reprod 2004;19:2377–84.

Bocca SM, Abuhamad AZ. Use of 3-dimensional sonography to assess uterine anomalies. J Ultrasound Med 2013;32:1–6.

Raga F, Bauset C, Remohi J, Bonilla-Musoles F, Simon C, Pellicer A. Reproductive impact of congenital Mullerian anomalies. Hum Reprod 1997;12:2277–81.

Rock JA, Schlaff WD. The obstetric consequences of uterovaginal anomalies. Fertil Steril 1985;43:681–92.

77
Oligospermia

A 28-year-old nulligravid woman comes to your office with primary infertility. She has regular menstrual cycles. Hysterosalpingography is normal. Her partner's repeat semen analysis confirms oligospermia. Physical examination demonstrates palpable varicoceles when standing and with Valsalva maneuver. The best next step in management for the male partner is

  (A) observation
  (B) testosterone therapy
* (C) varicocelectomy
  (D) clomiphene citrate therapy

Varicoceles, or abnormally dilated veins in the pampiniform plexus, have long been associated with male factor infertility. The link was made because of the increased incidence of varicoceles observed among infertile men and associated abnormalities found in semen analyses. Varicoceles have an incidence of approximately 4–23% in the general population, 21–41% in men with primary infertility, and 75–81% in men with secondary infertility. Untreated varicoceles may worsen over time.

Varicoceles can be diagnosed by several means; physical examination and scrotal ultrasonography are the most commonly used methods. The condition is graded at the time of the initial physical examination from 1 to 3 by means of the Dubin grading system: grade 3 is visible while the patient is standing; grade 2 is palpable without Valsalva maneuver; and grade 1 cannot be visualized and is only palpable with Valsalva maneuver. Clinical varicocele refers to those detectable by physical examination, either by palpation or visual inspection.

Most men with varicoceles are able to father children. Evidence shows that varicoceles are detrimental to male fertility and surgical correction offers an improvement in a couple's chances of obtaining a pregnancy, either spontaneously or through assisted reproductive technology.

To date, little research has been done to assess the utility of varicocele repair in the treatment of male factor infertility. The American Urological Association and the American Society for Reproductive Medicine have issued practice guidelines on varicocele and infertility. These organizations recommend that varicocele repair should be offered to infertile men with palpable lesions and one or more abnormal semen parameters. Therefore, the described patient should be offered varicocelectomy and not conservative management. Although pregnancies have been reported in women impregnated by men with varicocele, waiting on surgical treatment may increase the length of time to pregnancy. Repeat semen analysis may enable the patient and physician to assess if the abnormal semen parameter is consistent. However, given that the patient is infertile, he is a candidate for surgical repair. Whether or not the patient is in pain, surgical repair is indicated.

A recent meta-analysis done by the American Society of Andrology and the European Academy of Andrology demonstrated that men with idiopathic male infertility may benefit from therapy with clomiphene citrate, a selective estrogen receptor modulator. Clomiphene may increase spontaneous pregnancy rate and improve sperm concentration and motility. If clomiphene therapy is chosen, it is advisable to prescribe such therapy under the care of a urologist. Treatment for male infertility is based on identifying reversible causes of infertility and addressing them with appropriate intervention to achieve a pregnancy. The described male patient has a structural condition that may be significantly contributing to his abnormal semen parameters. Therefore, clomiphene is not the best treatment option for him. Testosterone therapy does not provide efficacious treatment for varicocele.

Jarow JP, Sharlip ID, Belker AM, Lipshultz LI, Sigman M, Thomas AJ, et al. Best practice policies for male infertility. Male Infertility Best Practice Policy Committee of the American Urological Association Inc. J Urol 2002;167:2138–44.

Kroese AC, de Lange NM, Collins J, Evers JLH. Surgery or embolization for varicoceles in subfertile men. Cochrane Database of Systematic Reviews 2012, Issue 10. Art. No.: CD000479. DOI: 10.1002/14651858. CD000479.pub5.

Report on varicocele and infertility. Male Infertility Best Practice Policy Committee of the American Urological Association and the Practice Committee of the American Society for Reproductive Medicine. Fertil Steril 2004;82(suppl):S142–5.

Report on varicocele and infertility. Practice Committee of the American Society for Reproductive Medicine. Fertil Steril 2006;86(suppl):S93–5.

78

Obesity and insulin resistance

A 27-year-old nulliparous woman is referred to you for evaluation of polycystic ovary syndrome (PCOS). She has oligomenorrhea and has a body mass index (BMI) of 38 (calculated as weight in kilograms divided by height in meters squared). Her hirsutism is minimal and is managed with laser treatment. Several of her relatives have type 2 diabetes mellitus. She wants to become pregnant in the near future. The most appropriate next step for this patient is

* (A) a 2-hour oral glucose tolerance test
* (B) nutrition consultation
* (C) metformin hydrochloride
* (D) bariatric surgery

Polycystic ovary syndrome is the most common endocrine disorder in women, with prevalence averaging 6–15%. Clinical manifestations include polycystic ovaries, hyperandrogenism (either clinical or biochemical), and oligo-ovulation or anovulation. More than 60% of PCOS patients are obese. Increases in BMI and abdominal (visceral) adiposity are associated with PCOS and worsening insulin resistance. Approximately 35% of women with PCOS will develop impaired glucose tolerance and 7–10% will develop type 2 diabetes mellitus. Increased hepatic gluconeogenesis and decreased peripheral utilization of glucose cause compensatory increases in insulin secretion. Elevated insulin levels contribute to the risk of hypertension, cardiovascular disease, and type 2 diabetes. In addition, hyperinsulinemia is one of the main promoters of hyperandrogenism seen in PCOS patients.

This patient has PCOS because she is oligomenorrheic, hyperandrogenic, and obese. Sixty percent of PCOS women are hirsute. The diagnosis of PCOS necessitates screening for impaired glucose tolerance and type 2 diabetes with a 2-hour glucose level after a 75-g glucose load (Box 78-1). This is the most appropriate next step in this patient. A hemoglobin A_{1c} test may be a potential substitute for the 2-hour glucose test; however, to date, the efficacy of using a hemoglobin A_{1c} test in patients with PCOS has not been completely proven.

The described patient has primary infertility but more importantly is obese with PCOS. Obese pregnant women are at increased risk of maternal and fetal complications. Maternal complications include gestational diabetes, hypertension, preeclampsia, and fetal macrosomia. Fetal complications include prematurity, stillbirth, spontaneous abortion, congenital anomalies, macrosomia, and childhood and adolescent obesity. Based on these complications, weight loss is recommended before treating her infertility.

Because obese women are at increased risk of maternal and fetal complications of pregnancy, obese women should be encouraged by their obstetricians to initiate a weight-reduction and exercise program before attempting pregnancy. A consultation with a nutritionist will provide the framework and knowledge for healthier eating habits. Substituting healthy foods for unhealthy ones and learning which snacks to incorporate into a dietary plan all contribute to weight loss. In addition, 150 minutes a week of moderate physical activity (ie, five 30-minute sessions) will produce a beneficial effect on glycemic control. Loss of 5% of body weight can lower circulating androgens and induce regular ovulatory cycles while allowing patients to attempt pregnancy without ovulation induction agents. It is important as a first step to determine whether the patient has diabetes because the presence or absence of the disorder will influence nutritional counseling and treatment.

Metformin hydrochloride is a biguanide used in the treatment and prevention of type 2 diabetes mellitus. It is an insulin-sensitizing agent that reduces peripheral insulin levels. In a large meta-analysis of patients at risk of diabetes, metformin reduced BMI by 5.3%, increased high-density lipoprotein, decreased low-density lipoprotein, decreased fasting insulin and glucose, and reduced new-onset diabetes by 40%. Metformin typically is reserved for patients who are diagnosed with impaired glucose tolerance or type 2 diabetes when lifestyle interventions have failed.

Bariatric surgery is indicated when lifestyle modifications and pharmacotherapy are unsuccessful. Comorbidities are reduced or eliminated with an improvement in quality of life. The most common bariatric procedures are Roux-en-Y (restrictive and malabsorptive) and adjustable gastric banding (restrictive). Most experts reserve surgery for patients with a BMI greater than 40 or patients with a BMI greater than 35 with comorbidities. When a bariatric patient becomes pregnant, it is prudent to consult her nutritionist and surgeon and incorporate them into the obstetric team.

> **BOX 78-1**
>
> **Suggested Evaluation for Patients With Polycystic Ovary Syndrome**
>
> *Physical*
> - Blood pressure
> - BMI (weight in kg divided by height in m^2)
>   — 25–30 = overweight, greater than 30 = obese
> - Waist circumference to determine body fat distribution
>   — Value greater than 35 inches = abnormal
> - Presence of stigmata of hyperandrogenism and insulin resistance
>   — Acne, hirsutism, androgenic alopecia, acanthosis nigricans
>
> *Laboratory*
> - Documentation of biochemical hyperandrogenemia
>   — Total testosterone and sex hormone-binding globulin or bioavailable and free testosterone
> - Exclusion of other causes of hyperandrogenism
>   — Thyroid-stimulating hormone levels (thyroid dysfunction)
>   — Prolactin (hyperprolactinemia)
>   — 17-hydroxyprogesterone (nonclassical congenital adrenal hyperplasia due to 21-hydroxylase deficiency)
>   Random normal level less than 4 ng/mL or morning fasting level less than 2 ng/mL
>   — Consider screening for Cushing syndrome and other rare disorders such as acromegaly
> - Evaluation for metabolic abnormalities
>   — 2-hour oral glucose tolerance test (fasting glucose less than 110 mg/dL = normal, 110–125 mg/dL = impaired, greater than 126 mg/dL = type 2 diabetes) followed by 75-g oral glucose ingestion and then 2-hour glucose level (less than 140 mg/dL = normal glucose tolerance, 140–199 mg/dL = impaired glucose tolerance, greater than 200 mg/dL = type 2 diabetes)
> - Fasting lipid and lipoprotein level (total cholesterol, high-density lipoproteins less than 50 mg/dL abnormal, triglycerides greater than 150 mg/dL abnormal [low-density lipoproteins usually calculated by Friedewald equation])
>
> *Ultrasonographic Examination*
> - Determination of polycystic ovaries: in one or both ovaries, either 12 or more follicles measuring 2–9 mm in diameter or increased ovarian volume (greater than 10 cm^3). If there is a follicle greater than 10 mm in diameter, the scan should be repeated at a time of ovarian quiescence in order to calculate volume and area. The presence of one polycystic ovary is sufficient to provide the diagnosis.
> - Identification of endometrial abnormalities
>
> *Optional Tests to Consider*
> - Gonadotropin determinations to determine cause of amenorrhea
> - Fasting insulin levels in younger women, those with severe stigmata of insulin resistance and hyperandrogenism, or those undergoing ovulation induction
> - 24-hour urinary free-cortisol excretion test or a low-dose dexamethasone suppression test in women with late onset of polycystic ovary syndrome symptoms or stigmata of Cushing syndrome

Diamanti-Kandarakis E, Dunaif A. Insulin resistance and the polycystic ovary syndrome revisited: an update on mechanisms and implications. Endocr Rev 2012;33:981–1030.

Knowler WC, Barrett-Connor E, Fowler SE, Hamman RF, Lachin JM, Walker EA, et al. Reduction in the incidence of type 2 diabetes with lifestyle intervention or metformin. Diabetes Prevention Program Research Group. N Engl J Med 2002;346:393–403.

Lerchbaum E, Schwetz V, Giuliani A, Obermayer-Pietsch B. Assessment of glucose metabolism in polycystic ovary syndrome: HbA1c or fasting glucose compared with the oral glucose tolerance test as a screening method. Hum Reprod 2013;28:2537–44.

Obesity in pregnancy. Committee Opinion No. 549. American College of Obstetricians and Gynecologists. Obstet Gynecol 2013;121:213–7.

Polycystic ovary syndrome. ACOG Practice Bulletin No. 108. American College of Obstetricians and Gynecologists. Obstet Gynecol 2009; 114:936–49.

Salpeter SR, Buckley NS, Kahn JA, Salpeter EE. Meta-analysis: metformin treatment in persons at risk for diabetes mellitus. Am J Med 2008;121:149–57.

79

Salpingitis isthmica nodosa

A 30-year-old nulligravid woman and her husband visit your clinic for primary infertility. Your diagnostic evaluation reveals a normal semen analysis and evidence of ovulatory cycles. Hysterosalpingography demonstrates bilateral proximal tubal occlusion. The situation in which microsurgical segmental tubal resection and reanastomosis is less successful is

    (A) luminal fibrosis
* (B) salpingitis isthmica nodosa
    (C) modified Pomeroy tubal ligation
    (D) tubal ligation via clip

Hysterosalpingography is considered a diagnostic test to assess the uterine cavity and fallopian tube patency in infertile patients. Proximal fallopian tube occlusion is demonstrated by failure of the contrast agent to enter the fallopian tubes. The failure of the dye to enter the fallopian tubes could be a result of a tubal spasm that has prevented the dye from entering the tubes. Alternatively, the proximal tubal occlusion may be diagnostic of a true fallopian tube abnormality.

In vitro fertilization (IVF) is an infertility treatment that can bypass the fallopian tubes. However, treatment with IVF may not always be necessary or accessible to the patient. Therefore, surgical treatment of proximal tubal occlusion is a reasonable alternative. There are two methodologies associated with proximal tubal repair: 1) hysteroscopic proximal tubal catheterization and 2) microsurgical tubal resection. Both techniques require specialized skill for microsurgery and advanced hysteroscopic techniques.

Salpingitis isthmica nodosa is a fallopian tube abnormality typified by diverticula (outpouchings) of the fallopian tube. The tubal mucosa penetrates in the myosalpinx and leads to a diverticular appearance of the fallopian tubes on hysterosalpingography. This condition is largely associated with infertility.

True proximal tubal occlusion is most often the result of luminal tubal fibrosis. Chronic inflammation and intratubal endometriosis also can cause blockage of the proximal fallopian tube. Intraluminar fibrosis, intratubal endometriosis, and chronic inflammation may be amenable to microsurgical excision and repair. Salpingitis isthmica nodosa is not amenable to surgical correction given that the fallopian tube has inherent abnormalities. The true cause for the diverticular appearance of the tube is unknown. However, repair of the tube is fruitless given that tubal damage is multifocal and that women with salpingitis isthmica nodosa are at high risk for ectopic pregnancy. Such patients are good candidates for IVF and bypassing of the abnormal fallopian tubes. Neither modified Pomeroy tubal ligation nor tubal ligation by means of the clip have significant bearing on microsurgical segmental tubal resection and reanastomosis.

Das S, Nardo LG, Seif MW. Proximal tubal disease: the place for tubal cannulation. Reprod Biomed Online 2007;15:383–8.

Renbaum L, Ufberg D, Sammel M, Zhou L, Jabara S, Barnhart K. Reliability of clinicians versus radiologists for detecting abnormalities on hysterosalpingogram films. Fertil Steril 2002;78:614–8.

Schippert C, Soergel P, Staboulidou I, Bassler C, Gagalick S, Hillemanns P, et al. The risk of ectopic pregnancy following tubal reconstructive microsurgery and assisted reproductive technology procedures. Arch Gynecol Obstet 2012;285:863–71.

80
Heavy menstrual bleeding

After she has undergone tubal ligation, a 29-year-old woman comes to your office with a history of monthly heavy menstrual bleeding. Ultrasonographic evaluation of her uterus and thyroid-stimulating hormone level are normal. The best nonsurgical therapy for her is

* (A) levonorgestrel intrauterine device (IUD)
* (B) continuous combination hormonal contraceptives
* (C) luteal phase progestin
* (D) nonsteroidal antiinflammatory drugs (NSAIDs)
* (E) tranexamic acid

Heavy menstrual bleeding, defined as menstrual blood loss greater than 80 mL per cycle, affects up to 30% of reproductive-aged women and negatively affects their physical activities and quality of life. Idiopathic heavy menstrual bleeding is regular bleeding in the absence of recognizable pelvic pathology or a bleeding disorder. Leiomyomas have been found in up to 40% of women with heavy menstrual bleeding; however, approximately 50% of women who undergo hysterectomy for this condition are found to have a normal uterus. Appendix C shows the International Federation of Gynecology and Obstetrics classification system for causes of abnormal uterine bleeding (AUB).

Data from some studies suggest that unrestrained local inflammatory events with or without deficient repair processes in the endometrium may contribute to the onset of AUB. Medical therapy such as NSAIDs, antifibrinolytics, continuous combination hormonal contraceptives, progestins, and the levonorgestrel IUD represent attractive options to avoid unnecessary surgery, such as endometrial ablation and hysterectomy. The levonorgestrel IUD provides treatment for at least 5 years, is as effective as endometrial ablation, and is reversible.

A recent study was performed among 571 women with AUB who were randomly assigned to the levonorgestrel IUD or usual medical treatment (NSAIDs, tranexamic acid, continuous combination oral contraceptives, and a progestin-only oral contraceptive). Over a 2-year period, the levonorgestrel IUD was more effective than usual medical treatment in reducing the effect of AUB on quality of life. Therefore, the levonorgestrel IUD would provide the best therapy for the described patient.

Abu Hashim H. Medical treatment of idiopathic heavy menstrual bleeding. What is new? An evidence based approach. Arch Gynecol Obstet 2013;287:251–60.

Endrikat J, Vilos G, Muysers C, Fortier M, Solomayer E, Lukkari-Lax E. The levonorgestrel-releasing intrauterine system provides a reliable, long-term treatment option for women with idiopathic menorrhagia. Arch Gynecol Obstet 2012;285:117–21.

Gupta J, Kai J, Middleton L, Pattison H, Gray R, Daniels J. Levonorgestrel intrauterine system versus medical therapy for menorrhagia. ECLIPSE Trial Collaborative Group. N Engl J Med 2013;368:128–37.

81

Late-onset congenital adrenal hyperplasia

An Ashkenazi Jewish 23-year-old nulligravid woman visits your office with irregular menstrual cycles. She is interested in becoming pregnant. She has seen her dermatologist for acne and hirsutism and has been shaving her chin weekly for the past 10 years. On physical examination, she has acne on her face and mild hair growth on her upper lip and chin. She shows no signs or symptoms of virilization. Her thyroid-stimulating hormone level is normal. The best next diagnostic step is to test for the level of

(A) total testosterone
* (B) morning follicular 17α-hydroxyprogesterone
(C) morning luteal 17α-hydroxyprogesterone
(D) dehydroepiandrosterone sulfate (DHEAS)
(E) prolactin

Late-onset congenital adrenal hyperplasia or nonclassic adrenal hyperplasia (CAH) affects approximately 1 in 1,000 non-Jewish whites and 1 in 30 Ashkenazi Jews. The disease is also prevalent in Alaska Native American Inuits, Hispanics, Italians, and Yugoslavs. The condition predominates in women who have a family history of nonclassic CAH or who have onset of hirsutism around menarche. It is an autosomal recessive condition that results in 50–80% loss of 21-hydroxylase activity (P450c21 enzyme) due to mutations in the CYP21A2 gene. This results in excessive adrenal androgen production. Enzyme defects in 11β-hydroxylase and 3β-hydroxysteroid dehydrogenase are rare causes of adrenal androgen overproduction. In patients with nonclassic CAH, the pathophysiology is characterized by inadequate cortisol synthesis, which stimulates pituitary adrenocorticotropic hormone (ACTH) secretion. Androgens proximal to the enzymatic blockade increase, causing hyperandrogenism.

The described patient has hirsutism, acne, and oligomenorrhea. These are the three most common symptoms in women with nonclassic CAH. Patients with nonclassic CAH have normal external genitalia, experience precocious pubarche, or may present as adolescents or adults with hyperandrogenism. This is in contrast to classic CAH (salt-wasting and virilizing forms), which manifests at birth with ambiguous genitalia.

Approximately 10% of all women are affected by androgen excess. The differential diagnosis includes polycystic ovary syndrome, nonclassic CAH, and androgen-secreting tumors. Differentiating between these disorders can present a diagnostic challenge. Polycystic ovary syndrome patients have a higher frequency of oligomenorrhea and amenorrhea with ovaries that are polycystic in appearance compared with the ovaries of patients with androgen-secreting tumors and nonclassic CAH. Laboratory evaluation with basal 17α-hydroxyprogesterone level helps to differentiate between nonclassic CAH and polycystic ovary syndrome. Most patients with nonclassic CAH will have a basal 17α-hydroxyprogesterone level greater than 200 ng/dL.

The best diagnostic test to assess for nonclassic CAH is a morning follicular serum 17α-hydroxyprogesterone level. Levels peak in the morning and reach a low later in the day. Morning follicular 17α-hydroxyprogesterone levels are higher in women with nonclassic CAH than in controls. By obtaining levels in the follicular phase and not in the luteal phase, false-positive levels are greatly reduced. Levels less than 200 ng/dL exclude nonclassic CAH, whereas levels greater than 800 ng/dL are virtually diagnostic for nonclassic CAH. The criterion standard for confirming the diagnosis of nonclassic CAH is the ACTH (cosyntropin) stimulation test. Baseline 17α-hydroxyprogesterone is drawn in the morning at any time during the menstrual cycle; ACTH (cosyntropin) 250 micrograms is administered intravenously as a bolus over a minute. A blood sample for 17α-hydroxyprogesterone is drawn 60 minutes after ACTH is given. In most patients with nonclassic CAH, stimulated 17α-hydroxyprogesterone levels will be in excess of 1,500 ng/dL at 1 hour. In clinical practice, it is not realistic to perform an ACTH stimulation test on all hyperandrogenic (hirsute) women who may have nonclassic CAH, given that it is a rare condition.

Total testosterone level would not be the best next diagnostic test. The described patient has risk factors for nonclassic CAH that include her ethnicity, perimenarchal onset of hirsutism, irregular cycles, and no virilization. If she was virilized, had rapid onset, or exhibited severe hirsutism, a total testosterone level would be indicated.

A total testosterone level greater than 200 ng/dL will identify most androgen-producing tumors. If the level is less than 200 ng/dL and the patient has rapid virilization or progressive hirsutism, an androgen-secreting tumor still should be assumed and further evaluation would be warranted.

Dehydroepiandrosterone sulfate is secreted primarily from the adrenal gland. It provides substrate for testosterone and dihydrotestosterone conversion; DHEAS levels are not elevated in nonclassic CAH. Women with androgen-secreting adrenal tumors can have elevated DHEAS levels in addition to elevated testosterone. However, a DHEAS level has little clinical value in a hirsute patient who is not virilized.

Hyperprolactinemia does not cause mild hirsutism by stimulating adrenal or ovarian androgen production. In addition, hyperprolactinemia can cause menstrual irregularity and galactorrhea in some women. For the described patient, testing for the prolactin level would not be the next best diagnostic test.

Auchus RJ. Congenital adrenal hyperplasia in adults. Curr Opin Endocrinol Diabetes Obes 2010;17:210–6.

Fritz MA, Speroff L. Hirsutism. In: Fritz MA, Speroff L, editors. Clinical gynecologic and infertility. 8th ed. Philadelphia (PA): Lippincott Williams & Wilkins; 2011. p. 533–65.

Pall M, Azziz R, Beires J, Pignatelli D. The phenotype of hirsute women: a comparison of polycystic ovary syndrome and 21-hydroxylase-deficient nonclassic adrenal hyperplasia. Fertil Steril 2010;94: 684–9.

Trapp CM, Oberfield SE. Recommendations for treatment of nonclassic congenital adrenal hyperplasia (NCCAH): an update. Steroids 2012;77:34–6.

Witchel SF. Nonclassic congenital adrenal hyperplasia. Curr Opin Endocrinol Diabetes Obes 2012;19:151–8.

82

Precocious puberty

You examine a 5-year-old Asian girl with a 9-month history of breast development and two episodes of vaginal bleeding. Physical examination demonstrates Tanner stage 3 breast development and Tanner stage 2 pubic hair development. Laboratory studies show a follicle-stimulating hormone (FSH) level of 4.3 mU/L, luteinizing hormone (LH) level of 3.6 mU/L, thyroid-stimulating hormone (TSH) level of 0.98 mU/L, prolactin level of 7.0 ng/mL, and estradiol level of 75 pg/mL. Bone age is advanced by 3 years with a predicted height of less than 1.52 m (60 in.). Cranial magnetic resonance imaging is normal. The most appropriate medical therapy is

    (A) oral ketoconazole
*   (B) injectable gonadotropin-releasing hormone (GnRH) agonist
    (C) depot medroxyprogesterone acetate
    (D) aromatase inhibitor

Precocious puberty is defined as early breast development, early pubic hair development, or both with menses before age 8 years. Because of earlier pubertal milestones among different ethnicities, some experts have recommended that these guidelines be changed to thelarche before age 7 years in white girls and before age 6 years in African American girls. Observations that suggest pathology and that indicate the need for evaluation at a borderline age are rapid development with bone age greater than 2 years and a predicted height of less than 1.5 m (59 in.) or 2 standard deviations below the target height, central nervous system symptoms, and symptoms that affect the emotional health of the girl.

Precocious puberty is early sexual development in which the normal sequence of events usually is preserved with shortened time intervals. Estrogen-induced accelerated growth, advanced bone age, and early closure of bone epiphyseal plates result in short stature. The younger the child, the more likely she will have a pathologic cause for premature sexual development. The etiologies of precocious puberty are commonly defined as central (GnRH-dependent) causes and peripheral (GnRH-independent) causes (Table 82-1).

Central precocious puberty consists of early activation of the hypothalamic–pituitary–ovarian axis and is idiopathic in greater than 50% of girls with central precocious puberty. Other cases involve intracranial processes that activate GnRH secretion through mostly unknown mechanisms. The most common tumor that causes central precocious puberty is a hypothalamic hamartoma, in which astroglial cells within the tumor release transforming growth factor-beta protein to induce GnRH release from the hypothalamus.

Adoption of children from less developed countries may be associated with central precocious puberty, most likely due to weight gain from rapid improvement in

TABLE 82-1. Causes of Isosexual Female Precocious Puberty

Central (GnRH-Dependent) Causes	Peripheral (GnRH-Independent) Causes
Idiopathic	Estrogen-producing estrogen tumor
International adoption	Simple ovarian cyst
Hypothalamic hamartoma	Hypothyroidism
Brain tumors	Estrogen-producing adrenal tumor
Hydrocephalus	McCune–Albright syndrome
Previous encephalitis, meningitis, head trauma, granulomatous disease	Exogenous estrogen
Subarachnoid cyst	Peutz–Jeghers syndrome
Neurofibromatosis 1	Silver–Russell syndrome
Sturge–Weber syndrome	

Abbreviation: GnRH, gonadotropin-releasing hormone.

nutrition. Primary hypothyroidism sometimes is considered a central cause but does not initially involve GnRH activation. Most likely, high TSH levels cross-react with FSH receptors on the pituitary. Hypothyroidism is the only cause of precocious puberty that does not result in an advanced bone age.

Peripheral precocious puberty is due to independent secretion of estrogen from the ovary or to an iatrogenic etiology from exposure to an estrogen or estrogen-like compound. Ovarian tumors, such as granulosa–theca cell, teratomas, or dysgerminoma, may directly secrete estrogen. The mechanism with other tumors appears to be induction of surrounding theca cell proliferation leading to increased hormone production. Cases of unilateral ovarian cysts causing peripheral precocious puberty have been reported, most likely with a similar mechanism. This is not to be confused with multiple ovarian cysts, which are the consequence of gonadotropin stimulation of the ovaries.

McCune–Albright syndrome is a unique cause of peripheral precocious puberty involving autonomous activation of gonadotropin receptors. Two G protein complexes, one stimulatory (Gsα) and one inhibitory (Giα), are responsible for cyclic-adenosine monophosphate activation after binding of a ligand (such as FSH and LH) to a G protein-coupled receptor. McCune–Albright syndrome is caused by a postfertilization mutation of the Gsα complex resulting in ovarian production of estrogen in the absence of gonadotropins. Although McCune–Albright syndrome can result in numerous abnormalities, the hallmarks include the triad of a unique bone lesion called polyostotic fibrous dysplasia, large pigmented skin lesions (café au lait patches), and peripheral precocious puberty. The peripheral precocious puberty seen with McCune–Albright syndrome typically has a very early onset in infancy.

The evaluation of isosexual precocious puberty includes a history with close attention to possible estrogen exposure, review of growth records looking for an increase in height percentile, and a physical examination to determine Tanner stages and presence of skin lesions. Initial testing includes a bone age test and blood analysis for estradiol, FSH, LH, and TSH. Bone age is increased in all cases of precocious puberty except when caused by primary hypothyroidism. A high estradiol level with a low gonadotropin level suggests an ovarian tumor with the next test being pelvic ultrasonography. A high gonadotropin level for age is consistent with central precocious puberty and the next step is magnetic resonance imaging. Cases without obvious diagnosis after initial testing may require GnRH stimulation testing. Administration of GnRH (or a GnRH agonist) results in a specific pattern of FSH and LH response depending on the diagnosis. In normal children, no significant increase in FSH or LH is seen. Children with premature thelarche have an FSH-dominant response whereas an LH-dominant response is diagnostic of central precocious puberty.

Treatment of precocious puberty depends on the etiology and the goals of therapy. Obviously, sources such as hypothyroidism and ovarian tumors require correction. Simple ovarian cysts often will resolve with time and observation. The decision to treat idiopathic or noncorrectable causes of central precocious puberty is dependent on the age of the child, the predicted final height, and the severity of symptoms. The treatment of choice is a GnRH agonist, with the most convenient being a 1-year implant of histrelin acetate. Before the discovery of GnRH agonist, precocious puberty was commonly treated with depot medroxyprogesterone acetate. This drug would decrease breast development and cease menses but would have little effect on bone maturation. Treatment for McCune–Albright syndrome requires the use of enzyme inhibitors given the autonomous action of the gonadotropin receptors independent of ligand. Options include the use of ketoconazole, which blocks production of sex steroids primarily through inhibition of 17α-hydroxylase activity; letrozole through inhibition of aromatase; and fulvestrant through antagonism of the estrogen receptor.

The most appropriate medical therapy for the described patient is injectable GnRH agonist. The long-term prognosis is positive for girls treated with a GnRH agonist for idiopathic central precocious puberty. Compared with controls, girls treated with GnRH agonist will have no difference in menstrual cycles or fertility.

Carel JC, Leger J. Clinical practice. Precocious puberty. N Engl J Med 2008;358:2366–77.

Fuqua JS. Treatment and outcomes of precocious puberty: an update. J Clin Endocrinol Metab 2013;98:2198–207.

Mason P, Narad C. Long-term growth and puberty concerns in international adoptees. Pediatr Clin North Am 2005;52:1351–8, vii.

83
Premature ovarian failure

A 28-year-old healthy nulligravid woman stopped taking oral contraceptives 1 year ago with the goal of pregnancy. However, she did not resume spontaneous menstrual cycles. Her primary care provider ordered tests that showed a follicle-stimulating hormone (FSH) level of 45 mIU/mL, estradiol level of 12 pg/mL, thyroid-stimulating hormone level of 1.37 mIU/mL, and a prolactin level of 9 ng/mL. The most appropriate next test or screen in her evaluation is

(A) bone density
* (B) karyotype
(C) *FMR1* gene mutation
(D) antimüllerian hormone
(E) antiovarian antibody

Primary ovarian insufficiency, also known as premature ovarian failure or hypergonadotropic hypogonadism, is defined as ovarian failure before age 40 years. Approximately 1% of women will experience primary ovarian insufficiency. The diagnosis generally includes amenorrhea, although some women with primary ovarian insufficiency may experience an occasional ovulatory cycle. Approximately 5–10% of women with primary ovarian insufficiency may conceive without assistance and have an uncomplicated pregnancy. Generally, patients have FSH levels above 30 mIU/mL. Some women with primary ovarian insufficiency will experience bleeding after a progestin withdrawal test. As a result, caution should be employed when using this test as part of the diagnostic evaluation of amenorrhea.

Premature ovarian failure is caused by a heterogenous group of disorders, although for most women no cause is identified. Women who were previously treated with gonadotoxic chemotherapy for a malignancy or autoimmune disease, such as lupus nephritis, may have transient or permanent damage to the ovarian function. Some of these women have immediate ovarian failure, whereas others will continue to menstruate but undergo early menopause. An autoimmune cause for primary ovarian insufficiency may be found in some women, and such individuals often have antiadrenal antibodies.

A number of genetic causes for primary ovarian insufficiency have been identified. Karyotype abnormalities, single-gene defects, and multifactorial polygenic conditions all can result in primary ovarian insufficiency. Turner syndrome (45,X) usually presents as primary amenorrhea and has a classic phenotype that includes streak ovaries. Other structural abnormalities of the X chromosome can result in primary ovarian insufficiency, such as Turner syndrome mosaic, mutations in regions of the X chromosome, trisomy X, and mutations that have a 46,XY karyotype (Swyer syndrome).

As a first step in the evaluation of women with primary ovarian insufficiency, such as the described patient, it is important to obtain a karyotype to evaluate for a missing or abnormal X chromosome and for the presence of any portion of a Y chromosome. If there is evidence of Turner syndrome, even in the mosaic form, the patient may need evaluation for serious sequelae of this syndrome, such as aortic rupture. If any portion of a Y chromosome is found, the patient is at risk of malignancy and should consider oophorectomy.

Evidence exists of an association between carriers of fragile X mutation and women with primary ovarian insufficiency. Fragile X mutations involve a dynamic trinucleotide repeat (CGG) sequence mutation in the X-linked *FMR1* gene at the terminal end of the long arm of the X chromosome. The normal *FMR1* gene has approximately 30 CGG repeats. The fully expanded form of the condition is called fragile X syndrome and has more than 200 repeats. Fragile X syndrome is the most common known genetic cause of mental retardation and autism in males. Fragile X premutation carriers have between 55 repeats and 200 repeats and may present with no symptoms, premature ovarian failure (approximately 15% of women with the premutation), or fragile X-associated tremor–ataxia syndrome.

In women with primary ovarian insufficiency, testing will reveal a fragile X premutation in 12% of women with a family history of primary ovarian insufficiency and 3% of women without such a family history. Male offspring born to women who are carriers of fragile X mutations are at significant risk of having mental retardation. It is important to offer screening for fragile X mutations in all women with unexplained primary ovarian insufficiency after testing for karyotype abnormalities has been performed. All carriers should be offered formal genetic counseling.

Women with primary ovarian insufficiency are at increased risk of osteopenia and osteoporosis. Once the diagnosis of primary ovarian insufficiency is made, the clinician should discuss strategies for bone protection, such as hormone therapy and dietary supplementation with calcium and vitamin D. Bone density evaluation, with a technique such as a dual energy X-ray absorptiometry scan, is important in women with primary ovarian insufficiency. However, bone density testing is not an urgent first step in the evaluation of primary ovarian insufficiency.

No evidence exists that testing for the antimüllerian hormone level will help in the diagnosis of primary ovarian insufficiency, and an elevated FSH level and amenorrhea are adequate to make the diagnosis. Similarly, serum antiovarian antibody levels have not been shown to be useful in the diagnosis of primary ovarian insufficiency. Moreover, serum antiovarian antibody level testing is not currently available in many clinical settings.

Carrier screening for fragile X syndrome. Committee Opinion No. 469. American College of Obstetricians and Gynecologists. Obstet Gynecol 2010;116:1008–10.

Rebar RW. Premature ovarian failure. Obstet Gynecol 2009;113: 1355–63.

Rebar RW, Connolly HV. Clinical features of young women with hypergonadotropic amenorrhea. Fertil Steril 1990;53:804–10.

Speroff L, Fritz MA. Amenorrhea. In: Speroff L, Fritz MA, editors. Clinical gynecologic endocrinology and infertility. 8th ed. Philadelphia (PA): Lippincott Williams and Wilkins; 2011. p. 464–73.

84

Acne in an adolescent patient

A 17-year-old adolescent woman with irregular menses and acne comes to your office. She is being treated by her dermatologist with topical antibiotics and topical tretinoin. Her evaluation is consistent with polycystic ovary syndrome (PCOS). Her best treatment option is

      (A) oral finasteride
      (B) oral spironolactone
*   (C) combination oral contraceptive (OC)
      (D) progestin-only OC
      (E) metformin hydrochloride

Acne is a multifactorial disease consisting of skin inflammation with exacerbation by systemic androgens. The major factors in acne include

- inflammation
- colonization of the pilosebaceous unit with the bacterium *Propionibacterium acnes*
- altered follicular growth and differentiation
- sebaceous gland hyperplasia with seborrhea

The main action of androgens is the increased production of sebum by the sebaceous gland. Sebum acts as a nutrient source for *P acnes* and enhances growth.

Acne is not normal in females and may require evaluation for syndromes of androgen excess such as PCOS, congenital adrenal hyperplasia, and very rarely Cushing syndrome. Most females with acne will have biochemical evidence of elevated dehydroepiandrosterone sulfate with or without elevated free testosterone and decreased levels of sex hormone-binding globulin (SHBG). Although numbers vary, approximately one in three women with acne will meet the clinical diagnosis of PCOS. The severity of acne does not correlate with androgen levels. The pubertal increase in insulin-like growth factor 1 (IGF-1) also plays a role in acne. The action of IGF-1 appears to increase active androgen receptors. The roles of the androgen receptor and IGF-1 in acne are made evident by clinical syndromes. Patients with complete androgen insensitivity syndrome or who lack IGF-1 (Laron dwarfism) do not have acne.

Females with obesity or a finding of acanthosis nigricans should have testing for glucose tolerance to evaluate for impaired glucose tolerance or type 2 diabetes mellitus. Ideally, this is done with a 2-hour oral glucose tolerance test. All women with acne need a referral to a dermatologist to help them manage the acne.

Treatment of acne is directed at decreasing inflammation and sebum production with topical and systemic therapy. Combination topical therapy is superior to monotherapy. Benzoyl peroxide is a mainstay of acne therapy because it has antiinflammatory, bactericidal, and antikeratinization properties. Topical antibiotics may offer benefits, but overuse has led to increased bacterial resistance. Retinoids are vitamin A derivatives that normalize abnormal growth of keratinocytes and decrease inflammation. Topical azelaic acid is a natural substance produced by a ubiquitous yeast and is bactericidal. Multiple combinations of these products are available.

Systemic treatment of acne incudes antibiotics, retinoic acid, and drugs to decrease androgens or block androgen action. The most commonly used antibiotic is tetracycline or minocycline. Oral isotretinoin is related to vitamin A and is highly effective for severe acne or cases refractory to standard therapy. However, it has the potential for severe adverse effects, including onset of psychiatric disorders, pseudotumor cerebri, severe hypertriglyceridemia with associated pancreatitis, hepatotoxicity, inflammatory bowel disease, vision changes, hearing loss, and teratogenicity. Use of the drug requires use of efficient birth control and monitoring of lipid levels and liver function tests.

Hormonal therapy for acne seeks to decrease free androgen levels, decrease conversion of testosterone to 5α-dihydrotestosterone, and block the androgen receptor. The most common therapy is a combination OC. Combination OCs decrease free androgens by increasing levels of SHBG and decrease ovarian androgen production. The magnitude of SHBG increase is dependent on the dose of ethinyl estradiol and the type of progestin in the pill. Third-generation progestins, including desogestrel and norgestimate, result in much higher levels of SHBG compared with first-generation norethindrone and second-generation levonorgestrel. Drospirenone, which is chemically related to spironolactone, yields an increase in SHBG similar to the third-generation progestins. The U.S. Food and Drug Administration (FDA) has approved three combination OCs for the treatment of acne: a 20-microgram ethinyl estradiol–drospirenone preparation, a 35-microgram ethinyl estradiol–norgestimate preparation in which the norgestimate dose increases from 0.18 mg to 0.215 mg to 0.25 mg, and an ethinyl estradiol–norethindrone combination in which the ethinyl estradiol dose increases from 20 micrograms to 30 micrograms to 35 micrograms. Note that FDA approval should not suggest that other combination OCs are not effective for acne treatment but more likely that the pharmaceutical company has chosen not to perform the necessary trials for this indication. Oral finasteride, oral spironolactone, and the progestin-only OC have not been approved by the FDA for the treatment of acne.

Type 2 5α-reductase inhibitor finasteride has been studied for treatment of acne and yields less response compared with combination OCs, spironolactone, and flutamide. This is most likely because of the lack of inhibition of type 1 5α-reductase that is found in the pilosebaceous unit. Dutasteride, an inhibitor of type 1 and type 2 5α-reductase, has yet to be studied.

Spironolactone has dual actions: blocking the androgen receptor and inhibiting 5α-reductase. The drug is used most commonly with a combination OC for several reasons. Spironolactone is potentially teratogenic and requires effective birth control. Spironolactone also may cause increased menstrual bleeding, which is prevented by a combination OC. The joint use of a combination OC and spironolactone is more effective than single therapy. Nonsteroidal pure antiandrogens, such as bicalutamide, flutamide, and hydroxyflutamide, also are effective for the treatment of acne. These drugs are not used frequently because of possible hepatotoxicity and the need for initial routine monitoring of liver function. They are also teratogenic.

Progestin-only contraceptives, such as oral norethindrone, implantable etonogestrel, injectable depot medroxyprogesterone acetate, and the levonorgestrel intrauterine device, do not increase SHBG and are not effective for the treatment of acne.

Metformin hydrochloride has a beneficial effect on acne in women diagnosed with PCOS. Because glucose utilization increases, insulin levels and circulating androgen levels decrease and SHBG levels increase. To date, no studies have been done to compare the efficacy of metformin with other treatment options for acne, and metformin is not approved by the FDA for the treatment of acne.

Oral corticosteroids are used only rarely for long-term treatment of acne because of the drugs' adverse effects. Oral corticosteroids work by decreasing inflammation and adrenal androgen production. Chronic therapy may be needed in women with adult-onset congenital adrenal hyperplasia who fail to respond to nonsteroidal therapies. Steroid adverse effects are dose dependent and numerous, the most significant adverse effects being diabetes mellitus, osteoporosis, and aseptic necrosis of the hip.

Dreno B, Layton A, Zouboulis CC, Lopez-Estebaranz JL, Zalewska-Janowska A, Bagatin E, et al. Adult female acne: a new paradigm. J Eur Acad Dermatol Venereol 2013;27:1063–70.

Katsambas AD, Dessinioti C. Hormonal therapy for acne: why not as first line therapy? facts and controversies. Clin Dermatol 2010;28: 17–23.

85

Psychologic effect of infertility

A 33-year-old nulligravid patient with a 2-year history of infertility has just completed her second cycle of in vitro fertilization (IVF) for severe male factor infertility. She had a negative serum β-hCG test result. After several weeks, her concerned husband calls your office and tells you that she has decreased energy, is tearful and anxious, and spends nearly all day in bed. The best next treatment option for the patient is

* (A) psychologic counseling
* (B) anxiolytic medication
* (C) antidepressants
* (D) acupuncture
* (E) immediate initiation of next treatment cycle

In 2005, the number of infertile individuals worldwide was estimated to be 60–80 million with a growth of approximately 2 million new cases per year. In developed countries, infertility is diagnosed in approximately 17–26% of reproductive-aged couples. The prevalence of infertility increases with age from 20% among women aged 35–39 years to 25–30% among women 40 years and older. Infertility treatment often is accompanied by psychologic disturbances that may significantly affect the relationship between the patient and her partner.

Numerous studies have shown that more than 50% of infertile couples consider infertility to be the most disappointing experience of their lives. Approximately 80% of infertile couples report infertility to be a stressful experience. Among infertile couples screened with the self-reported Beck Depression Inventory, symptoms of depression were observed in 37% of infertile women, and among those women, 19.4% had moderate or severe symptoms of depression. Depression among infertile women is the result of prolonged and severe normal psychologic response to the diagnosis, which comprises grief and mourning. Depression is highest in women who have been undergoing infertility treatment for the longest period. Among couples who are undergoing IVF, studies have demonstrated depression prevalence before IVF in 33% of women and 3.5% of men. After IVF failure, depression prevalence increased to 43% and 8% among women and men, respectively. It is believed that the incidence of depression in women is higher than in men because of a potentially higher propensity for women to express their feelings compared with men.

The best next treatment in the management of this patient is psychologic counseling. It may help to address the severity of her depression. Additionally, it may provide the patient or the couple with the support to endure additional treatment that is likely to result in pregnancy. Mental health practitioners experienced with this patient population would be preferable because they are sensitive to the nuances and complexities of infertile couples. Acupuncture is considered an alternative treatment to psychologic counseling, but its utility in this patient population remains controversial. There is no evidence that another course of infertility therapy would be beneficial for this couple. Many infertile couples who reinitiate treatment before their psychologic issues are addressed are likely to experience worsening of their effects and subsequent decline in their quality-of-life measures.

Alternative treatments, such as antidepressants, may be warranted. However, the patient must be evaluated thoroughly to assess the best treatment modality. Antidepressant medication may not provide the couple with the support required to proceed with additional treatments. Moreover, a number of anxiolytic medications are of the benzodiazepine class, which is contraindicated in pregnancy. Most patients would benefit from psychologic therapy to address the underlying problem and also cognitive–behavioral strategies to help address their stress.

Malcolm CE, Cumming DC. Follow-up of infertile couples who dropped out of a specialist fertility clinic. Fertil Steril 2004;81:269–70.

Pinborg A, Hougaard CO, Nyboe Andersen A, Molbo D, Schmidt L. Prospective longitudinal cohort study on cumulative 5-year delivery and adoption rates among 1338 couples initiating infertility treatment. Hum Reprod 2009;24:991–9.

Schmidt L. Psychosocial burden of infertility and assisted reproduction. Lancet 2006;367:379–80.

Verberg MF, Eijkemans MJ, Heijnen EM, Broekmans FJ, de Klerk C, Fauser BC, et al. Why do couples drop-out from IVF treatment? A prospective cohort study. Hum Reprod 2008;23:2050–5.

86
Primary amenorrhea due to gonadal dysgenesis

A 19-year-old nulligravid woman comes to your office with primary amenorrhea. She is 1.75 m (69 in.) tall with Tanner stage 1 breast development and Tanner stage 3 pubic hair. Her follicle-stimulating hormone level is 52 mIU/mL. Her endocrine screening is otherwise normal. Her karyotype is 46,XY. In addition to psychologic counseling, the next step in management should be

* (A) gonadectomy
* (B) Y-sequence analysis identification
* (C) combination oral contraceptives
* (D) growth hormone therapy
* (E) observation

The described patient has 46,XY gonadal dysgenesis, also known as Swyer syndrome. The classic presentation is a prepubertal phenotypic female with primary amenorrhea. Typically, there are no somatic abnormalities and patients will exhibit infantile uterus, cervix, and fallopian tubes as well as streak gonads. At the time of diagnosis, gonadectomy is recommended because of the significant risk (approximately 30%) of malignant transformation in the streak gonads. However, gonadectomy should be delayed until after puberty in women with androgen insensitivity syndrome. Gonadoblastoma is the most common tumor, followed by malignant transformation to a dysgerminoma. Choriocarcinoma may be seen rarely. Approximately 10–15% of cases of XY gonadal dysgenesis result from SRY mutations, another 10–15% arise from SRY deletions, and approximately 70% are of unknown genetic origin.

In diagnosing Swyer syndrome, the possibility of 45,X/46,XY mosaicism must be excluded. Patients with a 45,X cell line are at increased risk of the spectrum of cardiovascular issues associated with classic 45,X gonadal dysgenesis (Turner syndrome). Patients with a 45,X cell line may exhibit aortic dilation, coarctation, or a bicuspid aortic valve and are at increased risk of aortic dissection and rupture and possible death during pregnancy. Patients with nonmosaic pure 46,XY gonadal dysgenesis do not appear to be at increased risk of cardiovascular abnormalities or adverse obstetric outcomes.

Height may serve as one clue in distinguishing patients with pure 46,XY gonadal dysgenesis from patients with 45,X/46,XY mosaicism if classic features of Turner syndrome are not present. Patients with Swyer syndrome are typically of average or above-average height, whereas many patients with a 45,X cell line will exhibit short stature. A karyotype is essential in establishing the correct diagnosis.

After gonadectomy, exogenous hormone replacement is recommended to promote breast and uterine development. Estrogen alone should be given initially in a graduated fashion and cyclic progesterone or progestin should be added after the first episode of menstrual bleeding or typically after 2–3 years if no bleeding has occurred. Successful pregnancies have been achieved with oocyte donation in patients with 46,XY gonadal dysgenesis. The vast majority of reported pregnancies in patients with Swyer syndrome have required cesarean delivery for a variety of indications. The etiology of the high rate of cesarean delivery remains unclear but is probably multifactorial. The android shape of the pelvis in patients with a 46,XY karyotype may predispose these patients to labor abnormalities.

In the described patient, gonadectomy should be the next step in management. Because the karyotype clearly identifies a Y chromosome, Y-sequence analysis identification is not necessary. Testing for Y chromosome material would be necessary in a patient with gonadal dysgenesis if the karyotype did not identify Y chromosome material but perhaps showed a chromosomal fragment of unknown origin. Patients with cryptic Y chromosome sequences still should undergo gonadectomy, so the identification of even small amounts of Y chromosome material is important. Combination oral contraceptives are not appropriate for this patient at this time. Low-dose estrogen alone should be given initially for an extended period of months in order to try to mimic natural puberty and to facilitate normal breast and uterine development. Growth hormone therapy is not necessary in this patient because she is above-average height, although such therapy is prescribed frequently for girls with Turner syndrome. Observation is not appropriate given the need for gonadectomy and hormone therapy.

87

Intrauterine microinsert follow-up

A 35-year-old woman undergoes a hysteroscopic microinsert placement in her fallopian tubes without complication. After 3 months, follow-up hysterosalpingography (HSG) demonstrates unilateral tubal patency. The frequency with which patients will have at least one patent fallopian tube at 3 months after placement is

* (A) less than 5%
* (B) 5–9%
* (C) 10–14%
* (D) 15–19%
* (E) 20–24%

The hysteroscopically placed permanent contraceptive microinsert was approved by the U.S. Food and Drug Administration for female sterilization in 2002. The microinsert is placed within the tubal lumen with the assistance of a disposable delivery system in a wound-down state. After release from the delivery system, the insert expands and anchors itself within the tubal lumen, spanning the uterotubal junction. The microinsert has three components—a flexible stainless steel inner coil, a dynamic nickel–titanium expanding outer coil, and polyethylene terephthalate fibers. The polyethylene terephthalate fibers are wound in and around the inner coil. In its wound-down state, the insert is 4 cm long and 0.8 mm in diameter. When expanded, the outer coil is 1.5–2 mm in diameter and secures the insert within the proximal isthmic portion of the tube.

After placement of the microinsert, there is a local tissue response that results in a chronic inflammatory and fibrotic response to the polyethylene terephthalate fibers. The benign local tissue ingrowth against the polyethylene terephthalate fibers aids in device retention and prevention of pregnancy. After placement, the manufacturer recommends that 3–8 trailing outer coils be visible in the uterine cavity to ensure that the insert is placed distally enough to avoid expulsion but proximally enough to decrease the likelihood of insert perforation outside the tubal lumen. After the procedure, the patient is instructed that she is not protected from conception until follow-up HSG documents tubal occlusion. The documentation by HSG should be performed no sooner than 3 months after the placement of the microinsert.

In a retrospective report covering 7 years of experience of placement of the microinsert in 4,306 women in a single center, 4,108 (96.8%) women completed the standard 3-month follow-up protocol. In the 4,108 women, the rate of successful bilateral obstruction (including unilateral occlusion in women with only one tube) was 99.7% (4,095/4,108). Only 13 out of 4,108 women (0.3%) had one or both tubes patent at the 3-month follow-up visit. In support of this low tubal patency rate, only seven women (0.17%), three before the 3-month follow-up and four after the 3-month follow-up, became pregnant after the microinsert placement. One limitation of applying the aforementioned findings to others using the microinsert system may be the presence of a publication bias. All the procedures were performed (59%) or supervised (41%) by four gynecologists skilled in performing outpatient hysteroscopic procedures and very familiar with the sterilization technique. The aforementioned 3-month patency rate of 0.3% may not be achieved in all institutions. A recent review of the 3-month tubal patency rate after successful microinsert placement reported a rate of 3.5% tubal patency confirmed by HSG and subsequently a 0% patency rate at 6 months. The range of the tubal patency rates after successful microinsert placement appears to be less than 5% at 3 months after the initial procedure.

Benefits and risks of sterilization. Practice Bulletin No. 133. American College of Obstetricians and Gynecologists. Obstet Gynecol 2013;121:392–404.

Povedano B, Arjona JE, Velasco E, Monserrat JA, Lorente J, Castelo-Branco C. Complications of hysteroscopic Essure® sterilisation: report on 4306 procedures performed in a single centre. BJOG 2012;119:795–9.

88

Primary hypogonadotropic hypogonadism

A male partner of your fertility patient is found to be azoospermic on two consecutive semen analyses. Volume is normal but no sperm is identified. Postejaculatory urine also does not reveal any sperm. Laboratory results reveal a follicle-stimulating hormone (FSH) level of 0.2 mIU/mL, a luteinizing hormone (LH) level of 0.4 mIU/mL, a prolactin level of 3 micrograms/L, a thyroid-stimulating hormone level of 1.24 mIU/mL, and a total testosterone level of 162 ng/dL. The best next step is

* (A) magnetic resonance imaging (MRI) of anterior pituitary
* (B) testicular examination
* (C) counseling
* (D) peripheral karyotype
* (E) palpation of vas deferens

Primary hypogonadotropic hypogonadism is caused either by hypothalamic or pituitary disease and is characterized by low gonadotropin levels. By contrast, early andropause is accompanied by markedly elevated FSH and LH levels not seen in this clinical scenario. Pituitary imaging, preferably with MRI and gadolinium contrast, is necessary to evaluate a mass-occupying lesion in primary hypogonadotropic hypogonadism. A mass-occupying lesion, such as a hamartoma, craniopharyngioma, lymphoma, or pituitary tumor, would interrupt pulsatile gonadotropin-releasing hormone secretion. The failure of hormonal regulation can be determined easily. Endocrine deficiency leads to a lack of spermatogenesis and testosterone secretion due to decreased secretion patterns of LH and FSH. Standard treatment is human chorionic gonadotropin with later addition of human menopausal gonadotropin. These strategies can be effective in stimulating spermatogenesis. Only once pregnancy has been established should the patient return to testosterone substitution.

Secondary hypogonadotropic hypogonadism can be caused by obesity and anabolic steroids. As direct-to-consumer marketing becomes more common, many young men are developing secondary hypogonadotropic hypogonadism, azoospermia, and resultant infertility due to topical or injectable testosterone replacement.

Varicocele is a physical abnormality present in 2–22% of the adult male population and is more common in infertile men (25–40%). The exact effect of varicocele on fertility is unknown, but researchers suggest a relation to decreased testicular volume and decline in Leydig cell function. Treatment of varicocele to achieve pregnancy in infertile couples has been subject to criticism but may be considered, particularly if the defect is clinically palpable.

Cystic fibrosis is a fatal autosomal recessive disorder and is the most common genetic disease in white people. Men with cystic fibrosis are azoospermic because of congenital bilateral absence of the vas deferens. Men with isolated congenital bilateral absence of the vas deferens often carry mutations of the cystic fibrosis transmembrane conductance regulator gene. This gene is located in the short arm of chromosome 7 and encodes a membrane protein that functions as an ion channel and also influences formation of the ejaculatory duct, seminal vesicle, vas deferens, and distal two thirds of the epididymis. When a man has congenital bilateral absence of the vas deferens, it is important to test him and his partner for cystic fibrosis mutations. If the female partner is found to be negative for known mutations, her chance of being a carrier of unknown mutations is approximately 0.4%. The condition can be transmitted to offspring, and therefore genetic counseling and possibly preimplantation genetic diagnosis are indicated. Gonadotropin levels are typically normal with this condition. Low gonadotropin levels as described in this patient would not be seen in cases of varicocele, andropause, congenital bilateral absence of the vas deferens, or chromosomal abnormality.

Klinefelter syndrome and Y chromosome microdeletion often are associated with azoospermia and elevated gonadotropin levels. These conditions may be detected on peripheral blood testing. The best next step in diagnosis and evaluation of the described patient is MRI to evaluate the anterior pituitary. Testicular biopsy, at times used to confirm the diagnosis of primary testicular failure and nonobstructive azoospermia, is not indicated in this patient with low gonadotropin levels.

Dohle GR, Colpi GM, Hargreave TB, Papp GK, Jungwirth A, Weidner W. EAU guidelines on male infertility. EAU Working Group on Male Infertility. Eur Urol 2005;48:703–11.

Giannetta E, Gianfrilli D, Barbagallo F, Isidori AM, Lenzi A. Subclinical male hypogonadism. Best Pract Res Clin Endocrinol Metab 2012;26:539–50.

Nachtigall LB, Boepple PA, Pralong FP, Crowley WF Jr. Adult-onset idiopathic hypogonadotropic hypogonadism--a treatable form of male infertility. N Engl J Med 1997;336:410–5.

Oates RD, Amos JA. The genetic basis of congenital bilateral absence of the vas deferens and cystic fibrosis. J Androl 1994;15:1–8.

89

Hysterosalpingography complications

A 31-year-old woman, gravida 2, para 0, has been attempting conception without success for the past 5 years. She has a history of two ectopic pregnancies, 6 and 8 years ago. Both ectopic pregnancies were treated with methotrexate. She has regular menstrual cycles with significantly painful periods. Her husband has two children from a prior relationship. Among other tests, you perform hysterosalpingography (HSG) with the findings shown in Appendix B. The most appropriate immediate next step is

      (A) leuprolide acetate
*   (B) antibiotics
      (C) corticosteroids
      (D) bilateral salpingectomy
      (E) in vitro fertilization (IVF)

Hysterosalpingography is a radiologic procedure designed to evaluate the uterine cavity and the patency of the fallopian tubes. It involves inserting water-soluble radiopaque dye through a transcervical catheter and often is performed during evaluation for infertility or to confirm tubal obstruction after placement of a birth control implant for sterilization. Generally, HSG testing is performed within 5 days after menstrual bleeding ceases to avoid the possibility that the procedure could be done during the luteal phase after conception has occurred. To be conservative, a pregnancy test may be performed before the procedure. The technique for HSG involves using image-intensification fluoroscopy and usually requires only three to five basic films (a scout, an image evaluating the uterus and tubal patency, and an image to assess for pelvic areas of loculation). The total radiation exposure time is usually only 20–30 seconds. Many patients feel discomfort and uterine cramping during the procedure; pretreatment with a nonsteroidal antiinflammatory drug may mitigate those symptoms.

The described patient is at risk of tubal obstruction, given her history of two prior ectopic pregnancies. Although post-HSG pelvic inflammatory disease is uncommon (1.4–3.4%), it is a serious complication, especially in a patient population that strongly desires conception. If the tubes are dilated at the time of HSG, the risk of pelvic inflammatory disease after the procedure is as high as 10%. If there is any concern about active cervical or uterine infection on the day of the scheduled HSG, the test should be canceled and the patient should undergo evaluation and treatment for the infection. If the HSG demonstrates dilated, obstructed fallopian tubes, the patient should receive a postprocedure antibiotic, ie, doxycycline twice daily for 5 days. Treatment with antibiotics is the most appropriate next step in this case.

Leuprolide is a gonadotropin-releasing hormone agonist that often is used for treatment of endometriosis or leiomyomas as well as for IVF. The described patient has obstructed fallopian tubes, although the cause of the blockage is unknown. Common causes of tubal obstruction include prior pelvic infection or adhesions related to endometriosis or prior surgeries. Leuprolide would not be useful for the described patient.

In some rare instances, patients have a systemic reaction to the contrast dye used for HSG, with symptoms ranging from urticaria to bronchospasm and laryngeal edema. In patients with a history of prior sensitivity to iodinated contrast agents, it is recommended to pretreat with antihistamines and corticosteroids to prevent an allergic reaction. In the described patient, there is no evidence of an allergic reaction, so this treatment is not indicated. Another rare complication of HSG is uterine perforation. This may result in acute bleeding and require urgent surgery. Even in the extremely rare scenario of uterine perforation that requires surgery, it would be highly unlikely to require immediate bilateral salpingectomy.

This patient probably has significant tubal obstruction that is causing her infertility, and she will either require surgery or IVF to help her conceive. Distal tubal obstruction seen on HSG may require follow-up laparoscopy to better characterize the extent of disease and possibly repair the defect. In younger patients with mild pelvic disease,

90

Ovulation induction treatment options

A 30-year-old nulligravid woman requests counseling about how to become pregnant. She is currently taking a combination oral contraceptive as hormone therapy for Kallmann syndrome. Fifteen years ago, she was treated medically to develop secondary sexual characteristics. She was placed on subcutaneous recombinant follicle-stimulating hormone (FSH) therapy but did not respond to ovulation induction. The most appropriate next step in her management is

* (A) gonadotropin-releasing hormone (GnRH) agonist
* (B) GnRH antagonist
* (C) clomiphene citrate
* (D) human menopausal gonadotropins
* (E) aromatase inhibitor

Kallmann syndrome is a well-characterized disorder that features hypogonadotropic hypogonadism and anosmia. The syndrome affects 1 in 10,000 men and 1 in 50,000 women. Patients with this syndrome present with primary amenorrhea without development of secondary sexual characteristics. The failure of proper neuronal migration of the GnRH-secreting and olfactory neurons results in inadequate release of GnRH with a lack of production of FSH and luteinizing hormone (LH). This leads to a lack of folliculogenesis and steroid hormone production, which renders the patient infertile.

Initial treatment of Kallmann syndrome is aimed at restoring breast development and initiation of menstruation using estrogens and progestins. Pubic and axillary hair develops in response to adrenal androgens. Many case reports have observed the need for the use of menotropins (which carry LH and FSH activity) and a long ovarian stimulation for the development of a mature follicle. Although prolonged high doses of FSH alone can eventually induce folliculogenesis, there is always the risk of a multiple follicular response that requires cancellation of the cycle or of ovarian hyperstimulation syndrome. Reports have described a faster stimulation using lower doses of FSH with the addition of low-dose human chorionic gonadotropin or LH. Given the current understanding of folliculogenesis based on the two-cell theory, LH stimulates thecal cell androgen production, which serves as the substrate for aromatization in the follicle and release of estradiol, which is required for optimal follicular growth. The most appropriate next step in the management of the described patient would be to prescribe human menopausal gonadotropins.

In numerous case reports, pregnancy has been achieved using either a GnRH pump to mimic hypothalamic secretion or gonadotropin injections. Gonadotropin-releasing hormone agonists and GnRH antagonists can cause ovarian suppression and, thus, are ineffective for inducing folliculogenesis.

Clomiphene citrate is a selective estrogen receptor antagonist that is commonly used for ovulation induction in patients with an intact hypothalamic–gonadal–pituitary axis. Clomiphene has no utility in patients with Kallmann syndrome.

Aromatase inhibitors are a class of drugs that block the activity of aromatase, the enzyme responsible for the conversion of androgens into estradiol. Aromatase inhibitors cause a relative estrogen deficiency, which induces an increase of gonadotropin release in patients with a normal hypothalamic–gonadal–pituitary axis. However, they are not effective in patients with Kallmann syndrome.

Fechner A, Fong S, McGovern P. A review of Kallmann syndrome: genetics, pathophysiology, and clinical management. Obstet Gynecol Surv 2008;63:189–94.

Meczekalski B, Podfigurna-Stopa A, Smolarczyk R, Katulski K, Genazzani AR. Kallmann syndrome in women: from genes to diagnosis and treatment. Gynecol Endocrinol 2013;29:296–300.

Sipe CS, Van Voorhis BJ. Testosterone patch improves ovarian follicular response to gonadotrophins in a patient with Kallmann's syndrome: a case report. Hum Reprod 2007;22:1380–3.

91

Testing for insulin resistance in polycystic ovary syndrome

A 24-year-old nulligravid woman with periods of amenorrhea greater than 6 months may have polycystic ovary syndrome (PCOS). You notice darkening on the back of her neck that is velvety in appearance (Fig. 91-1; see color plate). The most accurate way to evaluate insulin resistance in patients with PCOS is

* (A) hyperinsulinemic euglycemic clamp
  (B) oral glucose tolerance test
  (C) fasting glucose level
  (D) fasting insulin level

Insulin resistance is a condition in which endogenous or exogenously administered insulin has less-than-normal effects on fat, muscle, and the liver. This insulin level results in increased hydrolysis of stored triglycerides and elevated circulating free fatty acid levels. In women with insulin resistance, there is increased hepatic gluconeogenesis resulting in an increased blood glucose concentration and a compensatory hyperinsulinemia. Increased circulating insulin levels contribute to hyperandrogenism by stimulating ovarian androgen production and inhibiting hepatic sex hormone-binding globulin production.

Women with insulin resistance often exhibit acanthosis nigricans (Fig. 91-1; see color plate), a hyperpigmented and velvety appearance of the skin commonly observed at the base of the neck, in the groin or axilla, or beneath the breasts. It strongly suggests insulin resistance and the possibility of diabetes mellitus. Insulin resistance and hyperinsulinemia are common among women with PCOS.

The criterion standard for assessment of insulin resistance is the hyperinsulinemic euglycemic clamp, which involves an intravenous infusion of insulin and glucose. The glucose infusion is increased until the infusion rate matches the rate of glucose uptake in tissues at a set insulin concentration. This rate is inversely proportional to the degree of insulin resistance. The lower the glucose concentration infused to maintain a normal fasting glucose level, the higher the degree of insulin resistance.

Euglycemic clamp studies are impractical in the busy clinical setting. Therefore, other surrogate markers are utilized to try to assess insulin resistance.

The standard glucose tolerance test is the most commonly utilized method to assess impaired glucose tolerance (IGT) and diabetes mellitus in patients with PCOS. The terms insulin resistance and glucose intolerance are erroneously used interchangeably. It is important to understand the difference between the information obtained from a euglycemic clamp and a 2-hour glucose tolerance test. A 2-hour glucose tolerance test involves ingestion of 75 g of glucose after fasting and assessment of the patient for fasting glucose and glucose level 2 hours after the load of glucose. Up to 35% of the PCOS population may exhibit IGT and 10% may exhibit obvious diabetes mellitus.

The prevalence of IGT and type 2 diabetes mellitus in U.S. women has been assessed in three large cross-sectional studies. The prevalence was 23–35% for IGT and 4–10% for type 2 diabetes mellitus. The prevalence rate of IGT in PCOS was three times higher than age-matched controls.

A fasting glucose level does not assess glucose control when challenged by food intake. It is considered inferior to test for glucose control after food intake. Fasting insulin levels can suggest insulin resistance. Fasting insulin levels greater than 20–30 mU/mL suggest insulin resistance but do not diagnose diabetes mellitus.

Historically, the diagnosis for diabetes mellitus was based on plasma glucose criteria and the 2-hour value on the 2-hour glucose tolerance test. In 2010, the American Diabetes Association and the Endocrine Society jointly recommended the use of the hemoglobin A_{1c} (HbA_{1c}) test to diagnose diabetes mellitus with a threshold of 6.5% or greater. The HbA_{1c} test has advantages over the 2-hour glucose tolerance test and fasting plasma glucose level criterion. The HbA_{1c} test is fast and does not require 2 hours of the patient's time. The convenience of the HbA_{1c} test may help health care providers to establish more diagnoses. Additionally, the HbA_{1c} test has little variability at times of stress or illness.

Ehrmann DA, Barnes RB, Rosenfield RL, Cavaghan MK, Imperial J. Prevalence of impaired glucose tolerance and diabetes in women with polycystic ovary syndrome. Diabetes Care 1999;22:141–6.

Ehrmann DA, Kasza K, Azziz R, Legro RS, Ghazzi MN. Effects of race and family history of type 2 diabetes on metabolic status of women with polycystic ovary syndrome. PCOS/Troglitazone Study Group. J Clin Endocrinol Metab 2005;90:66–71.

Legro RS, Kunselman AR, Dodson WC, Dunaif A. Prevalence and predictors of risk for type 2 diabetes mellitus and impaired glucose tolerance in polycystic ovary syndrome: a prospective, controlled study in 254 affected women. J Clin Endocrinol Metab 1999;84:165–9.

92

Abnormal uterine bleeding in an adolescent

A 14-year-old virginal girl comes to your office. She has had irregular heavy menses since menarche 1 year ago. She has menses that occur on an irregular basis every 2–5 months, and her bleeding lasts for 10–15 days, sometimes with spotting and other times with the need to change her pad every 2–3 hours. She has felt no lightheadedness or dizziness and has no other medical problems. The most likely explanation for her bleeding pattern is

* (A) endometriosis
* (B) oligo-ovulation
* (C) pregnancy
* (D) coagulopathy

Heavy menstrual bleeding is classically defined as bleeding that lasts longer than 7 days or blood loss in excess of 80 mL per cycle. Research has shown, however, that the patient's perception of light, moderate, or heavy bleeding corresponds poorly with her actual blood loss. This was clearly demonstrated in a classic Swedish study of menstrual blood loss. The investigators reported that 30% of women with light-to-average menstrual bleeding self-reported their menstrual periods to be heavy, whereas 40% of women with blood loss greater than 80 mL per cycle considered their menstrual periods to be only of moderate intensity. Consequently, heavy menstrual bleeding or "menorrhagia" is a relative term that cannot be defined solely by quantitative blood volume loss or days of bleeding. The physician's assessment of heavy menstrual bleeding must take into account the patient's perception of blood loss and the degree to which it affects her quality of life.

The PALM–COEIN classification system for causes of abnormal uterine bleeding was developed in 2011 by the International Federation of Gynecology and Obstetrics (Appendix C). The etiologies of abnormal uterine bleeding fall into two categories:

1. Structural causes, ie, polyp, adenomyosis, leiomyomata, malignancy (PALM)
2. Nonstructural causes, ie, coagulopathy, ovulatory dysfunction, endometrial, iatrogenic, and "not yet classified" (COEIN)

The abnormal bleeding in the described patient is likely to be related to anovulation or oligo-ovulation.

With anovulatory bleeding patterns, the history is often adequate to establish the diagnosis, and further testing can confirm the suspicion. In this case, the patient has infrequent, unpredictable bleeding that varies in intensity and has been occurring in this pattern since menarche. Such infrequent, unpredictable bleeding occurs most frequently in adolescents, perimenopausal women, obese women, and women with polycystic ovary syndrome. In adolescents, the condition can be related to immaturity of the hypothalamic–pituitary–ovarian axis, and it is common for the first few years after menarche. Although this is the most likely scenario in this case, other more serious pathologies must be ruled out.

Endometriosis is a common, benign condition that often presents with chronic pelvic pain, dysmenorrhea, dyspareunia, and infertility. Although this condition can occur in adolescents, it is more common in women aged 25–35 years. Usually women with endometriosis have regular menstrual cycles, making this an unlikely cause of the symptoms in this case.

Anatomic causes of menstrual bleeding are common, eg, benign lesions, such as polyps or leiomyomas, and malignancies (cervical or uterine). Uterine fibroids are the most common pelvic tumor in women with presenting symptoms of abnormal bleeding, pelvic pain or pressure, infertility, or recurrent pregnancy loss. Leiomyomas are common in reproductive-aged women and are found in approximately 80% of surgically removed uteri. Leiomyomas and polyps are uncommon in adolescents. Pregnancy should always be considered as a cause of abnormal uterine bleeding, especially in a case where the current bleeding pattern is a relatively sudden departure from the prior pattern. Adolescents may be unwilling to reveal their sexual history, so a pregnancy test should always be performed, even in girls who deny sexual activity. However, given that the timeframe for bleeding in this case is 1 year, it is unlikely that pregnancy-related complications are the cause of her symptoms.

It is important to rule out a coagulopathy in adolescents with abnormal bleeding, especially in adolescents who have excessive bleeding. Although the most common cause of abnormal bleeding in this age group is anovulation, coagulopathies can occur up to one third of the time. Coagulopathies may have significant consequences if they remain untreated. The most common coagulopathy is von Willebrand disease, but other coagulopathies, such as idiopathic thrombocytopenic purpura, platelet dysfunction, and thrombocytopenia due to malignancy, are possible. Most patients with coagulopathy exhibit regular bleeding patterns but have excessive bleeding that can necessitate changing a pad more frequently than every hour. For the described patient, testing for coagulopathy would be important, but coagulopathy is not the most likely diagnosis.

Deligeoroglou E, Karountzos V, Creatsas G. Abnormal uterine bleeding and dysfunctional uterine bleeding in pediatric and adolescent gynecology. Gynecol Endocrinol 2013;29:74–8.

Diagnosis of abnormal uterine bleeding in reproductive-aged women. Practice Bulletin No. 128. American College of Obstetricians and Gynecologists. Obstet Gynecol 2012;120:197–206.

Hallberg L, Hogdahl AM, Nilsson L, Rybo G. Menstrual blood loss--a population study. Variation at different ages and attempts to define normality. Acta Obstet Gynecol Scand 1966;45:320–51.

Munro MG, Critchley HO, Broder MS, Fraser IS. FIGO classification system (PALM-COEIN) for causes of abnormal uterine bleeding in nongravid women of reproductive age. FIGO Working Group on Menstrual Disorders. Int J Gynaecol Obstet 2011;113:3–13.

93

Elevated dehydroepiandrosterone sulfate

A 23-year-old nulligravid woman has had a history of irregular menses and hirsutism since she underwent menarche. She is suspected to have polycystic ovary syndrome. During the evaluation, her dehydroepiandrosterone sulfate (DHEAS) level is found to be 760 micrograms/dL. In order to further evaluate her abnormally elevated DHEAS level, you order imaging of the

(A) pelvis
(B) brain
* (C) abdomen
(D) chest

Compared with other steroids, DHEAS circulates in higher concentrations and is almost exclusively derived from the adrenal glands. The upper limit of normal is 350 micrograms/dL but can vary between laboratories. Mild elevations in DHEAS levels (350–700 micrograms/dL) have been noted in women with polycystic ovary syndrome and hirsutism. Elevation in DHEAS contributes to hirsutism because it acts as a prehormone for androgen synthesis in the hair follicle. Elevated levels of DHEAS also have been noted in Cushing syndrome, adrenal adenomas, adrenal carcinomas, and congenital adrenal hyperplasia resulting from *CYP11B*, *CYP21*, and 3β-HSD gene deficiencies. Levels of DHEAS are low in *CYP17* gene deficiency.

Elevated testosterone levels may originate from either the adrenal glands or the ovaries; however, a DHEAS level of 700 micrograms/dL or higher may indicate excessive adrenal secretion of this hormone and the possibility of an adrenal tumor. Malignancy cannot be diagnosed based on an elevated DHEAS level because approximately one half of patients with adrenal tumors and an elevated DHEAS level have a benign adrenal tumor. Adrenal malignant tumors are rare tumors with a higher occurrence in females than in males. The evaluation should include an abdominal computed tomography scan with specific evaluation of the adrenal glands for tumors. Abdominal magnetic resonance imaging provides an alternative method to evaluate elevated DHEAS levels. It may be preferable in patients who cannot tolerate intravenous iodinated contrast dye or if hemorrhage is suspected. Pelvic ultrasonography may be appropriate if an ovarian tumor is suspected. Brain magnetic resonance imaging may be appropriate for pituitary or hypothalamic abnormalities. A chest X-ray would not be useful in the care of this patient at this time.

Azziz R, Sanchez LA, Knochenhauer ES, Moran C, Lazenby J, Stephens KC, et al. Androgen excess in women: experience with over 1000 consecutive patients. J Clin Endocrinol Metab 2004;89:453–62.

Derksen J, Nagesser SK, Meinders AE, Haak HR, van de Velde CJ. Identification of virilizing adrenal tumors in hirsute women. N Engl J Med 1994;331:968–73.

McKenna TJ. Screening for sinister causes of hirsutism. N Engl J Med 1994;331:1015–6.

Stewart PM. The adrenal cortex. In: Williams textbook of endocrinology. In: Kronenberg HM, Melmed S, Polonsky KS, Larsen PR, eds. Williams textbook of endocrinology. 11th ed. Philadelphia (PA): WW Saunders; 2008. p. 445–58.

94

Low bone mass and bone physiology

A 62-year-old woman experienced a Colles (wrist) fracture in the past year. She has not used hormone therapy since she went through menopause 10 years ago. She currently takes dietary supplements of 400 international units/d of vitamin D and 1,500 mg/d of calcium. She is 1.52 m (5 ft) tall and weighs 68 kg (150 lb). Dual-energy X-ray absorptiometry (DXA) scans of the lumbar spine and hip were done (Fig. 94-1 and Fig. 94-2). The most appropriate next step in management is to prescribe

* (A) bisphosphonates
* (B) increased vitamin D and calcium supplements
* (C) teriparatide
* (D) estrogen therapy
* (E) calcitonin

Most postmenopausal women have low bone mass compared with the young adult reference population. The diagnostic thresholds for osteoporosis for any specific individual are based on the patient's bone mineral density test results compared with individuals of comparable age and sex. However, osteoporosis generally has no clinical symptoms until a fracture occurs, and ideally the disease should be identified and treated before a devastating fracture occurs. The most common type of fracture with osteoporosis is vertebral fracture; such fractures are asymptomatic two thirds of the time and may be diagnosed as an incidental finding on chest or abdominal X-rays. Hip fractures also are common and can occur suddenly with a fall or silently with subsequent destructive secondary osteoarthritis of the hip. Distal radial fractures (Colles fractures) frequently occur in the years after menopause. One meta-analysis of more than 40,000 women found that a prior fracture was associated with a significant risk for another fracture (relative risk = 1.86, 95% confidence interval, 1.75–1.98).

Screening for low bone mass should occur at age 65 years for all women and earlier for women with one or more risk factors. Risk factors include tobacco use, body mass index greater than 21 (calculated as weight in kilograms divided by height in meters squared), alcohol use, and parental fracture history.

Screening for low bone mass involves evaluation with a DXA scan of the spine and hip. In addition, in 2008, the World Health Organization task force developed a Fracture Risk Assessment Tool, which estimates the 10-year probability of a hip or other major fracture (Fig. 94-3).

Name.	Sex: Female	Height: 60.0 in
Patient ID:	Ethnicity: White	Weight: 150.4 lb
DOB:	Menopause Age: 50	Age: 56

Referring Physician:

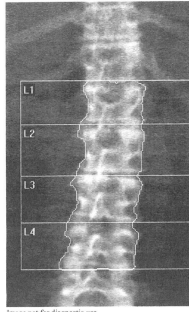

Image not for diagnostic use
k = 1.114, d0 = 49.3
116 x 126

Scan Information:
Scan Date: ID: A09171306
Scan Type: a Lumbar Spine
Analysis: September 17, 2013 10:53 Version 12.5:7
         Lumbar Spine
Operator: ER
Model: Discovery W (S/N 82365)
Comment:

DXA Results Summary:

Region	Area (cm²)	BMC (g)	BMD (g/cm²)	T-score	Z-score
L1	11.31	9.17	0.811	-1.0	-0.0
L2	14.32	12.57	0.878	-1.4	-0.2
L3	13.77	12.77	0.927	-1.4	-0.2
L4	15.52	13.86	0.893	-2.0	-0.8
Total	**54.92**	**48.37**	**0.881**	**-1.5**	**-0.4**

Total BMD CV 1.0%, ACF = 1.038, BCF = 1.010, TH = 7.138
WHO Classification: Osteopenia
Fracture Risk: Increased

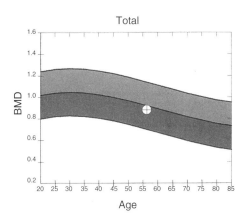

Physician's Comment:

T-score versus white female; z-score versus white female

FIG. 94-1. Dual energy X-ray absorptiometry (DXA) scan of the spine. (Courtesy of Janet Rubin, MD, University of North Carolina at Chapel Hill School of Medicine.)

Name:	Sex: Female	Height: 60.0 in
Patient ID:	Ethnicity: White	Weight: 150.4 lb
DOB:	Menopause Age: 50	Age: 56

Referring Physician:

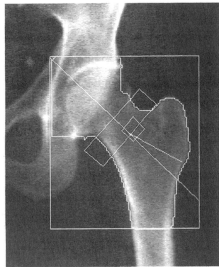

Image not for diagnostic use
k = 1.117, d0 = 51.9
89 x 99
NECK: -49 x 15

Scan Information:
Scan Date: ID: A09171305
Scan Type: a Left Hip
Analysis: September 17, 2013 10:47 Version 12.5:7
Left Hip
Operator: ER
Model: Discovery W (S/N 82365)
Comment:

DXA Results Summary:

Region	Area (cm²)	BMC (g)	BMD (g/cm²)	T-score	Z-score
Neck	4.55	2.30	0.504	-3.1	-2.0
Total	27.93	17.67	0.633	-2.5	-1.8

Total BMD CV 1.0%, ACF = 1.038, BCF = 1.010, TH = 6.388
WHO Classification: Osteoporosis
Fracture Risk: High

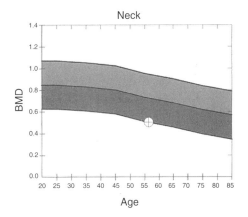

Physician's Comment:

T-score versus white female; z-score versus white female

FIG. 94-2. Dual energy X-ray absorptiometry (DXA) scan of the hip. (Courtesy of Janet Rubin, MD, University of North Carolina at Chapel Hill School of Medicine.)

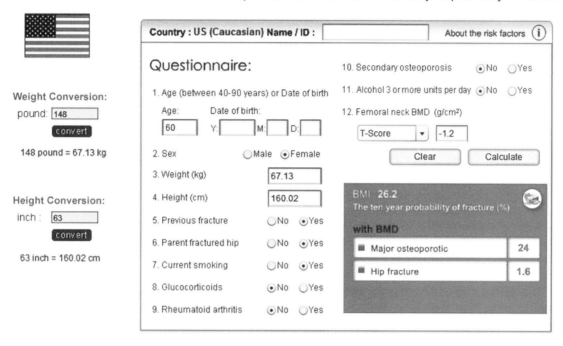

FIG. 94-3. FRAX® World Health Organization Fracture Risk Assessment Tool. Available at: http://www.shef.ac.uk/FRAX/. Retrieved June 21, 2013.

The National Osteoporosis Foundation recommends treatment if the patient has either

- a personal history of hip or vertebral fracture,
- a DXA scan with a T-score less than –2.5, or
- a bone mass in the osteopenic range (T-score of –1.0 to –2.5) and a 10-year probability of a hip fracture of 3% or more or a 10-year probability of any fracture of 20% or more based on Fracture Risk Assessment Tool estimates.

Traditionally, vitamin D and calcium supplementation have been standard treatments to reduce bone loss during menopause. However, a 2013 report from the U.S. Preventive Services Task Force concluded that there was insufficient evidence to confirm that supplementation with 1,000 mg or more of calcium or 400 international units of vitamin D has benefits in terms of primary prevention of fractures in postmenopausal women.

Bisphosphonates are a class of compounds that inhibit osteoclastic action and therefore reduce bone resorption. Most bisphosphonates available in the United States have been shown to reduce the incidence of spine and hip fractures and are considered the first-line treatment for postmenopausal women at risk for fracture. All bisphosphonates share a common P–C–P backbone structure, where C is carbon and each P is a phosphonate group; the two phosphonate groups act as a "bone hook" and are essential for binding to hydroxyapatite, a complex phosphate of calcium that is the chief structural element of vertebrate bone. Commonly used bisphosphonates in the United States include alendronate sodium, ibandronate sodium, risedronate sodium, and zoledronic acid. To help reduce the adverse effect of gastric irritation, these oral formulations must be taken on an empty stomach with a glass of water (no other liquid), and the patient must remain upright for 30–60 minutes. Case reports show that osteonecrosis of the jaw is associated with bisphosphonate use. In general, most of these cases occur in association with dental procedures in patients with poor dentition and established malignancies.

Other approved treatments to prevent fracture include estrogen, calcitonin, parathyroid hormone, and raloxifene hydrochloride. Teriparatide is a parathyroid hormone that is injected subcutaneously each day. It has been shown to increase bone density by stimulating new bone formation. However, given the cost, route of administration, and availability of other options, parathyroid hormone is not considered first-line therapy for treatment or prevention of osteoporosis.

Estrogen is a very effective antiresorptive agent for the treatment of osteoporosis. In the Women's Health Initiative study, estrogen was observed to reduce fracture risk by 40% in a low-risk population of menopausal women compared with placebo. However, the risks associated with estrogen use (eg, risk of breast cancer and cardiovascular disease) generally are considered significant enough to warrant use of other treatments, such as bisphosphonates, as first-line options for fracture prevention.

Calcitonin is administered in a daily nasal spray and has a relatively modest effect on bone mineral density compared with other treatment options. For this reason, it is rarely used as first-line treatment and instead is reserved for patients who cannot tolerate other more efficacious options.

Kanis JA, Johnell O, De Laet C, Johansson H, Oden A, Delmas P, et al. A meta-analysis of previous fracture and subsequent fracture risk. Bone 2004;35:375–82.

Moyer VA; U.S. Preventive Services Task Force. Vitamin D and calcium supplementation to prevent fractures in adults: U.S. Preventive Services Task Force recommendation statement. Ann Intern Med 2013;158:691–6.

National Osteoporosis Foundation. Clinician's guide to prevention and treatment of osteoporosis. Washington, DC: National Osteoporosis Foundation; 2010. Available at: http://nof.org/files/nof/public/content/file/344/upload/159.pdf. Retrieved July 12, 2013.

World Health Organization Collaborating Centre for Metabolic Bone Diseases. WHO Fracture Risk Assessment Tool (FRAX). Sheffield, United Kingdom: University of Sheffield; 2013. Available at: http://www.shef.ac.uk/FRAX/. Retrieved June 21, 2013.

95

Testing for ovarian reserve

A 30-year-old nulligravid woman comes to your office with a 1-year history of infertility. On evaluation, her partner has a normal semen analysis. Hysterosalpingography is normal. The earliest test that will identify a diminished ovarian reserve is

(A) follicle-stimulating hormone (FSH) level
(B) clomiphene citrate challenge test
* (C) antimüllerian hormone level
(D) inhibin B level

Ovarian reserve testing is an important aspect in the evaluation of fertility. It enables identification of women who may need more aggressive treatment. Ovarian reserve testing should be performed in all women older than 35 years, women with unexplained infertility, women with a prior history of ovarian surgery, women with a prior history of gonadotoxic treatment (chemotherapy or radiotherapy), and women who smoke.

It is important to realize that ovarian reserve testing only allows for quantitative evaluation of oocytes. Qualitative assessment of oocytes is unavailable and depends upon the age of the female partner. In addition, ovarian reserve testing does not predict pregnancy. Rather, it predicts the response of the patient to gonadotropin therapy.

The gonadotropin level will increase as the patient approaches menopause. The FSH level will typically increase first. The luteinizing hormone level may remain normal and increase later. An increase in FSH level is secondary to quantitative loss of follicles from the ovaries. With the loss of granulosa cells, inhibin production drops, and FSH is no longer suppressed because inhibin B inhibits FSH. This test can be performed at any time during cycle days 2–4, typically in conjunction with testing for estradiol level. Estradiol elevation in the early follicular phase may be related to the advanced follicular recruitment seen in ovarian aging.

If, on cycle days 2–4, the level of inhibin B (a product of granulosa cell secretion in the follicular phase) is low (below 45 pg/mL), there will be a poorer response to ovulation stimulation. Women with low inhibin B levels will be less likely to conceive than women with high inhibin B levels. However, because of the low sensitivity of this test, it has been largely abandoned.

The clomiphene citrate challenge test involves the administration of 100 mg clomiphene on days 5–9 of the menstrual cycle with FSH levels obtained on day 3 (before the administration of clomiphene) and on day 10. In patients with normal ovarian reserve, the clomiphene-dependent rise in FSH will be suppressed by inhibin B produced by the follicles. An abnormal test is defined by an abnormally high FSH on days 2–4 or day 10 (usually greater than 10 mIU/mL). Because of the low sensitivity of the clomiphene challenge test, it has been largely replaced by antimüllerian hormone level.

In women, antimüllerian hormone is produced by granulosa cells of preantral and small antral follicles. It is believed that antimüllerian hormone plays a role in the transition from resting primordial to growing follicles and in the recruitment of FSH-sensitive follicles. Because antimüllerian hormone declines with advancing age, it can be used as a marker of ovarian reserve. This decline takes place ahead of changes in FSH and luteinizing hormone levels. Normal antimüllerian hormone, which will vary between assays, is greater than 1.5 ng/mL. Thus, in testing diminished ovarian reserve, the earliest test to show an abnormal result is antimüllerian hormone level rather than the clomiphene challenge test, FSH level, or inhibin B level. Note that women with polycystic ovary syndrome may have a higher antimüllerian hormone level as a result of the presence of many small antral follicles in their ovaries (greater than 3.5 ng/mL).

Bukulmez O, Arici A. Assessment of ovarian reserve. Curr Opin Obstet Gynecol 2004;16:231–7.

Grynnerup AG, Lindhard A, Sorensen S. The role of anti-Müllerian hormone in female fertility and infertility—an overview. Acta Obstet Gynecol Scand 2012;91:1252–60.

La Marca A, Sighinolfi G, Radi D, Argento C, Baraldi E, Artenisio AC, et al. Anti-Müllerian hormone (AMH) as a predictive marker in assisted reproductive technology (ART). Hum Reprod Update 2010;16:113–30.

La Marca A, Volpe A. Anti-Müllerian hormone (AMH) in female reproduction: is measurement of circulating AMH a useful tool? Clin Endocrinol (Oxf) 2006;64:603–10.

Toner JP, Seifer DB. Why we may abandon basal follicle-stimulating hormone testing: a sea change in determining ovarian reserve using antimüllerian hormone. Fertil Steril 2013;99:1825–30.

Tremellen KP, Kolo M, Gilmore A, Lekamge DN. Anti-müllerian hormone as a marker of ovarian reserve. Aust N Z J Obstet Gynaecol 2005;45:20–4.

96
Chronic pelvic pain

You perform a laparoscopic resection of endometriosis in a 27-year-old woman, gravida 2, para 1, for Stage II endometriosis. She continues to experience cyclic, chronic pelvic pain. She has a history of deep vein thrombosis and is positive for factor V Leiden mutation. During her postoperative visit, she inquires about postsurgical options to minimize recurrence of her symptoms but not increase her thrombosis risk. You advise her that the best therapy to reduce her long-term chronic pelvic pain is a

(A) monthly nonsteroidal antiinflammatory drug (NSAID)
(B) continuous combination oral contraceptive (OC)
* (C) levonorgestrel intrauterine device (IUD)
(D) gonadotropin-releasing hormone (GnRH) analog
(E) aromatase inhibitor

Endometriosis affects 10–20% of women of reproductive age. Approximately 70–90% of women who experience chronic pelvic pain, dysmenorrhea, dyspareunia, infertility, and menstrual disturbances have endometriosis. The most common drugs used for the conservative treatment of endometriosis include NSAIDs, progestins such as depot medroxyprogesterone acetate, GnRH analogs to induce pseudomenopause, androgen derivatives, and continuous combination OCs. Safety concerns may limit the use of OCs in smokers, women older than 35 years, and patients with known coagulopathy. Depot medroxyprogesterone acetate and GnRH analogs require frequent visits for administration and may lead to hypoestrogenic symptoms and, in the long term, decreased adherence rates. Although GnRH analogs are effective for pain control, they should be limited to short-term use because of the risk of osteoporosis. Similarly, although NSAIDs might be beneficial for pain in the short term, they do not inhibit menses and do have long-term adverse effects. Aromatase inhibitors are a newer option and to date have not been proven to be efficacious in large randomized trials.

A 2003 randomized controlled trial monitored women with Stage I–IV endometriosis postsurgically who were either under observation or using the levonorgestrel IUD. Bleeding and dysmenorrhea were reduced in the levonorgestrel IUD group with additional reduction in nonmenstrual pain and dyspareunia. The reduction in relative risk of recurrence of dysmenorrhea was 78%. A 2005 randomized controlled trial compared women with Stage I–IV endometriosis who used either the levonorgestrel IUD or a GnRH analog. Both treatments reduced bleeding and improved pain; the quality of life assessment was similar in both groups.

The release rate of levonorgestrel from the levonorgestrel IUD is 20 micrograms/d during the first year and slowly decreases over the 5 years of use. As a 19-nortestosterone derivative, levonorgestrel exhibits progestational activity, inducing profound effects on the endometrium, which becomes atrophic and inactive. Therefore, based on the induced amenorrhea and endometrial atrophy, it is possible to speculate that the levonorgestrel IUD may be effective in controlling pain for the same period. Several hypotheses have been proposed to explain the mechanism of action of levonorgestrel in endometriosis. One hypothesis is that levonorgestrel acts directly on the endometriotic lesions in the peritoneum; however, the concentration of levonorgestrel is high only within the endometrium. Another hypothesis is that the level of levonorgestrel in peritoneal fluid is high and that the steroid acts directly on the endometriotic lesions in the peritoneum. Other theories suggest that endometrial production of estradiol-induced growth factors or growth factor-binding protein from the endometrium is inhibited, resulting in an antiproliferative effect, glandular atrophy, and decidualization.

Although endometriosis often is associated with infertility, not all patients with endometriosis wish to conceive. In patients who have completed childbearing or who wish to postpone pregnancy, the levonorgestrel IUD provides a good method for control of pain associated with endometriosis. The affordable cost and minimal adverse effects of the levonorgestrel IUD make it the best option for the described patient.

Bahamondes L, Petta CA, Fernandes A, Monteiro I. Use of the levonorgestrel-releasing intrauterine system in women with endometriosis, chronic pelvic pain and dysmenorrhea. Contraception 2007;75 (suppl):S134–9.

Vercellini P, Frontino G, De Giorgi O, Aimi G, Zaina B, Crosignani PG. Comparison of a levonorgestrel-releasing intrauterine device versus expectant management after conservative surgery for symptomatic endometriosis: a pilot study. Fertil Steril 2003;80:305–9.

97

Amenorrhea and galactorrhea

A 22-year-old woman who is not taking any medications comes to your office with a lack of menses for 9 months. She desires contraception. Physical examination is remarkable for nonspontaneous multiduct bilateral galactorrhea, atrophic-appearing vaginal mucosa, and absent cervical mucus. Laboratory testing shows a normal thyroid-stimulating hormone level and a prolactin level of 80 ng/dL. Magnetic resonance imaging (MRI) indicates a possible 3-mm anterior pituitary adenoma. In addition to repeat MRI and prolactin level in 1 year, the most appropriate therapeutic recommendation is

* (A) combination hormonal contraceptives
* (B) clinical observation
* (C) progestin-only oral contraceptives
* (D) bromocriptine mesylate

Management of hyperprolactinemia is dependent on the etiology of the condition and the prevention of associated consequences of estrogen deficiency. The pathologic causes of hyperprolactinemia are varied and may include pituitary tumors that secrete prolactin and other central nervous system (CNS) abnormalities that alter the inhibition by dopamine on pituitary lactotrophs. Such CNS abnormalities may include other pituitary lesions such as acromegaly, Cushing syndrome, and empty sella syndrome. Additional CNS abnormalities include tumors, infiltrating pituitary disorders such as sarcoidosis and histiocytosis X, encephalitis, head trauma, hydrocephalus, Rathke pouch cyst, pinealoma, pseudotumor cerebri, and metastatic disease. Primary hypothyroidism increases prolactin secretion by direct action of thyroid-releasing hormone on the pituitary. Circumstances that result in increased estrogen levels, such as development of polycystic ovary syndrome or liver dysfunction (due to decreased metabolism) and use of estrogen-containing drugs, also increase prolactin level. Because prolactin is renally excreted without metabolism, kidney failure causes hyperprolactinemia. In rare instances, tumors, such as ovarian teratomas and bronchogenic carcinomas, may secrete prolactin. Breast and chest wall abnormalities, such as mammoplasty, mastectomy, chest trauma, thoracic burns, and thoracic herpes zoster, may cause elevated prolactin levels by stimulating the serotonin-mediated neural reflex, mimicking suckling. Multiple drugs will increase prolactin, including dopamine antagonists (phenothiazines, metoclopramide, pimozide, sulpiride), dopamine reuptake inhibitors (amphetamines, tricyclic antidepressants), and catecholamine-depleting drugs (reserpine, methyldopa).

Generally, prolactin elevations result in suppression of GnRH amplitude and pulsatility and subsequent derangement of the hypothalamic–pituitary–gonadal axis. The degree of estrogen suppression correlates with the level of prolactin and determines the presenting symptoms. Mild elevations result in mild ovulatory dysfunction with luteal phase insufficiency, moderate levels with irregular cycles and episodic anovulation, and high levels with amenorrhea. These symptoms correlate with increasing levels of estrogen deficiency. Galactorrhea may be found in up to 80% of women with hyperprolactinemia but is not specific to the disorder. Up to 40–50% of women with galactorrhea do not have elevated prolactin levels. Such women do not need further evaluation. Spontaneous resolution with avoidance of nipple stimulation is common, and short-duration dopamine agonist may be helpful. Effects from pituitary tumors are correlated with size and commonly include CNS findings such as headaches and visual disturbances.

Women with non-tumor-related hyperprolactinemia or a microadenoma who lack bothersome galactorrhea or have no desire for pregnancy may be treated with combination hormonal contraceptives, which provide birth control and prevent bone loss associated with estrogen deficiency. A combination hormonal contraceptive, as opposed to a progestin-only oral contraceptive, should be used for better bone protection. The best recommendation for the described patient is a combination hormonal contraceptive with a repeat MRI and prolactin level test in 1 year.

Because of the varied presentation of hyperprolactinemia, prolactin level testing should be performed in women with menstrual irregularity, galactorrhea, and infertility. Blood should be drawn in the morning, fasting, and during the presumed follicular phase in women who have regular menses. The patient should avoid strenuous exercise or sexual activity before testing. If the patient has a borderline elevated prolactin level of 20–40 ng/mL, the test should be repeated before further evaluation. In cases without an obvious diagnosis from history or physical examination, the next line of testing is a thyroid-stimulating hormone level. Subsequent testing includes head MRI to evaluate for a CNS lesion. Prolactin levels correlate with the incidence of pituitary prolactinoma but not with other lesions. There is no threshold prolactin level for excluding a pituitary or CNS lesion, but lesions are less likely with mildly elevated prolactin levels. For this reason, other causes of mild hyperprolactinemia should be excluded before considering MRI.

Women with a pituitary microadenoma (ie, size less than 10 mm) (Fig. 97-1) do not require additional testing. The presence of a pituitary macroadenoma (ie, size greater than 10 mm) requires additional evaluation with formal visual field testing and anterior pituitary function testing consisting of testing for levels of insulin-like growth factor, morning and evening cortisol, follicle-stimulating hormone, luteinizing hormone, free thyroxine, and free α-subunit. This testing serves to determine anterior pituitary function and exclude production or secretion of another hormone by the adenoma.

Medical therapy with a dopamine agonist (bromocriptine, cabergoline) is indicated for women who do not desire birth control and who have non-tumor-related hyperprolactinemia, microadenoma, or bothersome galactorrhea. Women with a macroadenoma require immediate medical therapy and should use adequate contraceptives until the tumor is smaller than 10 mm. Consultation with a neurosurgeon may be obtained for women with substantial visual field defects or other CNS symptoms.

Patients with an initial negative MRI do not require repeat imaging unless new symptoms develop. Women with a microadenoma who are not undergoing dopamine agonist therapy require repeat imaging because approximately 7% will progress to a macroadenoma. No consensus exists on the time interval of the first follow-up scan, but 1 year seems prudent. In patients who are taking dopamine agonist therapy, normalization of the prolactin level correlates with absence of growth of a microprolactinoma. However, because there is no way to prove a microadenoma is secreting prolactin (especially with only mildly elevated blood prolactin), a repeat MRI

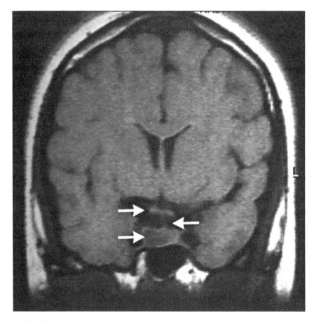

FIG. 97-1. Magnetic resonance imaging scan of a patient with hyperprolactinemia showing a pituitary microadenoma. The top arrow shows the hypothalamus, the middle arrow shows the infundibulum, and the bottom arrow shows the microadenoma.

at 1 year often is performed to establish a correlation of prolactin normalization and tumor shrinkage.

The most common dopamine agonists used for treatment are bromocriptine and cabergoline. Common adverse effects include nausea, vomiting, postural hypotension, and headache. Cabergoline is better tolerated and more efficacious in regard to tumor shrinkage and normalization of prolactin levels. Cabergoline does have one unique adverse effect: heart valve abnormalities leading to insufficiency. This has been seen only with high doses of cabergoline used for the treatment of Parkinson's disease. Studies suggest no increase in clinically significant valvular disease in women treated with cabergoline for hyperprolactinemia.

Ben-Jonathan N, Hnasko R. Dopamine as a prolactin (PRL) inhibitor. Endocrine Rev 2001;22:724–63.

Gillam MP, Molitch ME, Lombardi G, Colao A. Advances in the treatment of prolactinomas. Endocrine Rev 2006;27:485–534.

Steffensen C, Maegbaek ML, Laurberg P, Andersen M, Kistorp CM, Norrelund H, et al. Heart valve disease among patients with hyperprolactinemia: a nationwide population-based cohort study. J Clin Endocrinol Metab 2012;97:1629–34.

98

Ovarian androgen-secreting tumor

A nulliparous 45-year-old woman with amenorrhea visits your office. She reports a gradual increase in hirsutism over the past 6 months. On clinical examination, excessive hair growth is observed on her face, chest, lower back, thighs, and upper arms, with a Ferriman–Gallwey score of 23. Her body mass index is 32 (calculated as weight in kilograms divided by height in meters squared). She has a serum testosterone level of 350 pg/mL, sex hormone-binding globulin level of 35 pg/mL, and dehydroepiandrosterone sulfate (DHEAS) level of 85 micrograms/dL. Serum cortisol and 17-hydroxyprogesterone levels are normal. Transvaginal ultrasonography reveals a 2-cm solid mass in her left ovary. The best next step in management is

    (A) overnight dexamethasone suppression test
    (B) adrenocorticotropic hormone stimulation test
    (C) computerized tomography scan of the adrenal gland
*   (D) laparoscopic oophorectomy
    (E) pelvic magnetic resonance imaging (MRI)

When a patient presents with features of atypical hirsutism or polycystic ovary syndrome (PCOS), an extensive evaluation should be undertaken to exclude the 0.5% of patients who have other causes of hyperandrogenism, such as congenital adrenal hyperplasia, androgen-secreting tumors of the ovary and adrenal gland, Cushing syndrome, or hyperthecosis. Some of the unusual signs or symptoms associated with these atypical presentations include the onset of symptoms before age 15 years or a new diagnosis of PCOS in the third decade of life or later, rapid progression of symptoms, and development of virilization or a serum testosterone level in excess of twice the upper limit of the reference range (usually greater than 200 pg/mL). In women, the most common androgen-secreting tumors are ovarian, whereas in premenopausal women, the most common tumors are Sertoli–Leydig cell tumors, although lipoid and hilus cell tumors also may be observed. In postmenopausal women, hilus cell tumors are the most common androgen-secreting tumors and are quite often small. Any ovarian tumor has the ability to stimulate the production of androgens indirectly by causing hyperplasia of the surrounding ovarian stroma, such as ovarian teratomas or Brenner tumors.

Most androgen-secreting tumors are benign and in some rare instances require full staging. Less frequently, catheterization and sampling from adrenal and ovarian veins are required to elucidate the diagnosis or in cases in which no masses are identified by any type of imaging modality. Although this is a technically difficult procedure with potential complications, it may provide definitive diagnostic information. Virilization or severe hyperandrogenemia are rarely associated with PCOS. Clinical examination and a detailed history may be adequate to make the diagnosis of PCOS. However, measurement of serum testosterone, total testosterone, or an index of free testosterone is useful to screen for the rare patient with severe hyperandrogenism, prompting evaluation for diagnoses other than PCOS.

Androgen-secreting ovarian tumors are frequently small, solid masses with nonspecific appearance. To delineate the mass, it is best to perform a combination of transabdominal and transvaginal ultrasonography as part of the evaluation. An MRI scan also can be helpful; however, this modality is most useful in differentiating an ovarian tumor from a uterine myoma or other pelvic mass and during pregnancy. For the described patient, laparoscopic oophorectomy is the appropriate next step, given her rapid-onset hirsutism and high testosterone level, and should prove curative in most cases. The final pathology will dictate if staging is necessary. If necessary, staging can be performed as a separate procedure.

An overnight dexamethasone suppression test is indicated when there is suspicion for cortisol excess or as part of an evaluation for hirsutism to exclude Cushing syndrome from the differential diagnosis of PCOS. This patient does exhibit some of the clinical signs of cortisol excess, but the combination of a normal serum cortisol level and the very high testosterone level makes it unlikely that the correct diagnosis is Cushing syndrome.

An adrenocorticotropic hormone stimulation test would be indicated to confirm the diagnosis of adrenal hyperplasia if elevated serum 17-hydroxyprogesterone was identified as part of the evaluation for hirsutism or PCOS. Given this patient's normal 17-hydroxyprogesterone level, further evaluation is unnecessary.

A computerized tomography scan of the adrenal gland is indicated as part of the evaluation of an androgen-secreting adrenal tumor. This patient's normal DHEAS level likely excludes this diagnosis because virtually all adrenal tumors associated with excess testosterone production have concurrent significant DHEAS elevation. It is also important to emphasize that doing unnecessary adrenal gland imaging may result in unindicated interventions. Nonfunctioning adrenal tumors are common, can be found in up to 2% of the general population, and do not secrete hormones.

Ultrasonographic examination is the procedure of choice when imaging ovaries in the evaluation of hyperandrogenemia, whether in the diagnosis of polycystic ovaries or tumors. The described patient underwent pelvic ultrasonography that showed a solid ovarian mass that was clearly delineated; therefore, pelvic MRI will not add information that is useful to the differential diagnosis.

Dennedy MC, Smith D, O'Shea D, McKenna TJ. Investigation of patients with atypical or severe hyperandrogenaemia including androgen-secreting ovarian teratoma. Eur J Endocrinol 2010;162:213–20.

Unluhizarci K, Kaltsas G, Kelestimur F. Non polycystic ovary syndrome-related endocrine disorders associated with hirsutism. Eur J Clin Invest 2012;42:86–94.

99
Ectopic pregnancy

A 24-year-old woman, gravida 2, para 0, comes to your office with an unknown last menstrual period and a positive home pregnancy test. She reports postcoital spotting but no pelvic pain. Vital signs obtained during her visit are within normal limits. Abdominal and pelvic examinations reveal no tenderness, rebound, or guarding. The quantitative serum human chorionic gonadotropin (hCG) level is 2,800 mIU/mL. Transvaginal ultrasonography confirms thickened endometrium without evidence of an intrauterine gestational sac. A right paraovarian soft tissue mass is noted. Scant free fluid is noted in the posterior cul-de-sac. On review of her electronic medical records, you note that in an emergency department visit 2 nights before for nausea, her quantitative serum hCG level was 3,400 mIU/mL. The next step in management of this patient is

(A) prescribe methotrexate
(B) prescribe misoprostol
* (C) serial quantitative serum hCG level
(D) suction dilation and curettage (D&C)
(E) laparoscopy

Early pregnancy failure is the most common complication of pregnancy. Approximately 25% of recognized pregnancies end in miscarriage and approximately 1–2% will be ectopic pregnancies. Undiagnosed ectopic pregnancy can lead to maternal morbidity and mortality, so accurate and expeditious diagnosis is paramount. Ectopic pregnancy can compromise a woman's health and future fertility and remains a leading cause of maternal morbidity and mortality, accounting for up to 6% of deaths of women during pregnancy.

Diagnosis of women who initially present with pregnancy of unknown location requires multiple visits for blood tests, ultrasonography, and possible laparoscopy or D&C. During this diagnostic period, an unidentified ectopic pregnancy could rupture, necessitating emergency intervention or contributing to maternal morbidity. Conversely, a small percentage of pregnancies of unknown location and ectopic pregnancies may resolve spontaneously and not require medical or surgical intervention.

Modern management has resulted in a new set of pitfalls, such as overinterpretation of a single ultrasound, misunderstanding of the utility of serial hCG values, and inappropriate use of methotrexate that can result in iatrogenic complications. An error can result in false reassurance that a woman does not have an ectopic pregnancy or, conversely, interruption of a desired intrauterine pregnancy during diagnosis and management.

A dramatic change in management of ectopic pregnancy has occurred, with a shift from open surgical procedures transitioning through laparoscopic procedures toward a current predominance of medical management. Modern management should focus on ensuring that an intrauterine pregnancy is not interrupted as the result of diagnosis and treatment of a presumed ectopic pregnancy.

The ultrasonographic results obtained for the described patient would suggest an ectopic pregnancy. However, the findings must be put into clinical context with evaluation of the serial quantitative serum hCG level with consideration given to the patient's clinical circumstances. The described patient's serum hCG concentrations are falling and, therefore, close observation with serial serum hCG levels without immediate medical or surgical intervention is warranted for this stable patient. Suction D&C and misoprostol would be appropriate only in the case of a failed intrauterine pregnancy. Laparoscopy is not indicated at this time. Close follow-up is warranted.

Autry AM. Medical treatment of ectopic pregnancy: is there something new? Obstet Gynecol 2013;122:733–4.

Barnhart KT. Early pregnancy failure: beware of the pitfalls of modern management. Fertil Steril 2012;98:1061–5.

Senapati S, Barnhart KT. Biomarkers for ectopic pregnancy and pregnancy of unknown location. Fertil Steril 2013;99:1107–16.

100

Maternal virilization in hyperreactio luteinalis

A 28-year-old woman, gravida 1, comes to your office at 14 weeks of gestation with excessive facial hair growth. To date, her pregnancy has been free of complications. Ultrasonography for fetal organ evaluation shows adequate fetal growth and no fetal abnormalities. She has bilateral ovarian enlargement 10 cm in diameter and a multicystic appearance to the ovaries. The most likely diagnosis is

    (A) luteoma of pregnancy
* (B) hyperreactio luteinalis
    (C) granulosa cell tumor
    (D) ovarian hyperstimulation syndrome

The incidence of hyperandrogenism in pregnancy is low. Affected women can have a variety of symptoms, including hirsutism, acne, temporal balding, clitoromegaly, and deepening of the voice. The two most common causes of gestational hyperandrogenism are 1) luteomas of pregnancy and 2) hyperreactio luteinalis. The most likely diagnosis in the described patient is hyperreactio luteinalis, a condition characterized by moderate or marked cystic enlargement of the ovaries due to the formation of multiple benign theca-lutein cysts. Hyperreactio luteinalis is a complication of pregnancy usually distinguished by elevated serum human chorionic gonadotropin (hCG) levels in conditions such as hydatidiform mole and choriocarcinoma, although it may occur without markedly elevated hCG levels and has been reported in singleton pregnancies.

Hyperreactio luteinalis is thought to represent an exaggerated ovarian response to hCG. It may occur at any time in pregnancy, although most cases (54%) are noted in the third trimester. It is often bilateral but unilateral cases have been described. Differentiation from malignant conditions can be performed by radiologic studies, given that magnetic resonance imaging and ultrasonography show the ovaries to have a characteristic spoke-wheel appearance, representing compressed echogenic stroma surrounding multiple simple or minimally hemorrhagic cysts without solid elements (Fig. 100-1; see color plate). As hCG levels regress after delivery of the pregnancy, the ovaries slowly return to normal size without the need for surgical intervention. Fetal virilization during pregnancy does not occur because of placental and fetal aromatase activity.

The morphologic appearance of hyperreactio luteinalis on ultrasonography is nearly identical to that of ovarian hyperstimulation syndrome, a condition that occurs in the first trimester, and is associated nearly exclusively with ovulation stimulation with gonadotropins. This condition is associated with multiple corpus luteal cysts from multiple ovulations and, when severe, can be associated with ascites and pleural effusions. The condition typically regresses after the first trimester, and virilization is not seen with ovarian hyperstimulation syndrome. The degree of hirsutism in the hyperreactio luteinalis patient is measured by the Ferriman–Gallwey score through the use of representative images (Appendix D). This scoring system is used mainly for research purposes and is not a routinely used clinical tool.

Pregnancy luteomas are benign, solid, multinodular, nonneoplastic lesions characterized by bilateral ovarian enlargement during pregnancy that can simulate a tumor. The normal ovarian parenchyma is replaced by solid proliferation of luteinized stromal cells under the influence of hCG. Because stromal cells produce androgen, virilization of the mother can occur. Luteoma of pregnancy is most often seen in African American multiparous women in the third or fourth decade of life. It is often asymptomatic and discovered incidentally during cesarean delivery.

Granulosa cell tumors are rare tumors that arise from granulosa cells. Although granulosa cells normally occur only in the ovary, the peak age at which they occur is 50–55 years.

Mature cystic teratomas (dermoid cysts) constitute the most common tumor of the ovary to be identified in a pregnant patient. Dermoids often display a characteristic appearance as a uniformly solid and echogenic lesion or as a complex cystic lesion with a focal echogenic nodule that causes distal acoustic shadowing. They may be bilateral in a minority of cases. They are of germ cell origin and do not produce androgens.

Ovarian hyperstimulation is a complication of ovarian stimulation characterized by abdominal pain, bloating, and enlarged ovaries. If it presents in pregnancy, it occurs early in the first few weeks and typically resolves soon thereafter. Hirsutism is not a presenting sign of ovarian hyperstimulation syndrome.

Angioni S, Portoghese E, Milano F, Melis GB, Fulghesu AM. Hirsutism and hyperandrogenism associated with hyperreactio luteinalis in a singleton pregnancy: a case report. Gynecol Endocrinol 2007;23:248–51.

Langer JE, Coleman BG. Case 1: Diagnosis: hyperreactio luteinalis complicating a normal pregnancy. Ultrasound Q 2007;23:63–6.

Phelan N, Conway GS. Management of ovarian disease in pregnancy. Best Pract Res Clin Endocrinol Metab 2011;25:985–92.

Yacobozzi M, Nguyen D, Rakita D. Adnexal masses in pregnancy. Semin Ultrasound CT MR 2012;33:55–64.

101

Osteoporosis and celiac disease

A postmenopausal woman reports to you with a diagnosis of osteoporosis for which she is taking oral bisphosphonate. She reports a history of chronic diarrhea and celiac disease. In spite of her bisphosphonate compliance, her bone mineral density (BMD) has worsened in comparison with her previous screening. The best next step to optimize her BMD is

* (A) adherence to a gluten-free diet
* (B) weight-bearing exercise
* (C) supplemental calcium and vitamin D
* (D) parathyroid hormone

Celiac disease is a gluten-sensitive enteropathy characterized by reversible small-bowel mucosal atrophy in a genetically predisposed person resulting from an inappropriate immune response to dietary gliadin, a component of wheat proteins. Celiac disease has long been associated with metabolic bone disease. It also has been associated with infertility and pregnancy loss. Low BMD can be detected even in children and adolescents with newly diagnosed celiac disease. The need to consider celiac disease as a pathologic factor in individuals with osteoporosis applies particularly to osteoporosis patients who fail to respond to standard treatment.

The American Gastroenterological Association recently reviewed studies of osteoporosis in celiac disease according to standard levels of evidence. All such studies have shown low mean BMD around the time of diagnosis of untreated celiac disease. A pooled analysis indicated very low bone mass in approximately 40% of individuals in the spine and 15% of individuals at the hip.

When patients with symptomatic celiac disease were compared with asymptomatic patients, symptomatic patients were found to have significantly lower BMD.

Prospective studies have shown significant improvement in BMD after introduction of a gluten-free diet. For the described patient, a gluten-free diet would decrease her diarrhea, improve her gastrointestinal absorption, and have a greater effect on her BMD. It is paramount to institute a gluten-free diet first before initiating other medical treatments. Thus, prescribing intravenous bisphosphonate would not be indicated as a first step. However, weight-bearing exercise along with supplemental calcium and vitamin D should be encouraged. Calcitonin can be considered if bone loss continues in spite of adherence to the gluten-free diet.

Bernstein CN, Leslie WD, Leboff MS. AGA technical review on osteoporosis in gastrointestinal diseases. Gastroenterology 2003;124:795–841.

Geraci A. Osteoporosis and celiac disease: is it useful to a new guideline [editorial]? J Gastrointest Digest Syst 2011;1:e103.

Mazure R, Vazquez H, Gonzalez D, Mautalen C, Pedreira S, Boerr L, et al. Bone mineral affection in asymptomatic adult patients with celiac disease. Am J Gastroenterol 1994;89:2130–4.

Miller PD. Monitoring osteoporosis therapies. Curr Osteoporos Rep 2007;5:38–43.

102

Intrauterine device complications with infection

A 35-year-old woman, gravida 2, para 2, visits your clinic for contraceptive counseling. She is a healthy, monogamous woman who currently does not take any medications. She has had no prior surgeries, abnormal Pap tests, pelvic infections, or other gynecologic problems. After discussing several options, she expresses interest in an intrauterine device (IUD). When considering insertion of an IUD in this patient, the most important way to prevent pelvic infection is

(A) oral antibiotics on the day of insertion
(B) screening for gonorrhea and chlamydial infection
(C) pelvic ultrasonography before insertion
* (D) use of aseptic techniques

The IUD is an extremely safe and effective method of contraception. In addition to its high efficacy for pregnancy prevention, it is long acting and rapidly reversible, with minimal risks and adverse effects. It is a safe option for most women, and women who have contraindications to the use of oral contraceptives should consider an IUD. Although the IUD is the most commonly used method of reversible contraception worldwide, its usage is relatively low in the United States, where approximately 2% of women used IUD contraception in 2002 compared with 7.7% in 2009. This percentage, however, pales in comparison to contraception use in other parts of the world (eg, it is estimated that in parts of Asia, 50% of women use IUDs). A possible explanation of this low level of use in the United States is a history of negative publicity about IUDs, including inaccurate information about risks, fears of litigation, and concerns about the mechanism of action to prevent pregnancy. A type of IUD from the 1970s was shown to be linked to an increased incidence of pelvic inflammatory disease (PID). However, the IUDs that are currently available have not been associated with an increased risk of infection.

An IUD generally is inserted in the office setting without sedation. The patient must be informed of the risks and benefits of the IUD and read the package brochure that comes with the device, then sign a written consent before insertion of the IUD. Reported complications of insertion of an IUD include uterine perforation and vasovagal response during insertion. If an IUD is inserted during an early, unrecognized pregnancy, it may result in a miscarriage. For this reason, an IUD is ideally inserted in the follicular phase, after cessation of menses, or in women who are using another reliable method of contraception, are abstinent, or who desire emergency contraception within 5 days of unprotected intercourse, which is a U.S. Food and Drug Administration-approved indication for a copper-containing IUD.

An approximate 10% risk of developing PID exists if an IUD is inserted during an active cervical infection with *Neisseria gonorrhoeae* or *Chlamydia trachomatis*. For this reason, a patient's medical history should be evaluated before insertion of an IUD. If a patient is at high risk of infection or if cervicitis is noted on speculum examination, she should be screened for gonorrhea and chlamydial infection, and IUD insertion should be delayed. For women in whom the risk is unclear, screening for gonorrhea and chlamydial infection at the time of insertion with subsequent treatment if results are positive is appropriate. However, in patients who are at low risk for cervicitis, such as the described patient, the risk of PID is relatively low. A study of 57,000 IUD insertions between 2005 and 2009 found that only 0.5% of patients developed a significant pelvic infection. No difference in infection risk was observed for women who were screened for gonorrhea and chlamydial infection before IUD insertion compared with those who were not screened. Therefore, it is not recommended to routinely screen low-risk women for subclinical cervical infection before IUD insertion. Nonetheless, IUDs should always be inserted with the use of aseptic techniques, taking care to adequately clean the cervix with povidone-iodine and to maintain sterility for the procedure itself.

Antibiotic prophylaxis during IUD insertion is not recommended for low-risk asymptomatic women. A Cochrane review summarized data from four large randomized controlled trials and concluded that there was no significant decrease in the occurrence of PID in patients who did or did not receive antibiotic prophylaxis (odds ratio = 0.89, 95% confidence interval, 0.53–1.51).

After insertion of the IUD, if there is any question about appropriate placement of the IUD, ultrasonography can help confirm the location of the device. For example, if there is severe discomfort or bleeding during or after placement of an IUD, or if the IUD string is missing,

103

Androgen disorders

A 19-year-old woman with primary amenorrhea and lack of breast development visits your clinic. She is 1.73 m (68 in.) tall and weighs 63.5 kg (140 lb). On physical examination, she has Tanner stage 1 breast development and Tanner stage 1 pubic hair. Genital examination reveals a cervix. Her follicle-stimulating hormone and luteinizing hormone levels are 76 mIU/mL and 64 mIU/mL, respectively. Her estradiol level is less than 20 pg/mL. The most likely diagnosis is

(A) androgen insensitivity syndrome
(B) Kallmann syndrome
(C) Turner syndrome
* (D) Swyer syndrome

Primary amenorrhea is the lack of menstrual cycles after age 15 years and lack of development of secondary sexual characteristics. Amenorrhea can be secondary to several etiologies, such as hypothalamic–pituitary causes, ovarian failure, anovulation secondary to polycystic ovary syndrome, and anatomic abnormalities that affect the outflow tract.

Swyer syndrome is a form of gonadal dysgenesis. The genetic sex of the affected individual is 46,XY, and at birth the neonate is phenotypically female and does not appear to have any sexual ambiguity. Swyer syndrome patients are raised as females and typically are seen in puberty with primary amenorrhea and lack of development of secondary sexual characteristics. Approximately 10–20% of cases are a result of *SRY* gene mutations. Other genes, such as *SF1*, *WT1*, and *DAX1*, are possible candidates in the interference of sex determination. The remaining cases are of unknown origin. Because the *SRY* gene is responsible for development of the testes, and later the Sertoli cells which produce antimüllerian hormone, the mutations in *SRY* lead to the continued development of the müllerian ducts (uterus, fallopian tubes, cervix, and upper vagina). The lack of testicular development and testosterone production does not allow for further wolffian duct development or virilization of the external genitalia.

Therefore, these individuals will be born phenotypically female with female internal genitalia. In addition, these individuals do not have a functional gonad but rather have a streak of fibrous tissue called a streak gonad. Because of the lack of functional gonads, the gonadotropin levels are elevated and there are undetectable ovarian steroid hormone (estradiol) levels. The diagnosis is made when a patient with this presentation has a 46,XY karyotype.

Treatment of these individuals starts with psychologic counseling as well as pelvic ultrasonography to exclude growing gonadal tumors. The presence of Y chromosome material places the patient at risk of gonadal tumors such as gonadoblastomas and gonadal dysgerminomas. The risk of tumor development can be as high as 75%. For this reason, once diagnosis is made, the patient should undergo gonadectomy. The gonads should be sent for pathology in case tumors are noted. Secondary sexual development should be addressed using estrogen replacement with initial low-dose estradiol or conjugated equine estrogen with interval increases in the dose until full breast development is achieved, at which time progesterone should be added or combination oral contraceptives can be used. Fertility still can be achieved with the use of donor oocytes.

The described patient most likely has Swyer syndrome. In androgen insensitivity syndrome, physical examination will reveal breast development, a blind-ending vagina, and a lack of female internal genitalia. Individuals with Kallmann syndrome typically have low gonadotropin levels. Turner syndrome patients are typically short and may have other physical stigmata, such as a webbed neck, a shield chest, and a wide-carrying angle. Some patients with mosaic Turner syndrome may achieve normal height. Mosaic Turner syndrome could be considered in the differential diagnosis of the patient under discussion.

Jorgensen PB, Kjartansdottir KR, Fedder J. Care of women with XY karyotype: a clinical practice guideline. Fertil Steril 2010;94:105–13.

Lim HN, Freestone SH, Romero D, Kwok C, Hughes IA, Hawkins JR. Candidate genes in complete and partial XY sex reversal: mutation analysis of *SRY*, *SRY*-related genes and *FTZ-F1*. Mol Cell Endocrinol 1998;140:51–8.

McElreavey K, Fellous M. Sex determination and the Y chromosome. Am J Med Genet 1999;89:176–85.

Zielinska D, Zajaczek S, Rzepka-Gorska I. Tumors of dysgenetic gonads in Swyer syndrome. J Pediatr Surg 2007;42:1721–4.

104
Metabolic syndrome

A 36-year-old woman, gravida 2, para 1, aborta 1, has gained a significant amount of weight since she had a 10-week miscarriage 1 year ago; her weight gain was 17.2 kg (38 lb) over the past 15 months. She is 1.65 m (65 in.) tall and now weighs 88.4 kg (198 lb). Her body mass index is 32.9 (calculated as weight in kilograms divided by height in meters squared). She has recently noticed malaise and the need to urinate more frequently. She has normal menstrual cycles with an interval of 27–29 days and no gynecologic complaints. Her blood pressure is 142/92 mm Hg. Her waist circumference is 99 cm (39 in.). She has acanthosis nigricans present in her axilla and at the nape of her neck. She does not have abdominal striae or hyperpigmentation. Her fasting blood glucose level is 116 mg/dL, her thyroid-stimulating hormone level is 1.1 mU/mL, her plasma triglyceride level is 171 mg/dL, and her testosterone level is 30 ng/dL. The most likely diagnosis is

(A) autoimmune thyroiditis
(B) Cushing syndrome
* (C) metabolic syndrome
(D) late-onset congenital adrenal hyperplasia
(E) polycystic ovary syndrome (PCOS)

Metabolic syndrome, also known as syndrome X or insulin resistance syndrome, is associated with a cluster of risk factors for atherosclerotic cardiovascular disease. When first described, the syndrome constellation included obesity, type 2 diabetes mellitus, hypertension, and atherogenic dyslipidemia. Over the past 30 years, however, several definitions for metabolic syndrome have been proposed (Table 104-1).

Although the American Diabetes Association has chosen to drop the term "metabolic syndrome," the American College of Cardiology continues to use the term and believes that it helps frame and conceptualize patient management. The central question is whether there is a unique pathophysiology for this "syndrome" and whether metabolic syndrome better estimates individual risk beyond combining the individual components using a risk assessment tool, such as the Framingham Point Score for cardiovascular disease. Moreover, individuals who do not have the exact combination of features to merit a diagnosis of metabolic syndrome still may be at very high risk of cardiovascular disease. The aberrations attributed to metabolic syndrome are also risk factors for cardiovascular disease. "Insulin resistance syndrome" is also an unsatisfactory term because insulin-mediated glucose uptake varies greatly. Individuals may have elevated insulin without hyperglycemia or else hyperglycemia without hyperinsulinemia. Moreover, persons with normal insulin and glucose levels may be at significant risk of cardiovascular disease. Not all insulin-resistant hyperinsulinemic individuals show the other features of metabolic syndrome.

The described patient has four of the five features needed for diagnosis of metabolic syndrome according to the National Cholesterol Education Program: 1) waist circumference greater than 88 cm (35 in.) in women, 2) plasma triglycerides elevated over 150 mg/dL, 3) blood pressure over 130/85 mm Hg, and 4) fasting blood glucose over 110 mg/dL. The fifth criterion (which does not apply to this case) is a high-density lipoprotein cholesterol

TABLE 104-1. Criteria for the Diagnosis of Metabolic Syndrome

Agency	Blood Pressure	Lipids	Central Obesity	Glucose	Other	Diagnostic Requirements
WHO	On antihypertensive therapy or BP 140/90 mm Hg or more	Plasma triglycerides over 150 mg/dL and/or HDL-C less than 35 mg/dL in men or less than 39 mg/dL in women	Body mass index* 30 or more and/or waist–hip ratio more than 0.9 or higher in men or greater than 0.85 in women	IGT or as in type 2 diabetes mellitus	Microalbuminuria UAE 20 micrograms/min or more or A:Cr 30 mg/g or more	Type 2 diabetes mellitus or IGT and two additional abnormalities; if GT normal, three required
NCEP ATP III	BP 130/85 mm Hg or greater	Plasma triglycerides over 150 mg/dL and/or HDL-C less than 40 mg/dL in men or less than 50 mg/dL in women	Waist circumference greater than 102 cm (40 in.) (men), greater than 88 cm (35 in.) (women)	FBG 110 mg/dL or more		Three or more of criteria listed
AACE	Hypertension	Plasma triglycerides over 150 mg/dL and/or HDL-C less than 35 mg/dL in men or less than 45 mg/dL in women	Waist circumference 102 cm (40 in.) or greater in men, 88 cm (35 in.) or greater in women	IFG or as in type 2 diabetes mellitus	Insulin resistance, acanthosis nigricans, hyperuricemia. Minor criteria: hypercoagulability, CHD, PCOS, vascular endothelial dysfunction, microalbuminuria	Not specified
IDF	BP 130/85 mm Hg or less	Plasma triglycerides over 150 mg/dL and/or HDL-C less than 40 mg/dL in men or less than 50 mg/dL in women	Waist circumference greater than 102 cm (40 in.) (men), greater than 88 cm (35 in.) (women)	FBG 110 mg/dL or greater		

Abbreviations: AACE, American Association of Clinical Endocrinologists; A:Cr, albumin:creatine; BP, blood pressure; CHD, coronary heart disease; FBG, fasting blood glucose; HDL-C, high-density lipoprotein cholesterol; IDF, International Diabetes Foundation; IFG, impaired fasting glucose; IGT, impaired glucose tolerance; NCEP ATP III, National Cholesterol Education Program Adult Treatment Panel III; PCOS, polycystic ovary syndrome; UAE, urinary albumin excretion; WHO, World Health Organization.

*Body mass index, weight in kilograms divided by height in meters squared (kg/m^2).

level of less than 50 mg/dL in women. These criteria identify a patient with obesity, hypertension, diabetes mellitus, and dyslipidemia. This constellation of symptoms comprises metabolic syndrome.

Late-onset congenital adrenal hyperplasia is due to a defect and insufficiency in 21α-hydroxylase, 11β-hydroxylase, or 3β-hydroxysteroid (type II) dehydrogenase enzymes with resultant excessive androgen production. Although obesity and fatigue may be present, the hallmark is severe hirsutism and varying degrees of virilization.

Polycystic ovary syndrome is seen in women with varying degrees of anovulation and androgen excess. Hyperinsulinemia or insulin resistance and hyperandrogenism also are recognized features. Patients are often obese, with acanthosis nigricans. Hirsutism is rarely as severe as that seen in congenital adrenal hyperplasia. Because menstrual irregularities are common in PCOS and the woman described has normal cycles, PCOS is a less likely diagnosis.

Cushing syndrome is the persistent overproduction of cortisol, usually because of excess adrenocorticotropic hormone or adrenocorticotropic hormone-secreting tumors. The features of Cushing syndrome include central obesity and fat deposits in the cheeks, dorsocervical area (buffalo hump), and supraclavicular regions. The limbs are thin and wasted, and hirsutism is rarely a feature. Hyperpigmentation is confined to skin creases and pressure points. The striae from Cushing syndrome are violaceous, whereas those from simple obesity are silvery.

Autoimmune thyroiditis commonly presents with weight gain, fatigue, facial puffiness, and peripheral edema. Hirsutism is rarely a feature, and indeed hair loss often is reported with the condition.

Hyperlipidemia may be present with all of the conditions listed. The most likely diagnosis in the described patient is metabolic syndrome.

Executive Summary of The Third Report of The National Cholesterol Education Program (NCEP) Expert Panel on Detection, Evaluation, And Treatment of High Blood Cholesterol In Adults (Adult Treatment Panel III). Expert Panel on Detection, Evaluation, and Treatment of High Blood Cholesterol in Adults. JAMA 2001;285:2486–97.

Gabir MM, Hanson RL, Dabelea D, Imperatore G, Roumain J, Bennett PH, et al. The 1997 American Diabetes Association and 1999 World Health Organization criteria for hyperglycemia in the diagnosis and prediction of diabetes. Diabetes Care 2000;23:1108–12.

Grundy SM, Cleeman JI, Daniels SR, Donato KA, Eckel RH, Franklin BA, et al. Diagnosis and management of the metabolic syndrome: an American Heart Association/National Heart, Lung, and Blood Institute Scientific Statement. American Heart Association and the National Heart, Lung, and Blood Institute [published errata appear in Circulation 2005;112:e298. Circulation 2005;112:e297]. Circulation 2005;112:2735–52.

Hanley AJ, Karter AJ, Williams K, Festa A, D'Agostino RB Jr, Wagenknecht LE, et al. Prediction of type 2 diabetes mellitus with alternative definitions of the metabolic syndrome: the Insulin Resistance Atherosclerosis Study. Circulation 2005;112:3713–21.

Kahn R, Buse J, Ferrannini E, Stern M. The metabolic syndrome: time for a critical appraisal: joint statement from the American Diabetes Association and the European Association for the Study of Diabetes. American Diabetes Association and the European Association for the Study of Diabetes. Diabetes Care 2005;28:2289–304.

Liu J, Grundy SM, Wang W, Smith SC Jr, Vega GL, Wu Z, et al. Ten-year risk of cardiovascular incidence related to diabetes, prediabetes, and the metabolic syndrome. Am Heart J 2007;153:552–8.

Mitka M. Does the metabolic syndrome really exist? Diabetes and heart disease groups spar over issue. JAMA 2005;294:2010–3.

Orchard TJ, Temprosa M, Goldberg R, Haffner S, Ratner R, Marcovina S, et al. The effect of metformin and intensive lifestyle intervention on the metabolic syndrome: the Diabetes Prevention Program randomized trial. Diabetes Prevention Program Research Group. Ann Intern Med 2005;142:611–9.

105

Complications of robotic-assisted surgery

A 45-year-old multiparous woman comes to your office for a preoperative visit in preparation for a robotic-assisted total hysterectomy for symptomatic uterine fibroids. She has heard that hysterectomy can be associated with small-bowel obstruction, bladder injury, increased blood loss, vaginal cuff dehiscence, and brachial plexus injury. As part of informed consent, you inform her that of all of the complications named, relative to laparoscopic-assisted surgery, robotic-assisted surgery increases her risk of

  (A) vaginal cuff dehiscence
  (B) bladder injury
* (C) small-bowel obstruction
  (D) increased blood loss
  (E) brachial plexus injury

Hysterectomy for benign gynecologic disease is perhaps one of the most commonly performed gynecologic surgical procedures; approximately one in nine women in the United States will ultimately have this procedure performed. Hysterectomies were originally performed via laparotomy or vaginally. In the late 1980s, reports linking laparoscopic procedures with shorter hospital stays and faster recoveries led to the inclusion of this surgical technique among the possible surgical options. More recently, robotic-assisted hysterectomy has been introduced as an alternative minimally invasive approach to traditional hysterectomy.

The robotic platform was originally developed by the Stanford Research Institute, the U.S. Department of Defense, and the National Aeronautics and Space Administration in an attempt to facilitate telesurgery for wounded soldiers. The intent was to position the operating surgeon remote to the battlefield to perform surgery in the battlefield operating room through robotic assistance. Although telesurgery was found to be technically possible, its practical use was limited by several factors, including telecommunication bandwidth requirements. Robotic-assisted laparoscopic surgery was further developed, commercialized, and approved by the U.S. Food and Drug

Administration in 2001 for urologic procedures, in 2002 for thoracic surgery, and in 2005 for gynecologic procedures. Since Food and Drug Administration approval in 2005, thousands of gynecologic surgeons have received training in this alternative technique. Supporters of the procedure argue that robotic assistance allows physicians to be more comfortable in applying a minimally invasive technique, whereas in the past a laparotomy would have been performed.

A recent cohort study of 264,758 women who underwent hysterectomy for benign gynecologic disorders at 441 hospitals across the United States in 2007–2010 showed that the use of robotic-assisted hysterectomy increased from 0.5% to 9.5% of all hysterectomies in 2010 with a concurrent decrease in the rate of abdominal hysterectomy. Although patients who underwent a robotic-assisted hysterectomy had shorter stays, the transfusion requirements, death rates, and rates of discharge to a nursing facility were similar. Total costs associated with robotic-assisted hysterectomy were higher than for laparoscopic hysterectomy.

The rate of vaginal cuff dehiscence is comparable with other gynecologic techniques for hysterectomy. There are no reports of an increased risk of brachial plexus injury with robotic-assisted procedures despite the positioning of the patient on the surgical table. Intraoperative injuries to the urinary system are equally common in laparoscopic and robotic procedures. A recent Cochrane review of randomized controlled trials of robotic surgery for benign gynecologic conditions found no differences in intraoperative complications between robotic and laparoscopic surgery. However, robotic surgery led to a higher incidence of postoperative complications (odds ratio =5.44, 95% confidence interval, 1.57–18.82; P=.008), including urinary tract infection, small-bowel obstruction, wound infection, erosion, abdominal wall pain necessitating trigger point injection, and abscess. The authors concluded that robotic surgery was not associated with improved effectiveness or safety but that it was associated with substantially increased costs.

Kashani S, Gallo T, Sargent A, Elsahwi K, Silasi DA, Azodi M. Vaginal cuff dehiscence in robotic-assisted total hysterectomy. JSLS 2012;16:530–6.

Liu H, Lu D, Wang L, Shi G, Song H, Clarke J. Robotic surgery for benign gynaecological disease. Cochrane Database of Systematic Reviews 2012, Issue 2. Art. No.: CD008978. DOI: 10.1002/14651858.CD008978.pub2.

Wright JD, Ananth CV, Lewin SN, Burke WM, Lu YS, Neugut AI, et al. Robotically assisted vs laparoscopic hysterectomy among women with benign gynecologic disease. JAMA 2013;20:689–98.

106
Heavy menstrual bleeding

A 37-year-old woman, gravida 3, para 3, comes to your office with a 1-year history of heavy menstrual bleeding. The most recent episode started 2 weeks ago and ended last night. She describes the bleeding as heavy with passage of small clots. On examination, her uterus is normal in size and no bleeding from the cervix is noted. Pelvic ultrasonography is normal and she has a negative urine pregnancy test. Her thyroid-stimulating hormone level is normal and her hemoglobin level is 10 g/dL. She reports no orthostatic symptoms. The most effective nonsurgical management to reduce the effects of vaginal bleeding on her quality of life is

* (A) levonorgestrel intrauterine device (IUD)
* (B) tranexamic acid
* (C) progestin-only oral contraceptives (OCs)
* (D) combination hormonal OCs

Heavy menstrual bleeding is blood loss during menses in excess of 80 mL. It is a common problem that can significantly affect a woman's quality of life. In the United States, heavy menstrual flow accounts for approximately 18% of visits to the gynecologist. The quantification of blood loss during menses can be difficult to assess, given that it relies on the patient's perception of the amount of blood lost and how she feels about its effects on her quality of life. Appendix C shows the differential diagnosis of abnormal uterine bleeding. After an evaluation is completed, treatment of the underlying cause of excessive bleeding is performed.

Heavy menstrual bleeding may be addressed with either medical or surgical therapy. Surgery can be performed in cases in which removal of anatomic lesions is necessary or, if the patient does not desire future fertility, she may choose either endometrial ablation or hysterectomy. Medical treatment is less invasive and can provide the patient with fertility-sparing treatment. Some patients are poor surgical candidates and may have to rely on medical interventions only.

Medical management can be hormonal or nonhormonal. Hormonal therapies include combination hormonal OCs, progestin-only OCs, and the levonorgestrel IUD. The disadvantages of OCs are that they require daily administration, carry the risks of development of hypercoagulability and deep vein thrombosis, and may not completely control the patient's bleeding. Nonhormonal methods include the use of tranexamic acid and nonsteroidal antiinflammatory drugs, eg, naproxen and ibuprofen. Tranexamic acid, although effective, is expensive and requires patient compliance for effectiveness.

Although initially developed as a contraceptive method, the levonorgestrel IUD can provide treatment for heavy menstrual bleeding. The levonorgestrel IUD has been shown to cause greater reduction in menstrual blood loss compared with other hormonal and nonhormonal treatments and is more effective in reducing the effects of heavy menstrual bleeding on the patient's quality of life. In addition to its effects on quality of life, the levonorgestrel IUD is an effective reversible contraceptive that lacks any negative metabolic effects, does not require daily use, and has effects that are not dependent on patient adherence.

Endrikat J, Vilos G, Muysers C, Fortier M, Solomayer E, Lukkari-Lax E. The levonorgestrel-releasing intrauterine system provides a reliable, long-term treatment option for women with idiopathic menorrhagia. Arch Gynecol Obstet 2012;285:117–21.

Gupta J, Kai J, Middleton L, Pattison H, Gray R, Daniels J; ECLIPSE Trial Collaborative Group. Levonorgestrel intrauterine system versus medical therapy for menorrhagia. N Engl J Med 2013;368:128–37.

Higham JM, Shaw RW. Clinical associations with objective menstrual blood volume. Eur J Obstet Gynecol Reprod Biol 1999;82:73–6.

Munro MG, Critchley HO, Broder MS, Fraser IS; FIGO Working Group on Menstrual Disorders. FIGO classification system (PALM–COEIN) for causes of abnormal uterine bleeding in nongravid women of reproductive age. Int J Gynaecol Obstet 2011;113:3–13.

Nicholson WK, Ellison SA, Grason H, Powe NR. Patterns of ambulatory care use for gynecologic conditions: a national study. Am J Obstet Gynecol 2001;184:523–30.

Stewart A, Cummins C, Gold L, Jordan R, Phillips W. The effectiveness of the levonorgestrel-releasing intrauterine system in menorrhagia: a systematic review. BJOG 2001;108:74–86.

Whiteman MK, Kuklina E, Jamieson DJ, Hillis SD, Marchbanks PA. Inpatient hospitalization for gynecologic disorders in the United States. Am J Obstet Gynecol 2010;202:541.e1–6.

107
Virilization due to Sertoli–Leydig cell tumor

A 28-year-old woman has a 3-month history of temporal scalp balding and amenorrhea. Transvaginal ultrasonography reveals a 3-cm right solid ovarian mass, and a computed tomography scan shows normal adrenal glands. The hormone that is most likely to be elevated is

* (A) dehydroepiandrosterone sulfate (DHEAS)
* (B) 17α-hydroxyprogesterone
* (C) 17α-hydroxypregnenolone
* (D) testosterone

Androgen-producing tumors that lead to signs of virilization can be isolated to an ovarian origin by a combination of diagnostic imaging and an elevated serum testosterone level. The hallmark changes of virilization include temporal scalp balding, deepening of the patient's voice, and the presence of clitoromegaly (abnormally enlarged clitoris with a diameter greater than 10 mm). These changes are in addition to facial and body hirsutism. The findings raise concern for an adrenal or ovarian tumor. Usually, the virilizing changes are sudden in onset and represent a marked change noticed by the patient or family members. It often has been stated that one can diagnose the presence of an androgen-secreting tumor by patient history alone.

Imaging in combination with serum hormone testing is helpful in localizing the source of excess androgen secretion. Tumors within the adrenal glands are associated with an elevated serum DHEAS level of greater than 700 micrograms/dL. Because 95% of circulating DHEAS is produced by the adrenal glands, DHEAS makes an ideal marker for adrenal function. Tumors within an ovary are associated with an elevated serum testosterone level of greater than 200 ng/dL. The ovaries produce approximately 50% of circulating testosterone, with the sources almost equally divided between direct ovarian secretion and peripheral conversion of androstenedione to testosterone. Once the source of the androgen excess is localized to either the adrenal glands or ovaries, imaging can frequently narrow down the origin to a discrete mass. When the exact location of the tumor is in doubt, selective venous catheterization of the adrenal glands and the ovaries can be helpful in deciding which organ should be explored further. This is especially important in the case of an ovarian tumor when future fertility is desired. In contrast, if the androgen excess occurs in a postmenopausal woman without a discrete adrenal mass, bilateral oophorectomy would be preferable.

Ovarian Sertoli–Leydig cell tumors are rare, accounting for approximately 0.5% of all ovarian tumors. Because of the associated hormone production of Sertoli–Leydig cell tumors (primarily androgen secretion and sometimes estrogen secretion), the presence of characteristic endocrine-mediated symptoms contributes to the diagnosis of such tumors. Approximately two thirds of cases occur in women younger than 30 years, but these tumors can occur throughout a woman's lifetime. The tumors are typically unilateral with approximately 1.5% occurring bilaterally, and 90% are Stage 1 at diagnosis. The overall prognosis of Sertoli–Leydig cell tumors is good for Stage 1 well-differentiated neoplasms. Conservative surgery is an option for women with such tumors who wish to preserve their fertility. Some Sertoli–Leydig cell tumors have primarily estrogenic symptoms such as postmenopausal or anovulatory bleeding; other types of these tumors show no obvious endocrine manifestations. The Sertoli–Leydig cell tumors with minimal hormonal alterations are often larger tumors (greater than 10 cm in diameter), may be associated with tumor rupture, and are more likely to be poorly differentiated. When a tumor exhibits no endocrine changes, this finding may indicate the presence of a tumor with a more aggressive biologic behavior.

In the described patient, the presence of signs of virilization and an ovarian mass mean that testosterone will be the most elevated hormone. If the patient had an exclusive adrenal mass with normal-sized ovaries, the serum DHEAS level would be elevated. The presence of virilization (temporal scalp balding) makes other etiologies of general hirsutism such as adult-onset congenital adrenal hyperplasia and Cushing syndrome less likely. These two causes are associated with elevations in serum 17-hydroxyprogesterone and cortisol, respectively. A serum 17α-hydroxypregnenolone level would be elevated in comparison with a 17α-hydroxyprogesterone level in a patient with a 3β-hydroxysteroid dehydrogenase deficiency, which is a cause of adult-onset congenital adrenal hyperplasia, but not in this setting of virilization with an ovarian mass.

Gui T, Cao D, Shen K, Yang J, Zhang Y, Yu Q, et al. A clinico-pathological analysis of 40 cases of ovarian Sertoli-Leydig cell tumors. Gynecol Oncol 2012;127:384–9.

Petersons CJ, Burt MG. The utility of adrenal and ovarian venous sampling in the investigation of androgen-secreting tumours. Intern Med J 2011;41:69–70.

Weng CS, Chen MY, Wang TY, Tsai HW, Hung YC, Yu KJ, et al. Sertoli-Leydig cell tumors of the ovary: a Taiwanese Gynecologic Oncology Group study. Taiwan J Obstet Gynecol 2013;52:66–70.

108

Functional hypothalamic amenorrhea and osteoporosis

A 17-year-old nulligravid adolescent with secondary amenorrhea for 10 months comes to your office with her mother. She exercises for 2 hours daily and has been actively working on weight loss for the past 2 years, with a drop in weight from 54.4 kg to 43.5 kg (120 lb to 96 lb). Her height is 1.57 m (62 in.) and her current body mass index is 17.6 (calculated as weight in kilograms divided by height in meters squared). A dual-energy X-ray absorptiometry (DXA) scan shows that her bone mineral density is 3 standard deviations below the mean for her age group. In addition to nutritional and psychological counseling, the most appropriate medical option for this patient is

    (A) alendronate sodium
    (B) raloxifene hydrochloride
    (C) human parathyroid hormone therapy
*   (D) combination oral contraceptives (OCs)

Personal, behavioral, and environmental factors that can disrupt the normal function of the hypothalamic–pituitary–ovarian axis may cause functional amenorrhea. Hypothalamic function typically is impaired, although women with intact hypothalamic pathways may have missed or absent menses. The degree of disruption depends on the inherent stability of the hypothalamic–pituitary–ovarian axis and the severity of the imposed stress. Physical stress, emotional disturbances, illness, weight loss, and exercise have all been implicated as causes of hypothalamic hypogonadism. The term "female athlete triad" refers to the combination of inadequate caloric balance, amenorrhea, and osteoporosis and usually is seen in patients with self-imposed severe caloric restriction combined with extreme athletic output. Alterations in neuropeptides, neurotransmitters, and neurosteroids can cause gonadotropin-releasing hormone pulsatile secretion to cease. In the early phases of the female athlete triad, the patient may maintain normal estrogen and gonadotropin levels, but in the later phases, the hormone levels drop to subphysiologic levels.

Young women with the female athlete triad can have low estrogen levels and subsequently low bone mass, putting them at risk of pathologic stress fractures. In addition, young women with the female athlete triad have poor bone development at precisely the time they should be obtaining peak bone mass. It has been estimated that women obtain 95% of their total lifetime bone mass by age 18 years. Young athletes may alter their bone density in an unevenly distributed fashion, with normal bone density in weight-bearing sites and decreased bone density in non-weight-bearing sites. These young women may never obtain an optimal bone mass, predisposing them to significant osteoporosis later in life.

In premenopausal women, the z-score derived from a DXA scan should be used to evaluate bone mineral density at the lumbar spine and hip. An important clinical advantage of the DXA scan compared with other types of bone mineral density measurement is that its ability to identify patients at risk of fracture has been assessed and proven in a large number of epidemiologic studies. The z-score represents the bone mass compared with individuals of similar age and sex. A z-score is concerning when it is two or more standard deviations below the mean.

The first step in treatment of young women with the female athlete triad should be to restore normal energy balance. Higher caloric intake is likely needed to restore bone mass and should include vitamin D, calcium, trace minerals, and adequate protein intake. Nutritional counseling and often psychologic support and behavior therapy are essential to reversing this challenging disease.

Oral contraceptives can provide estrogen to these women to help improve bone density. Although the best strategy to foster better bone health is through improved nutritional status, this strategy can take time, and some patients are extremely resistant to diet and exercise modification. Hormonal therapy and OC use can stabilize bone loss but do not reverse prior bone loss. Extensive counseling should occur when prescribing OCs to these patients. Women with the female athlete triad must understand that OC use will result in cyclic bleeding, but that the resumption of bleeding is iatrogenic in origin and is not indicative of a return of their own hypothalamic function.

Bisphosphonates, such as alendronate sodium, should be avoided in patients with the female athlete triad. These medications have a long half-life and can be absorbed into bone and leach out later in life, possibly during a future pregnancy. These medications have demonstrated adverse effects in animal models and should be used with caution in reproductive-aged women.

The effects on bone density of selective estrogen receptor modulators, such as raloxifene hydrochloride, are inferior to the effects of estrogen. For younger women, these medications offer no advantage over exogenous estrogen or OCs. Parathyroid hormone has been used in one trial to treat bone loss in women with endometriosis who take gonadotropin-releasing hormone agonists. However, parathyroid hormone should be used with caution in young women who may have open epiphyses and would not be the first-line option for the described patient.

Rickenlund A, Carlstrom K, Ekblom B, Brismar TB, Von Schoultz B, Hirschberg AL. Effects of oral contraceptives on body composition and physical performance in female athletes. J Clin Endocrinol Metab 2004;89:4364–70.

Silveira LF, Latronico AC. Approach to the patient with hypogonadotropic hypogonadism. J Clin Endocrinol Metab 2013;98:1781–8.

Thein-Nissenbaum J. Long term consequences of the female athlete triad. Maturitas 2013;75:107–12.

Young N, Formica C, Szmukler G, Seeman E. Bone density at weight-bearing and nonweight-bearing sites in ballet dancers: the effects of exercise, hypogonadism, and body weight. J Clin Endocrinol Metab 1994;78:449–54.

109

Abnormal uterine bleeding

A 27-year-old nulligravid woman who has been diagnosed with polycystic ovary syndrome (PCOS) comes to your office with irregular bleeding. She reports no vaginal bleeding for 6 months followed by bleeding every 2 weeks. She has not experienced moliminal symptoms associated with her menstrual periods. An office urine pregnancy test is negative. Transvaginal ultrasonography identifies polycystic-appearing ovaries bilaterally and an endometrial stripe measuring 22 mm. Her serum reproductive hormone levels today will most likely be

	Estradiol (pg/mL)	Luteinizing Hormone (LH) (mIU/mL)	Progesterone (ng/mL)
(A)	Less than 32	0.2	0.4
(B)	Less than 32	12	0.4
* (C)	97	12	0.4
(D)	220	7	12

Polycystic ovary syndrome is a complex, multidimensional disorder resulting in defects in reproduction and metabolism. The prevalence of PCOS is estimated at 4–12% of reproductive-aged women. Menstrual abnormalities seen with chronic anovulation include secondary amenorrhea, oligomenorrhea, and heavy menstrual bleeding.

The pathophysiology of PCOS encompasses inherent ovarian dysfunction that is strongly influenced by external factors, such as disturbances of the hypothalamic–pituitary–ovarian axis. Exaggerated gonadotropin-releasing hormone (GnRH) pulsatility results in hypersecretion of LH, which has effects on ovarian androgen production and oocyte development. Patients may report false-positive urinary ovulation predictor tests due to chronic LH elevation. Persistently heightened GnRH pulsatility then alters the LH–follicle-stimulating hormone (FSH) ratio and contributes to increased ovarian androgen secretion and aberrant follicle maturation. Changes in the LH–FSH ratio, although a contributor to hyperandrogenism, are not used as criteria to diagnose PCOS.

The irregular bleeding reported by the described patient is most consistent with anovulatory bleeding and, therefore, progesterone values would be expected to be less than 3 ng/mL. Given her diagnosis of PCOS, we would expect altered GnRH pulsatility leading to LH hypersecretion, absence of ovulation, progesterone level below 3–5 ng/mL, and some FSH secretion from the anterior pituitary, causing granulosa cell production of detectable estradiol levels. This is emphasized in the clinical example in the unopposed endometrial proliferation observed with measurement of a thickened endometrial stripe. An estradiol level greater than 200 pg/mL would be consistent with a periovulatory follicle and would be unlikely in this patient. Bloodwork performed today most likely will reveal the hormonal level pattern of a prolonged follicular phase (option C).

Boutzios G, Karalaki M, Zapanti E. Common pathophysiological mechanisms involved in luteal phase deficiency and polycystic ovary syndrome. Impact on fertility. Endocrine 2013;43:314–7.

Cheung AP. Polycystic ovary syndrome: a contemporary view. J Obstet Gynaecol Can 2010;32:423–5, 426–8.

Johnston-MacAnanny EB, Park JK, Berga SL. Polycystic ovary syndrome. In: Falcone T, Hurd WW, eds. Clinical reproductive medicine and surgery: a practical guide. New York (NY): Springer-Verlag; 2013. p. 113–23.

110
Cushing syndrome

A 28-year-old nulligravid woman is referred to you for secondary amenorrhea. She tells you that she has been gaining weight and has fatigue and muscle weakness. On physical examination, her blood pressure is 165/95 mm Hg and her weight is 99.8 kg (220 lb). She has round facies, facial hirsutism, truncal obesity, and development of blue-tone abdominal striae. The best screening test to confirm the diagnosis is

* (A) overnight dexamethasone suppression test
* (B) 17-hydroxyprogesterone
* (C) urinary 17-ketosteroids
* (D) serum adrenocorticotropic hormone (ACTH)
* (E) morning serum cortisol

Cushing syndrome is defined as overexposure of the body's tissues to cortisol. It is a rare diagnosis; approximately one to five new cases occur per million people per year in the United States. Cushing syndrome should be distinguished from Cushing disease and implies an ACTH-secreting tumor. Identification of the clinical presentation of Cushing syndrome and its biochemical confirmation present a challenging clinical problem for physicians. Early diagnosis of Cushing syndrome is crucial because the natural history of the condition is marked by significant excess of mortality and morbidity. Serum ACTH-dependent causes account for 85% of Cushing syndrome cases. Patients with Cushing syndrome have pituitary oversecretion or ectopic production of ACTH or corticotrophin-releasing hormone; ACTH-independent causes account for the remaining 15% of cases, which arise from adrenal adenoma or carcinoma. In addition, some women develop the condition because of the exogenous use of corticosteroids.

The diagnostic process is divided into two sequential steps: first, the diagnosis and confirmation of hypersecretion of cortisol, and, second, the differential diagnosis of its etiology. Patients with Cushing syndrome are relatively insensitive to corticosteroid feedback and exhibit oversecretion of cortisol that is not circadian in nature. If any of the screening test results are positive, an additional screening test is performed to confirm the diagnosis. Once the diagnosis is established with two positive screening test results, the cause of Cushing syndrome is established in order to initiate treatment as soon as possible.

A recent meta-analysis commissioned by the Endocrine Society Diagnosis of Cushing Syndrome Task Force found similar accuracy for such tests as 24-hour urine cortisol, 1-mg overnight dexamethasone suppression test, serum or salivary midnight cortisol, and combined strategies based on these tests. An overnight dexamethasone suppression test offers the simplest way of screening for the condition. Dexamethasone at a dose of 1 mg is given orally at 11:00 PM and a plasma cortisol level is drawn at 8:00 AM the next morning. A value of less than 5 micrograms/dL rules out Cushing syndrome, and the number of patients with abnormal suppression is less than 1%. Although historically obese patients have a false-positive rate that is higher than the general population, a recent study of 86 obese patients where a cut-off for the suppression of serum cortisol was reduced to 3 micrograms/dL reduced the false-positive rate to 2.3%, minimizing any extensive follow-up studies. A 24-hour urine collection for free cortisol is a useful measurement in the basal state to detect Cushing syndrome; however, it is cumbersome and difficult to collect. Although a midnight plasma cortisol level of less than 15 micrograms/dL is helpful to detect Cushing syndrome, it is difficult to use in clinical practice because of the time for collection. Likewise, morning serum cortisol is not a suitable screening test for Cushing syndrome.

The current guidelines published by the Endocrine Society recommend against the use of tests for random serum cortisol or plasma ACTH levels and urinary 17-ketosteroids, pituitary and adrenal imaging, and the 8-mg dexamethasone suppression test as first-line tests for the diagnosis of Cushing syndrome. A high-dose dexamethasone suppression test is used to determine the etiology of Cushing syndrome, particularly when it is associated with an ACTH level. In the presence of a basal level ACTH less than 5 micrograms/mL, a high-dose dexamethasone suppression (taking 2 mg every 6 hours for 2 consecutive days), and no decrease in urinary steroids (17-hydroxysteroid and cortisol) by at least 40%, an adrenal tumor is likely. When ACTH is measurable in the blood (greater than 20 pg/mL), an ectopic ACTH-producing tumor is unlikely if the urinary steroids decrease by at least 40%. Cushing disease is present when an ACTH and a chest X-ray are normal and an abnormal sella is found on computed tomography scan of the pituitary gland. A low plasma ACTH level (less than 5 pg/mL) suggests an autonomous cortisol-secreting tumor, and the precise location of the abnormality would be identified with computed tomography or magnetic resonance imaging of the adrenal gland. A high plasma ACTH level (greater than 50 pg/mL) suggests ectopic secretion and should be differentiated from patients with Cushing disease. In cases where ACTH is elevated, intrapetrosal sinus catheterization is the only test able to differentiate Cushing disease from ectopic secretion of cortisol. A 17-hydroxyprogesterone level is used as a screening test for 21-hydroxylase deficiency and not Cushing syndrome, so it would not be appropriate in this case. Therefore, the best screening test to confirm the diagnosis is overnight dexamethasone suppression test.

Elamin MB, Murad MH, Mullan R, Erickson D, Harris K, Nadeem S, et al. Accuracy of diagnostic tests for Cushing's syndrome: a systematic review and metaanalyses. J Clin Endocrinol Metab 2008;93:1553–62.

Guignat L, Bertherat J. The diagnosis of Cushing's syndrome: an Endocrine Society Clinical Practice Guideline: commentary from a European perspective. Eur J Endocrinol 2010;163:9–13.

Nieman LK, Biller BM, Findling JW, Newell-Price J, Savage MO, Stewart PM, et al. The diagnosis of Cushing's syndrome: an Endocrine Society Clinical Practice Guideline. J Clin Endocrinol Metab 2008;93:1526–40.

111
Retrograde ejaculation

A 35-year-old male with diabetes mellitus type 1 comes to your office with his 31-year-old partner to be evaluated for infertility. His diabetes mellitus was poorly controlled for a number of years but is now well controlled. He reports that his urine is cloudy after intercourse. His semen analysis demonstrates low-volume oligospermia. A postejaculatory urinalysis demonstrates the presence of sperm. A repeat semen analysis performed with the use of postejaculatory urine after alkalinization is normal. The most appropriate initial treatment to help this couple achieve pregnancy is

    (A) donor sperm insemination
* (B) intrauterine insemination
    (C) in vitro fertilization (IVF)
    (D) surgical correction

Ejaculatory dysfunction is one the most common male sexual disorders and includes premature ejaculation, retrograde ejaculation, delayed ejaculation, anorgasmia, and aspermia (absence of sperm). Retrograde ejaculation can be categorized into two types: 1) functional and 2) anatomic. Functional disorders, such as multiple sclerosis, diabetic neuropathy, or spinal cord injury, may result from medical conditions that predispose the patient to retrograde ejaculation. Anatomic causes include urologic bladder neck surgery such as transurethral resection of the prostate.

Retrograde ejaculation causes semen to reflux back into the bladder rather than pass through the urethra. Sperm exposed to acidic urine may affect normal semen parameters and contribute to infertility. Correct diagnosis of retrograde ejaculation requires evaluation of postejaculatory urine, usually in patients who have a history of low semen volume (less than 1 mL).

In order to perform the postejaculatory urinalysis, the urine specimen should be centrifuged, followed by microscopic examination of the pellet. In men with azoospermia or aspermia, the presence of sperm on postejaculatory urinalysis is suggestive of retrograde ejaculation. Significant numbers of sperm observed in the urine confirm the diagnosis of retrograde ejaculation.

Before ejaculation, the urine should be alkalinized using oral sodium bicarbonate. After ejaculation, the urine should be obtained via voiding or catheterization. Sperm are isolated by centrifugation, washed, and then prepared for use in intrauterine insemination or IVF if the semen parameters are unfavorable. In some patients, antihistamines or ephedrine can be used to try to close the bladder neck to prevent retrograde ejaculation at the time of orgasm. Potential adverse effects include headache, hypertension, and urinary retention.

In the couple described, intrauterine insemination is the most appropriate initial treatment. Donor sperm insemination is not necessary for this couple given the male partner's normal semen analysis with the use of postejaculatory urine after alkalinization. At this time, IVF is not necessary given his normal semen analysis. In patients with an abnormal semen analysis or patients who have failed to conceive with intrauterine insemination, IVF should be considered. This patient's retrograde ejaculation is probably due to diabetic neuropathy and, thus, surgical correction is not indicated at this time.

Fedder J, Kaspersen MD, Brandslund I, Hojgaard A. Retrograde ejaculation and sexual dysfunction in men with diabetes mellitus: a prospective, controlled study. Andrology 2013;1:602–6.

Jefferys A, Siassakos D, Wardle P. The management of retrograde ejaculation: a systematic review and update. Fertil Steril 2012;97:306–12.

van der Linden PJ, Nan PM, te Velde ER, van Kooy RJ. Retrograde ejaculation: successful treatment with artificial insemination. Obstet Gynecol 1992;79:126–8.

112
Hirsutism

A 27-year-old woman presents to your office with hirsutism and irregular menses related to polycystic ovary syndrome (PCOS). She is otherwise healthy and is not trying to conceive at this time. You counsel her that, in addition to lifestyle modifications, the best initial pharmacologic treatment for her symptoms is

    (A) levonorgestrel intrauterine device
* (B) combination oral contraceptives (OCs)
    (C) metformin hydrochloride
    (D) spironolactone
    (E) flutamide

Hirsutism occurs when fine vellus hairs differentiate into dark, coarse terminal hairs. It is usually a clinical sign of another underlying disease, such as, in this patient, PCOS. The subjective perception of the patient and not the severity of the hirsutism should dictate the need for treatment. Management often will require a two-pronged approach that comprises

1. medical therapy to decrease androgen activity and
2. cosmetic therapy to remove terminal hairs that are present.

In the described patient, given her diagnosis of PCOS together with irregular menses and hirsutism, a combination estrogen–progestin OC should be recommended as the primary treatment. Combination OCs work through several mechanisms. The estrogen component increases sex hormone-binding globulin production in the liver, thereby decreasing circulating testosterone levels. The progestin component inhibits luteinizing hormone production, further decreasing testosterone levels. An additional benefit of OCs includes prevention of endometrial neoplasia.

When combination OCs alone are not sufficient for management of a patient's hirsutism, other modalities can be used along with OCs. Antiandrogens, particularly spironolactone, can be used to treat hyperandrogenism. Antiandrogens have multiple mechanisms of action, including competing with androgens to bind to the androgen receptor. All antiandrogens are teratogenic and should be used only in conjunction with very reliable contraception. Spironolactone can cause hyperkalemia and probably should be avoided in women with renal compromise. Flutamide and finasteride are other antiandrogen alternatives, but they do not appear to be superior to spironolactone. Additionally, flutamide carries the risk of hepatotoxicity. With all treatments for hyperandrogenism, it may take 6 months or more before the patient's symptoms improve.

Insulin-sensitizing agents, such as metformin, play an important role in addressing the metabolic sequelae of PCOS, but they have not demonstrated a clear benefit in the treatment of hirsutism. However, insulin-sensitizing agents do decrease circulating androgen levels, improve ovulation rates, and improve glucose tolerance. The levonorgestrel intrauterine device would provide endometrial protection for this anovulatory patient but would not improve her hirsutism. Topical eflornithine, an inhibitor of the enzyme ornithine decarboxylase, is another approved treatment for facial hirsutism. It requires twice-daily treatment and is effective while it is being used, but symptoms tend to recur once the therapy is stopped.

Mechanical hair removal techniques, such as shaving, plucking, waxing, depilatory cream, electrolysis, and laser, are employed frequently by women with hirsutism. In the medical literature, laser treatment has been studied to a greater extent than has electrolysis. Laser hair removal is effective in women with PCOS, particularly when utilized in conjunction with medical treatment to decrease circulating androgen levels. Classically, laser hair removal works best on women with dark hair and light skin because the laser can selectively focus on the target hair without damaging or discoloring the surrounding tissue. One randomized trial found that the combination of eflornithine and laser treatment was more effective than laser alone in treating hirsutism.

Escobar-Morreale HF, Carmina E, Dewailly D, Gambineri A, Kelestimur F, Moghetti P, et al. Epidemiology, diagnosis and management of hirsutism: a consensus statement by the Androgen Excess and Polycystic Ovary Syndrome Society. Hum Reprod Update 2012;18:146–70.

Polycystic ovary syndrome. ACOG Practice Bulletin No. 108. American College of Obstetricians and Gynecologists. Obstet Gynecol 2009;114:936–49.

Swiglo BA, Cosma M, Flynn DN, Kurtz DM, Labella ML, Mullan RJ, et al. Clinical review: Antiandrogens for the treatment of hirsutism: a systematic review and metaanalyses of randomized controlled trials. J Clin Endocrinol Metab 2008;93:1153–60.

113

Bioidentical hormones

A 52-year-old woman, gravida 2, para 2, has been amenorrheic for 9 months and has been experiencing frequent hot flushes. She has heard about bioidentical hormones and would like to learn more. You inform her that one such U.S. Food and Drug Administration (FDA)-approved bioidentical hormone prescribed for menopausal symptoms is

* (A) 17β-estradiol
  (B) ethinyl estradiol
  (C) biestrogen
  (D) triestrogen

Hormone therapy (HT) is very effective for the treatment of menopausal symptoms. Controversy about the appropriate use of HT has developed in recent years. In response to the Women's Health Initiative (WHI) study, patients with menopausal symptoms have begun to seek alternatives to traditional HT. Because of media coverage and direct-to-consumer advertising, compounded bioidentical hormones have experienced a surge in popularity. Patients are told that a natural bioidentical product can be created or compounded specifically for them. These compounded products are promoted as safe and effective. Available evidence does not support claims for the superiority of bioidentical products over conventional HT for the treatment of menopausal symptoms. Moreover, some compounded bioidentical products may pose additional safety risks due to variable purity and potency.

As a result of the findings of the WHI study, HT is no longer prescribed to improve a woman's long-term health; however, HT is appropriate to treat menopausal symptoms, particularly around the time of menopause. Such HT is available as commercial products or compounded formulations, in which an individualized preparation is created as a unique medication tailored to the needs of a specific patient.

The term "bioidentical hormones" refers to supplements that are chemically and structurally similar if not identical to hormones naturally produced by the body. Commercially available bioidentical hormones that have been approved by the FDA include estradiol and micronized progesterone. The term bioidentical hormones has become misleading because compounding pharmacies and the media use it to suggest that such preparations are more natural and safer than HT prescribed by physicians. Because of the confusion surrounding the term, the FDA and the Endocrine Society now recognize it to be a marketing term and not a medical term.

Frequently used compounded formulations include biestrogen and triestrogen, which are both made up on a milligram-per-milligram basis. Biestrogen consists of 20% estradiol and 80% estriol, and triestrogen is 10% estradiol, 10% estrone, and 80% estriol. Historically, the intent of compounding was to provide a medication with different ingredients, preservatives, or routes of administration than commercially available medications (eg, to avoid a specific preservative for a patient with an allergy). However, the practice of blending commercially available products to create a new custom-made compound for a specific patient is very different. A paucity of data exists on the efficacy and safety of most of these new compounds. In contrast, all FDA-approved hormonal products must provide to consumers information on the efficacy, adverse effects, and complications of the product and display a black box warning to address the findings of the WHI study. Compounded hormones are not approved or regulated by the FDA. As a result, such products are also exempt from displaying warnings or contraindications.

In general, HT should be used at the lowest possible dose to manage a woman's menopausal symptoms. Because symptom control is the most important therapeutic endpoint, testing of hormone levels in any fashion (serum, urine, or saliva) is typically unnecessary. Salivary testing, which is recommended by many compounding advocates, is fraught with limitations and is not biologically meaningful for measuring sex hormones.

For patients who are interested in treatment with naturally occurring hormones, 17β-estradiol (transdermal or oral) and micronized progesterone (oral or vaginal) are FDA-approved products. Ethinyl estradiol is a synthetic estrogen commonly used in oral contraceptives. The American College of Obstetricians and Gynecologists and the American Society for Reproductive Medicine agree that compounded hormones are not superior to conventional menopausal HT and that they pose potential risks, including variable purity and potency as well as possible underdosage or overdosage.

Compounded bioidentical menopausal hormone therapy. Committee Opinion No. 532. American College of Obstetricians and Gynecologists. Obstet Gynecol 2012;120:411–5.

Files JA, Ko MG, Pruthi S. Bioidentical hormone therapy. Mayo Clin Proc 2011;86:673–80, quiz 680.

Huntley AL. Compounded or confused? Bioidentical hormones and menopausal health. Menopause Int 2011;17:16–8.

114

Hypothyroidism

During a primary infertility evaluation, a 35-year-old healthy woman has a normal physical examination with a thyroid-stimulating hormone (TSH) level of 4.5 international units/L and positive thyroid microsomal antibodies. The best next test for this patient is

    (A) thyroid ultrasonography
    (B) iodine-131 uptake scan
* (C) repeat TSH level
    (D) serum iodine level

Primary subclinical hypothyroidism is defined as an elevated TSH level; in comparison, *clinical hypothyroidism* is defined as an elevated TSH level accompanied by decreased free and total thyroxine levels. As the name suggests, patients with subclinical hypothyroidism are most commonly asymptomatic. Primary hypothyroidism is caused by decreased thyroxine production in the thyroid gland with a compensatory increase in TSH (thyrotropin). Secondary hypothyroidism is caused by a lack of production of TSH by the pituitary and is much less common.

In developed countries, the most common cause of primary hypothyroidism is autoimmune destruction of the thyroid gland through Hashimoto thyroiditis, also known as autoimmune thyroiditis. Less common etiologies include iodine deficiency, iatrogenic causes due to surgery, irradiation, and drugs used to treat hyperthyroidism (eg, lithium); the infiltrating diseases sarcoidosis and amyloidosis; and TSH receptor mutations. Hashimoto thyroiditis is 15–20 times more common in women compared with men, with peak incidence between ages 30 years and 50 years. The disease may present in several ways, including euthyroid goiter, subclinical hypothyroidism with goiter, hypothyroidism without goiter, painless (silent) thyroiditis, and postpartum thyroiditis.

Thyroid disease in women includes subclinical hypothyroidism (approximately 7.5%), hypothyroidism (1.8%), thyroid peroxidase antibodies (10%), and thyroglobulin antibodies (3%). Thyroid peroxidase is responsible for converting iodide into iodine in the generation of thyroxine. Antibodies to thyroid peroxidase were first identified in thyroid microsomal protein preparations, hence the term antimicrosomal antibodies. Subclinical hypothyroidism with or without thyroid peroxidase antibodies increases the risk of hypothyroidism. The incidence of hypothyroidism in women with subclinical hypothyroidism and thyroid peroxidase antibodies is approximately 4.3% per year; incidence in women with subclinical hypothyroidism alone is 2.6% per year and that in women with thyroid peroxidase antibodies alone is 2.1% per year. Approximately 16–20% of pregnant women found to have thyroid peroxidase antibodies with normal TSH in the first trimester will have an elevated TSH by the third trimester. These women also have a much higher risk of postpartum thyroiditis. The presence of thyroglobulin antibodies does not substantially increase the risk of hypothyroidism. Unlike the thyroid-stimulating immunoglobulins found in Graves disease, in the first trimester thyroid peroxidase antibodies have no adverse effect on the fetus, and ultrasonographic monitoring of the fetal thyroid is not necessary.

The upper limit of normal TSH levels in euthyroid women remains controversial. Levels of TSH normally increase with obesity and age. For every 10 years of age after age 39 years, TSH increases 0.03 mU/L. In addition, TSH levels vary diurnally up to 50%, with the highest level occurring at the onset of sleep. Thus, it is prudent to repeat the TSH level test in patients with borderline-elevated TSH levels. In nonpregnant women who are not trying to conceive, the National Academy of Clinical Biochemistry reports an upper normal limit of TSH of

2.5 mU/L. Data from the National Health and Nutrition Examination Survey III support an upper normal limit of TSH of 4.12 mU/L. In addition, National Health and Nutrition Examination Survey III data support an upper normal TSH level in women who are planning pregnancy or in pregnant women in the first trimester of 2.5 mU/L, in the second trimester of 3.0 mU/L, and in the third trimester of 3.5 mU/L.

No universal agreement has been reached on screening recommendations for hypothyroidism. The American Thyroid Association supports screening all patients every 5 years starting at age 35 years. The American College of Physicians recommends screening women after age 50 years if they have symptoms or findings suggestive of thyroid disease. The American College of Obstetricians and Gynecologists does not recommend thyroid screening during pregnancy in asymptomatic women without a personal history of thyroid disease.

The appropriate screening test for primary hypothyroidism is a TSH level test. No other functional studies are needed, such as testing for thyroxine or triiodothyronine, if the TSH level is normal. In nonpregnant women who are not trying to conceive, routine assay for thyroid peroxidase antibodies remains controversial. A finding of thyroid peroxidase antibodies helps define the diagnosis of Hashimoto thyroiditis. A finding of positive antibodies in a woman with a normal TSH level also increases the risk of subsequent development of hypothyroidism and requires more frequent follow-up testing. In contrast, the finding of positive antibodies in a patient with an elevated TSH level does not affect the decision to treat. Evidence supports the testing of women with recurrent pregnancy loss for thyroid peroxidase antibodies because an association has been shown. Adverse pregnancy outcomes, such as preterm delivery before 34 weeks of gestation and neonatal respiratory distress, have been observed in women with thyroid peroxidase antibodies and normal TSH levels (2.5 mU/L or less). No current consensus exists regarding the benefit of treating women with normal TSH levels and elevated antibodies for recurrent pregnancy loss or to prevent pregnancy complications.

Thyroid ultrasonography is not necessary in women who have hypothyroidism and a normal thyroid examination. Ultrasonography should be performed if a thyroid nodule is suspected to determine the need for pathologic evaluation for cancer. Radioactive iodine uptake scanning is used to evaluate for hyperthyroidism and is not indicated in the evaluation of hypothyroidism. Iodine deficiency remains a cause of endemic goiter and hypothyroidism in underdeveloped countries but rarely in developed countries. Because of a high incidence of goiter in the Great Lakes region of the United States, iodine supplementation of salt was started in the 1920s. Iodine is found in seafood and many fruits and vegetables. Foods with the highest iodine content include haddock, yogurt, white bread, and milk. Iodine evaluation is not indicated in a patient with a normal TSH level and no goiter.

Spontaneous normalization of elevated TSH levels is not uncommon, especially in individuals with levels less than 10 mU/L and absent antibodies. For patients who have started taking thyroxine in this situation, it is reasonable to intermittently discontinue the drug and reassess the need for medication.

Subclinical hypothyroidism is associated with health issues. Meta-analyses have shown increased coronary heart mortality in subjects younger than 65 years who have a TSH level of 10 mU/L or higher. Contributing factors may include increased low-density lipoprotein cholesterol, increased insulin, impaired left ventricular function, increased peripheral resistance, and increased vascular inflammation. Subclinical hypothyroidism also is associated with adverse pregnancy outcomes, such as spontaneous abortion and fetal loss. An association with impaired fetal neurocognitive development also has been found in some, but not all, studies.

The U.S. Preventive Services Task Force supports treatment of all women (pregnant and nonpregnant) with subclinical hypothyroidism who have a TSH level of 10 mU/L or greater. Because of a lack of clinical trials, the U.S. Preventive Services Task Force has made no specific recommendation regarding treatment of women with TSH levels below 10 mU/L.

In accordance with current data and recommendations, the most common course of action is to treat women who desire pregnancy and pregnant women with a TSH greater than 2.5 mU/L regardless of thyroid peroxidase antibody status. In contrast, the treatment of women who desire pregnancy and of pregnant women with positive thyroid peroxidase antibodies but with normal TSH remains controversial.

Cooper DS. Clinical practice. Subclinical hypothyroidism. N Engl J Med 2001;345:260–5.

Garber JR, Cobin RH, Gharib H, Hennessey JV, Klein I, Mechanick JI, et al. Clinical practice guidelines for hypothyroidism in adults: cosponsored by the American Association of Clinical Endocrinologists and the American Thyroid Association. American Association Of Clinical Endocrinologists And American Thyroid Association Taskforce On Hypothyroidism In Adults [published errata appear in Thyroid 2013;23:251. Thyroid 2013;23:129]. Thyroid 2012;22:1200–35.

Huber G, Staub JJ, Meier C, Mitrache C, Guglielmetti M, Huber P, et al. Prospective study of the spontaneous course of subclinical hypothyroidism: prognostic value of thyrotropin, thyroid reserve, and thyroid antibodies. J Clin Endocrinol Metab 2002;87:3221–6.

Stagnaro-Green A, Abalovich M, Alexander E, Azizi F, Mestman J, Negro R, et al. Guidelines of the American Thyroid Association for the diagnosis and management of thyroid disease during pregnancy and postpartum. American Thyroid Association Taskforce on Thyroid Disease During Pregnancy and Postpartum. Thyroid 2011; 21:1081–125.

115
Ambiguous genitalia

You have been called to evaluate a neonate after delivery. On physical examination, the genitalia are ambiguous. You suspect congenital adrenal hyperplasia (CAH). The neonate's electrolytes are normal and you await results of a karyotype, pelvic ultrasonography, and a steroid hormone profile. The most appropriate step in the management of this neonate is to

- (A) perform a cosyntropin stimulation test
- (B) assign preliminary gender and meet with the family
- (C) consult with a pediatric surgeon
- * (D) use a multidisciplinary team to assist in gender assignment
- (E) start adrenal steroid replacement

The birth of an infant with sexual ambiguity is a medical and psychosocial emergency for the infant, parents, and health care team. Congenital adrenal hyperplasia is an autosomal recessive disorder that occurs in approximately 1 in 16,000 births. If not diagnosed promptly, CAH can be life threatening. More than 90% of cases are due to a 21-hydroxylase deficiency. This enzyme is needed for cortisol synthesis in the adrenal cortex; 21-hydroxylase catalyzes conversion of 17-hydroxyprogesterone to 11-deoxycortisol, which is a precursor for cortisol. In addition, 21-hydroxylase catalyzes conversion of progesterone to deoxycorticosterone, a precursor for aldosterone. Newborns with 21-hydroxylase deficiency cannot synthesize cortisol, which causes the adrenal gland to be stimulated by adrenocorticotropic hormone, driving excess cortisol precursors to produce excess androgens (dehydroepiandrosterone and androstenedione). These androgens cause ambiguous genitalia in the newborn. Approximately 95% of these cases result in masculinization of a 46,XX offspring but do not significantly affect male sexual differentiation.

A defect in aldosterone biosynthesis (salt wasting) causes excessive sodium excretion with the development of hypovolemia and hyperreninemia, resulting in hyperkalemia. Cortisol deficiency causes increased antidiuretic hormone secretion. These deficiencies together contribute to hyponatremic dehydration and shock if not diagnosed and treated quickly.

Initial newborn screening includes a physical examination, pelvic ultrasonography, and a karyotype, which can be completed in 24 hours. If CAH is suspected, electrolyte levels and a serum 17-hydroxyprogesterone level should be obtained quickly. Newborns in some locales can be screened for CAH using dried blood on filter paper to test for elevated 17-hydroxyprogesterone. If electrolytes are abnormal and a positive newborn screen is obtained, treatment should be initiated immediately.

In the case of a sick infant with abnormal electrolytes, most practitioners do not delay treatment to perform a cosyntropin stimulation test. The cosyntropin stimulation test is the criterion standard for diagnosing CAH. Cosyntropin stimulates the adrenal gland. A newborn with salt-wasting disease will have the highest 17-hydroxyprogesterone levels whereas simple virilizing newborns have lower 17-hydroxyprogesterone levels. The test takes an hour and, thus, treatment of a sick infant with salt-wasting CAH should not be delayed for the cosyntropin test.

The most appropriate management for this newborn is to use a multidisciplinary team to assist in gender assignment and work with the family. The multidisciplinary team and the family should work together to formulate the best strategy for clinical care. Ideally, a multidisciplinary team would include a pediatric endocrinologist, a pediatric surgeon, a pediatric urologist, a psychologist, a pediatrician, and a coordinator. A careful family history is needed to help the medical team formulate the proper diagnosis. This decision takes into account the parents' cultural background and desires when gender assignment is made. To assign the preliminary gender and meet with the family would not be the appropriate next step. Instead, the multidisciplinary team and family together should decide gender assignment.

Consultation with a pediatric surgeon would not be the best next step in the management of this newborn. Once a diagnosis is established and the salt-wasting form of CAH is ruled out, the multidisciplinary team, including the pediatric surgeon, should decide on a possible surgical correction. Genital surgery is performed to maximize anatomy and sexual function. Single-stage surgical correction between the ages of 2 months and 6 months is advocated by some because at that age tissues are estrogenized and pliable. Early correction can mitigate stigmatization and minimize family concerns. However, some advocate waiting until the child is old enough to give informed consent before surgery is performed.

In the newborn with CAH who has signs of adrenal crisis, such as vomiting, diarrhea, weight loss, hypertension, hyponatremia, hyperkalemia, or hypoglycemia, medical care must be initiated immediately. Electrolytes are replaced, and, if 17-hydroxyprogesterone is elevated, a stress dose of steroids is given followed by a regimen of steroids until the newborn is stable. Diagnosing and stabilizing this newborn is of primary concern and is managed by the pediatric team. Once the newborn has a proper diagnosis and is medically stable, it is up to the multidisciplinary team to assign gender.

Houk CP, Hughes IA, Ahmed SF, Lee PA. Summary of consensus statement on intersex disorders and their management. International Intersex Consensus Conference. Writing Committee for the International Intersex Consensus Conference Participants. Pediatrics 2006;118:753–7.

Sandberg DE, Gardner M, Cohen-Kettenis PT. Psychological aspects of the treatment of patients with disorders of sex development. Semin Reprod Med 2012;30:443–52.

White PC, Bachega TA. Congenital adrenal hyperplasia due to 21 hydroxylase deficiency: from birth to adulthood. Semin Reprod Med 2012;30:400–9.

Wilson JD, Rivarola MA, Mendonca BB, Warne GL, Josso N, Drop SL, et al. Advice on the management of ambiguous genitalia to a young endocrinologist from experienced clinicians. Semin Reprod Med 2012;30:339–50.

116
Multifetal pregnancy reduction

A 31-year-old woman, gravida 1, para 0, underwent infertility treatment with gonadotropins and intrauterine insemination and conceived a triplet pregnancy. She is very concerned about continuing such a high-risk pregnancy and is considering multifetal pregnancy reduction. You explain to her that the most frequent obstetric risk associated with a triplet pregnancy is

(A) miscarriage
* (B) preterm delivery
(C) fetal growth restriction
(D) preeclampsia

High-order multiple gestations, defined as pregnancies with three or more fetuses, are associated with an increased risk of multiple adverse obstetric outcomes. In particular, preterm delivery is almost guaranteed; approximately 75–100% of triplets are delivered preterm. The risk of spontaneous loss of the entire pregnancy is approximately 15% for triplets. The average gestational age at delivery for triplets is 32.2 weeks (versus 35.3 weeks for twins). Compared with singleton pregnancies, cerebral palsy occurs 17 times more often in triplet pregnancies. Approximately 50–60% of triplets exhibit growth restriction compared with 14–25% of twins. In addition, 75% of triplets and 25% of twins require admission to the neonatal intensive care unit. The risk of preeclampsia is increased significantly in multiple gestations and is more likely to occur earlier and to meet the criteria for severe or atypical preeclampsia. Women with multiple gestations are also at increased risk of gestational diabetes mellitus, which affects approximately 3–6% of twin pregnancies and up to 22–39% of triplet pregnancies. The risk of death in the first year of life is 20-fold higher in triplet pregnancies compared with singleton pregnancies. Multifetal gestations also are associated with increased maternal morbidity and health care costs, given that women with multiple gestations are six times more likely to be hospitalized with pregnancy complications. Significant psychosocial complications also can follow multiple births, including severe parenting stress, parental depression, child abuse, and divorce.

Fertility therapy has led to a dramatic increase in multifetal pregnancies. From 1980 to 2009, the twin rate increased by 76%. The high-order multifetal pregnancy rate increased by more than 400% between 1980 and 1998 but decreased by 29% from 1998 to 2009. Preventive strategies to limit multifetal pregnancies, particularly high-order multiples, are essential in order to address this issue. In recent years, there has been an increased emphasis by the American Society for Reproductive Medicine on limiting the number of embryos transferred at the time of in vitro fertilization. Ideally, single embryo transfer should be considered in select patients. However, the attempt to limit the incidence of multifetal pregnancies with the use of controlled ovarian hyperstimulation with gonadotropins has proved more challenging. The risk of high-order multiples with controlled ovarian hyperstimulation is approximately 9%, which is significantly higher than the risk with in vitro fertilization (approximately 1%). In the case of high-order multifetal pregnancies, counseling should include discussion of *multifetal pregnancy reduction,* defined as a first-trimester or early second-trimester procedure to reduce the total number of fetuses, by one or more, in a multifetal pregnancy.

In terms of the medical benefits of multifetal pregnancy reduction, the goal is to increase the probability of achieving at least one live birth and decrease the risk of spontaneous loss of the entire pregnancy. Estimates of perinatal outcomes after fetal reduction include a loss rate of 11–12% and a very early preterm delivery rate of 4.5%. The loss rate is dependent on multiple factors, including the experience of the operator, the number of fetuses present, and ease of access. Additionally, multifetal pregnancy reduction is associated with a decreased risk of preterm delivery, hypertension, preeclampsia, and gestational diabetes mellitus. Patients may opt to have the fetuses tested for aneuploidy or genetic anomalies before reduction is performed.

High-order multiple gestations present medical and ethical dilemmas. To maximize the safety of the pregnancy, a maternal–fetal medicine specialist often is consulted regarding selective reduction of a high-order multiple pregnancy. For many patients, moral, religious, social, cultural, and economic factors also may play a role in this difficult decision-making process.

Dodd JM, Crowther CA. Reduction of the number of fetuses for women with a multiple pregnancy. Cochrane Database of Systematic Reviews 2012, Issue 10. Art. No.: CD003932. DOI: 10.1002/14651858. CD003932.pub2.

Evans MI, Dommergues M, Wapner RJ, Goldberg JD, Lynch L, Zador IE, et al. International, collaborative experience of 1789 patients having multifetal pregnancy reduction: a plateauing of risks and outcomes. J Soc Gynecol Investig 1996;3:23–6.

Malizia BA, Dodge LE, Penzias AS, Hacker MR. The cumulative probability of liveborn multiples after in vitro fertilization: a cohort study of more than 10,000 women. Fertil Steril 2013;99:393–9.

Multifetal pregnancy reduction. Committee Opinion No. 553. American College of Obstetricians and Gynecologists. Obstet Gynecol 2013;121:405–10.

Multiple gestation: complicated twin, triplet, and high-order multifetal pregnancy. ACOG Practice Bulletin No. 56. American College of Obstetricians and Gynecologists. Obstet Gynecol 2004;104:869–83.

117

Hypothalamic amenorrhea

A 31-year-old nulligravid woman comes to your office with an 18-month history of amenorrhea after she stopped taking cyclical oral contraceptives (OCs). She runs 50 km (31.1 mi) a week and has a body mass index of 19 (calculated as weight in kilograms divided by height in meters squared). Her follicle-stimulating hormone (FSH) level is 5.0 mIU/mL, luteinizing hormone (LH) level is 2.5 mIU/mL, prolactin level is 12 ng/mL, thyroid-stimulating hormone level is 2.1 mU/mL, and estradiol level is 15 pg/mL. She has no signs of hyperandrogenism. You determine that the diagnosis is

(A) post-OC amenorrhea
* (B) functional hypothalamic amenorrhea
(C) polycystic ovary syndrome
(D) ovarian insufficiency

Secondary amenorrhea is defined as loss of menstruation for 6 months. The cessation of the described patient's menses is not normal. In general, most cases of amenorrhea are due to hypothalamic amenorrhea, polycystic ovary syndrome, ovarian insufficiency, or hyperprolactinemia. A thorough history, physical examination, and targeted hormone evaluation often identify the etiology of the amenorrhea. Initial hormone evaluation includes testing for levels of serum human chorionic gonadotropin, FSH, LH, thyroid-stimulating hormone, and prolactin.

An intact, functional hypothalamic–pituitary–ovarian axis and a genital outflow tract are prerequisites for normal menstrual cycles. Dysfunction at any level can cause a spectrum of menstrual disturbances.

The described patient was menstruating regularly on cyclical OCs for 5 years. Once she discontinued taking OCs, she had no menses. Women who have not resumed menstrual function within 6 months of stopping OC use should be evaluated for secondary amenorrhea. There is no known relation between OC use and secondary amenorrhea. Waiting more than 6 months before initiation of a formal evaluation would delay diagnosis and treatment of her amenorrhea.

Functional hypothalamic amenorrhea is due to suppression of gonadotropin-releasing hormone pulsatility with ultimate suppression of the hypothalamic–pituitary–ovarian axis. Pituitary gonadotropin secretion is at basal levels, causing low estrogen levels. Functional hypothalamic amenorrhea usually is caused by weight loss, exercise, or stress. Gonadotropin levels are low to normal with low estrogen levels in the presence of normal brain imaging (normal pituitary and sellar magnetic resonance imaging). This patient's hypogonadism is due to strenuous exercise. Most patients with hypothalamic amenorrhea produce low gonadotropins. The described patient's FSH and LH levels are normal. The differences in glycosylation of the gonadotropins can result in decreased bioactivity and, thus, hypogonadism, even when gonadotropin levels are normal. Given that functional hypothalamic amenorrhea is a diagnosis of exclusion, other causes of menstrual irregularity need to be ruled out. For example, hyperprolactinemia and thyroid dysfunction are correctable causes of irregular menses and amenorrhea.

Long-distance runners often have low body weight and lean body mass. Among competitive athletes, distance runners have the highest incidence of amenorrhea. Initial counseling for this patient should include increasing caloric intake and reducing her running intensity in conjunction with nutritional counseling. After taking these steps, menstrual function often will resume and patients can attempt pregnancy. If amenorrhea continues and pregnancy is desired, ovulation induction with gonadotropins is the treatment of choice. Clomiphene citrate in this patient is unlikely to induce ovulation without an intact hypothalamic–pituitary–ovarian axis.

Polycystic ovary syndrome is also a diagnosis of exclusion. Features include clinical or biochemical evidence of hyperandrogenism and chronic anovulation with or without polycystic-appearing ovaries.

Ovarian insufficiency (hypergonadotropic hypogonadism) in women younger than 40 years occurs in approximately 1% of women. Follicle-stimulating hormone levels above 20 mIU/mL and LH levels above 40 mIU/mL are diagnostic for ovarian insufficiency. The etiology of ovarian insufficiency in most women is unknown. Iatrogenic causes of ovarian insufficiency include chemotherapy and radiation for treatment of malignancies. In rare cases, a woman who discontinues OCs in order to start family building and will have secondary amenorrhea with elevated gonadotropins. In such a case, OC use masks ovarian insufficiency developed while the patient was taking OCs. This can be a devastating diagnosis, especially when OCs are discontinued in the hope of starting a family.

Fritz MA, Speroff L. Amenorrhea. In: Fritz MA, Speroff L, editors. Clinical gynecologic and infertility. 8th ed. Philadelphia (PA): Lippincott Williams & Wilkins; 2011. p. 435–93.

Gordon CM. Clinical practice. Functional hypothalamic amenorrhea. N Engl J Med 2010;363:365–71.

Practice Committee of the American Society for Reproductive Medicine. Current evaluation of amenorrhea. Fertil Steril 2008;90:S219–25.

Santoro N. Update in hyper- and hypogonadotropic amenorrhea. J Clin Endocrinol Metab 2011;96:3281–8.

118
Luteal phase deficiency

A 38-year-old woman, para 1, who has been trying to become pregnant visits your clinic. She is concerned that she has shorter cycles than her peers. Her thyroid-stimulating hormone and prolactin levels are normal. Her menstrual cycle intervals are typically 25–26 days. She has read about luteal phase deficiency and wants to know if low progesterone may lead to infertility. The most appropriate next step is

(A) basal body temperature charting
(B) midluteal progesterone level testing
(C) timed endometrial biopsy
* (D) telling her that no good test for luteal phase deficiency exists

Progesterone is vital to the process of implantation and early embryonic development. After ovulation, progesterone is produced first by the corpus luteum and then by the trophoblast. Progesterone maintains the integrity of the endometrium in the secretory phase. Withdrawal of progesterone leads to rapid shedding of the secretory endometrium. If pregnancy is established, embryonic growth and development is contingent upon adequate progesterone production from the corpus luteum.

The importance of ovarian progesterone in implantation and early embryonic growth has been demonstrated elegantly in several clinical models. If the corpus luteum is surgically removed before 7 weeks of gestation, a high risk of pregnancy loss exists, but if supplemental exogenous progesterone is given after luteectomy, a miscarriage before 7 weeks can be prevented. Exogenous progesterone has enabled pregnancy by means of oocyte donation in women who have ovarian failure. Mifepristone, a progesterone antagonist, will induce a high rate of pregnancy loss if it is used before 7 weeks of gestation.

Given the critical role that progesterone plays in nidation, it is logical to think that inadequate endogenous progesterone production would lead to insufficient endometrial development for the support of early pregnancy. Luteal phase deficiency was first described in 1949. Low circulating progesterone levels after ovulation were assumed to cause delayed endometrial maturation. A significant delay was thought to cause asynchrony between the embryo and endometrium, leading to infertility or early pregnancy wastage. Several means of diagnosing luteal phase deficiency have been proposed: charting basal body temperature, measuring serum progesterone levels in the luteal phase, monitoring luteal phase length from LH surge, and performing endometrial biopsy. Although measuring midluteal progesterone levels and charting basal body temperature are accurate in identifying recent ovulation, neither test is useful to judge the quality or adequacy of the luteal phase.

The endometrial biopsy was a central part of the basic infertility workup until recently. The secretory endometrium has specific histologic features that mark each day after ovulation. Endometrial biopsy allows experienced pathologists to assign a date to endometrial specimens that correspond to the number of days past ovulation. If the biopsy specimen shows a discrepancy of more than 2 days from menstrual dating, this is considered "out of phase." Two out-of-phase biopsies are considered proof of luteal phase deficiency. Until recently this test was accepted as the criterion standard for diagnosis of luteal phase deficiency.

Many flaws can be found in the interpretation of endometrial biopsy results. The standard endometrial dating criteria were derived from a population of infertile patients. This meant that the reference population was abnormal from the outset. Histologic dating shows significant interobserver and intraobserver variation, which raises questions about the validity of this test. Variations seen among individual patients, different cycles within the same patient, and different observers of the same histologic specimen were too large, showing that the endometrial biopsy lacked precision.

Controversy over the clinical relevance of luteal phase deficiency is due in part to the lack of a reliable test to diagnose this disorder. In a multicenter randomized controlled clinical trial of 847 women with regular cycles, abnormal histologic dating could not discriminate between infertile women and those with proven fertility. As an independent entity that causes infertility, luteal phase deficiency was disproven and endometrial biopsy has been abandoned for this indication. Endometrial biopsy remains valid for tissue confirmation of ovulation or evaluation of pathologic states such as endometrial cancer or hyperplasia. In terms of the most appropriate next step for the described patient, you should counsel her that there is no reliable test for luteal phase deficiency.

The clinical relevance of luteal phase deficiency: a committee opinion. Practice Committee of the American Society for Reproductive Medicine. Fertil Steril 2012;98:1112–7.

Coutifaris C, Myers ER, Guzick DS, Diamond MP, Carson SA, Legro RS, et al. Histological dating of timed endometrial biopsy tissue is not related to fertility status. NICHD National Cooperative Reproductive Medicine Network. Fertil Steril 2004;82:1264–72.

Murray MJ, Meyer WR, Zaino RJ, Lessey BA, Novotny DB, Ireland K, et al. A critical analysis of the accuracy, reproducibility, and clinical utility of histologic endometrial dating in fertile women. Fertil Steril 2004;81:1333–43.

Progesterone supplementation during the luteal phase and in early pregnancy in the treatment of infertility: an educational bulletin. Practice Committee of the American Society for Reproductive Medicine. Fertil Steril 2008;89:789–92.

Speroff L, Fritz MA. Female infertility. In: Speroff L, Fritz MA. Clinical gynecologic endocrinology and infertility. 8th ed. Philadelphia (PA): Lippincott Williams and Wilkins; 2011. p. 1164–6.

Speroff L, Fritz MA. Recurrent early pregnancy loss. In: Speroff L, Fritz MA. Clinical gynecologic endocrinology and infertility. 8th ed. Philadelphia (PA): Lippincott Williams and Wilkins; 2011. p. 1216–7.

119

Donor sperm use

A 30-year-old woman, gravida 1, para 0, has been trying to conceive for more than 1 year with her partner. As a teenager, she terminated a pregnancy she had conceived with a different partner. Her current partner's semen analysis reveals a normal volume of azoospermia on two separate centrifuged specimens. His urologic examination is notable for decreased testicular size. He has a total testosterone level of 250 ng/dL, a follicle-stimulating hormone (FSH) level of 28 mIU/mL, negative cystic fibrosis testing, 46,XY karyotype, and a Y chromosome *AZFa* microdeletion. In counseling the couple, you explain that their best chance for conception is with

(A) microsurgical testicular sperm extraction
(B) testicular biopsy
(C) epididymal sperm aspiration
(D) intracytoplasmic sperm injection
* (E) donor sperm insemination

Approximately 15% of couples who try to conceive for a 1-year period fail to do so. Up to 50% of those cases can be attributed to a male factor, sometimes manifested as complete absence of sperm in the ejaculate, or azoospermia. Azoospermia is confirmed when no sperm are found in at least two separate centrifuged semen samples.

An azoospermic man should be evaluated with a complete history and physical examination. In addition, serum total testosterone and FSH levels should be measured. Men with hypogonadotropic hypogonadism also should have a serum prolactin level test and hypothalamic–pituitary imaging to exclude pituitary tumors.

If the ejaculate volume is low, and testicular volume and FSH levels are normal, the patient may have obstructive azoospermia. In cases of suspected obstructive azoospermia or severe oligospermia, patients should be screened for cystic fibrosis. If at least one vas deferens is palpable, a diagnostic testicular biopsy with simultaneous testicular sperm extraction can be performed. Possible reconstruction may subsequently be an option.

In contrast to obstructive azoospermia, men with nonobstructive azoospermia often present with low testicular volume and an elevated FSH level. Approximately 12% of men with nonobstructive azoospermia will be found to have karyotypic abnormalities, and 18% of these will involve the Y chromosome. All men with azoospermia due to testicular failure should be offered genetic testing with karyotype and Y chromosome microdeletion testing. Testing for terminal microdeletions of the long arm of the Y chromosome can help predict the likelihood of finding sperm at the time of testicular exploration. Microdeletions of the Y chromosome will be found in approximately 10–15% of men with severe oligospermia or azoospermia. In particular, three deletions in nonoverlapping regions of the long arm of the Y chromosome (Yq11), called the azoospermia factor (*AZF*), have been identified: 1) *AZFa* (proximal), 2) *AZFb* (central), and 3) *AZFc* (distal). Deletions involving the entire *AZFa* or *AZFb* region predict a very poor prognosis for sperm retrieval, even with extensive testicular exploration. In men with *AZFc* deletions, reduced numbers of sperm may be present in the ejaculate. Even if the ejaculate is azoospermic in a patient with an *AZFc* deletion, sperm extraction by testicular biopsy still may be successful. In fact, men with a known *AZFc* deletion have a greater probability of successful sperm retrieval than do men with idiopathic azoospermia. If a man with an *AZFc* deletion successfully conceives, any sons will inherit the abnormality. Because the prognosis varies dramatically according to the specific microdeletion, testing for Y chromosome microdeletion is strongly recommended for genetic and preoperative counseling in these patients.

In the described patient, testicular biopsy for sperm procurement would be unsuccessful because of his *AZFa* microdeletion; therefore, microsurgical testicular sperm extraction, testicular biopsy, epididymal sperm aspiration, and intracytoplasmic sperm injection would all be futile. For this couple, donor sperm insemination would provide their best chance for conception. The couple also might consider adoption.

Hopps CV, Mielnik A, Goldstein M, Palermo GD, Rosenwaks Z, Schlegel PN. Detection of sperm in men with Y chromosome microdeletions of the *AZFa*, *AZFb* and *AZFc* regions. Hum Reprod 2003; 18:1660–5.

Practice Committee of American Society for Reproductive Medicine in collaboration with Society for Male Reproduction and Urology. Evaluation of the azoospermic male. Fertil Steril 2008;90(Suppl):S74–7.

Stahl PJ, Masson P, Mielnik A, Marean MB, Schlegel PN, Paduch DA. A decade of experience emphasizes that testing for Y microdeletions is essential in American men with azoospermia and severe oligozoospermia. Fertil Steril 2010;94:1753–6.

120

Normal menopausal transition

A 39-year-old woman requests counseling for irregular menstrual cycles. She has no menopausal symptoms at present but is concerned about the lengthening of her menstrual interval from 28 days to 33 days with occasional missed cycles. The urine pregnancy test result is negative. You tell her that if she develops irregular menses and menopausal symptoms before age 45 years, the most appropriate next step is to check

- (A) patient menstrual calendar
- (B) estradiol and inhibin B levels
- (C) follicle-stimulating hormone (FSH) level
* (D) thyroid-stimulating hormone (TSH) and prolactin levels

Estimates of the median age at menopause typically range from 50–52 years of age. The Massachusetts Women's Health Study (N=2,570) found the median age at menopause to be 51.3 years. The transition to menopause normally is associated with alterations in menstrual bleeding and the onset of vasomotor symptoms, such as night sweats.

Many of the early symptoms of the normal menopausal transition will overlap with other common conditions such as depression, mood swings, degenerative arthritis, and hypothyroidism. The vast majority of women enter menopause between ages 45 years and 56 years. Earlier onset of menopause is associated with smoking, lower socioeconomic status, malnutrition, and a maternal history of early menopause.

Discerning whether somatic and psychological symptoms in the fifth decade are a result of aging, depression, menopause, hypothyroidism, midlife stressors, or chronic illness can be clinically challenging. A normal TSH level can rule out hypothyroidism. Joint pain and stiffness can suggest degenerative arthritis, but this would be unlikely if the patient has a negative family history and a normal physical examination. Subclinical depression and midlife crisis are best assessed by careful attention to recent changes or stressors in the patient's personal, social, or professional life.

Menopausal symptoms that occur in a patient before age 45 years are likely to be caused by something other than perimenopause. A patient who develops irregular menses and menopausal symptoms before age 40 years should be investigated for primary ovarian insufficiency, but other causes, such as hypothyroidism, are much more common. When a patient has mild depression and fatigue before this age, hypothyroidism is a much more likely diagnosis.

The described patient does not have symptoms at this time but has asked what tests would be done if she were to develop irregular menses. You tell her that checking TSH and prolactin levels is the first step in the evaluation of ovulatory dysfunction. For the patient to keep a menstrual calendar to monitor for shortening of the intermenstrual interval would not be appropriate as the next step. Similarly, checking estradiol and inhibin B levels would constitute nonspecific tests in the detection of menopause. If the patient's TSH and prolactin levels are normal, the FSH level should be checked next. However, checking the FSH level would not be the first step. An FSH level in the menopausal range would identify primary ovarian insufficiency in the patient.

McKinlay SM, Brambilla DJ, Posner JG. The normal menopause transition. Maturitas 1992;14:103–15.

Randolph JF Jr, Crawford S, Dennerstein L, Cain K, Harlow SD, Little R, et al. The value of follicle-stimulating hormone concentration and clinical findings as markers of the late menopausal transition. J Clin Endocrinol Metab 2006;91:3034–40.

Speroff L, Fritz MA. Menopause and the perimenopausal transition. In: Speroff L, Fritz MA. Clinical gynecologic endocrinology and infertility. 8th ed. Philadelphia (PA): Lippincott Williams and Wilkins; 2011. p. 686–7.

Testing and interpreting measures of ovarian reserve: a committee opinion. Practice Committee of the American Society for Reproductive Medicine. Fertil Steril 2012;98:1407–14.

121
Window of fertilization

A 28-year-old woman tells you that she would like to have her intrauterine device removed so she can begin trying to conceive her first child. She is unsure of how to time intercourse and asks for your advice. She has read that there is an optimal time to conceive known as the "window of fertilization." She and her husband plan to have intercourse every other day starting after cessation of menstrual flow. You inform her that the time that the window of fertilization ends is

* (A) the day of ovulation
* (B) 48 hours before maximal production of cervical mucus
* (C) when the urinary luteinizing hormone kit turns positive
* (D) the day of peak progesterone levels

The luteinizing hormone (LH) surge typically occurs on day 13 in a 28-day menstrual cycle. Follicular rupture follows 36–40 hours after the onset of the LH surge or up to 24 hours after its peak. The usual reason to measure serum or urinary LH is to predict ovulation.

Data suggest that the fertile window opens before follicular rupture. The window begins the day before ovulation and then declines rapidly thereafter. The fertilizable life of an oocyte is 12–24 hours, meaning that the fertile window closes on the day of ovulation. Sperm, however, can survive in the reproductive tract for as long as 5 days, retaining the ability to fertilize an ovum even before ovulation occurs. The great majority of pregnancies occur when coitus takes place in the 3-day interval just before ovulation and the closure of the fertile window on the day of ovulation.

Many options exist for couples who are trying to conceive and who need to know when to plan intercourse. Basal body temperature monitoring, calendar timing, and ultrasonographic follicular monitoring, as well as serum LH testing and urinary LH testing, are options. Basal body temperature charting and calendar timing have the advantage of being low cost and noninvasive. For this reason, these methods are a logical starting place for patients. However, these methods are not very reliable for pinpointing the fertile window because of the normal variation in cycle length and the fact that the temperature rise accompanying ovulation usually is detected after the fact. Thus, the first day of temperature rise on the basal body temperature chart is not helpful to gauge the closing of the fertile window. When greater accuracy is needed, LH testing and ultrasonography may be employed.

Serum LH measurements and ultrasonography need to be performed serially and, as a result, are expensive. Therefore, these techniques usually are employed in conjunction with other fertility treatments, such as insemination or ovulation induction. Urinary LH kits are useful to predict ovulation when compared with the other tests available. Fertility medications, polycystic ovary syndrome, and the menopausal state can yield false-positive results with over-the-counter urinary LH kits. Patient self-assessment of cervical mucus changes provides another accurate test that has essentially no cost.

Peak production of cervical mucus occurs 2–3 days before ovulation and does not mark the end of the fertility window. Similarly, the first day of the LH surge would represent the midpoint of the fertile window, not the end. Serum progesterone levels are useful to confirm that ovulation has taken place but have no role in predicting ovulation.

Optimizing natural fertility. Practice Committee of American Society for Reproductive Medicine in collaboration with Society for Reproductive Endocrinology and Infertility. Fertil Steril 2008;90:S1–6.

Speroff L, Fritz MA. Regulation of the menstrual cycle. In: Speroff L, Fritz MA. Clinical gynecologic endocrinology and infertility. 8th ed. Philadelphia (PA): Lippincott Williams and Wilkins; 2011. p. 199–242.

Stanford JB, White GL, Hatasaka H. Timing intercourse to achieve pregnancy: current evidence. Obstet Gynecol 2002;100:1333–41.

Wilcox AJ, Weinberg CR, Baird DD. Timing of sexual intercourse in relation to ovulation. Effects on the probability of conception, survival of the pregnancy, and sex of the baby. N Engl J Med 1995;333:1517–21.

122

Use of stem cells in reproduction

The type of stem cells that presently hold the most promise for stem cell-based therapy for infertile patients is

  (A) hematopoietic
  (B) embryonal
  (C) embryonal carcinoma
* (D) induced pluripotent

The developing field of regenerative medicine attempts to regenerate and replace damaged tissue. Research studies are investigating the creation of functional gametes from stem cells. It is thought that in the future this technology may help infertile couples to utilize their own genetic material to produce gametes for reproduction.

Nearly every adult human tissue is believed to harbor a population of stem cells, which are involved in renewal (eg, replacement of blood cells), remodeling (eg, alterations to breast tissue before lactation), or repair (eg, wound healing). Many of these cells are multipotent, with the ability to differentiate into a number of cell types. Researchers have long sought to harness their power for the treatment of human disease or injury.

Hematopoietic stem cells (HSCs) isolated from bone marrow have been used successfully for the treatment of bone marrow and blood cancers for more than four decades. However, adult stem cells such as HSCs appear to have a strong bias toward making cell types of the same lineage. This has limited their utility in basic research and their therapeutic potential. Therefore, they are not typically seen as a viable option for generating gametes.

Embryonic stem cells (ESCs) offer greater promise because they are capable of differentiating into cells from all three germ layers: ectoderm, mesoderm, and endoderm. A number of studies also have shown that differentiation of mouse and human ESCs into germline cells also is possible in vitro. Publication of these studies was accompanied by speculation that ESC-derived gametes might have a role in reproductive medicine.

Failure to conceive through assisted reproductive technologies because of the poor quality of their gametes leaves patients with the option of utilizing donated sperm and oocytes. Several authors raised the possibility that ESCs might provide a convenient and inexhaustible source of oocytes. However, this approach would not allow patients to have biologically related offspring. It also creates an interesting ethical specter that a surplus embryo might become the "parent" of a multitude of offspring. Therefore, the use of ESCs as a therapy for infertility is not ideal.

Embryonal carcinoma cells are pluripotent cells derived from teratocarcinoma cells. Under the right conditions, they can be differentiated into germline stem cells. However, zygotes created from embryonal carcinoma cell-derived germ cells arrest at the 6–8-cell stage, presumably because of embryo imprinting abnormalities. Embryonal carcinoma cells also do not allow for generation of patient-specific therapies, and because of their teratocarcinoma origin, they may acquire additional undesirable genetic mutations. Therefore, they are not currently a reasonable source of gametes for infertile patients.

At present, induced pluripotent stem cells appear to be the type of stem cells that offer the most promising stem cell-based therapy for infertile patients. Groundbreaking studies have demonstrated that retroviral-induced expression of genes such as *Oct4*, *Sox2*, *c-Myc*, and *Klf4* in somatic cells is sufficient to convert them into a pluripotent state. It is possible to transform a terminally differentiated skin fibroblast into a cell that resembles an ESC by morphology, growth properties, and expression marker genes. This discovery won a Nobel Prize award in 2012. This newly characterized class of stem cells opens the door for individualized therapies for a host of diseases and other health problems, including infertility. Induced pluripotent stem cells can be differentiated into germ cells that are capable of producing apparently healthy offspring. One possible limitation of their utility is the fact that viral modification of the genome is required for their creation. These viral vectors are oncogenic and bring up safety concerns in this transgenic approach. Recent publications have suggested that small molecules may soon replace the use of viruses, opening the door to therapeutic possibilities.

Hayashi K, Ogushi S, Kurimoto K, Shimamoto S, Ohta H, Saitou M. Offspring from oocytes derived from in vitro primordial germ cell-like cells in mice. Science 2012;338:971–5.

Ichida JK, Blanchard J, Lam K, Son EY, Chung JE, Egli D, et al. A small-molecule inhibitor of Tgf-β signaling replaces $Sox2$ in reprogramming by inducing $Nanog$. Cell Stem Cell 2009;5:491–503.

Maherali N, Sridharan R, Xie W, Utikal J, Eminli S, Arnold K, et al. Directly reprogrammed fibroblasts show global epigenetic remodeling and widespread tissue contribution. Cell Stem Cell 2007;1:55–70.

Nayernia K, Li M, Jaroszynski L, Khusainov R, Wulf G, Schwandt I, et al. Stem cell based therapeutical approach of male infertility by teratocarcinoma derived germ cells. Hum Mol Genet 2004;13:1451–60.

Nicholas CR, Haston KM, Grewall AK, Longacre TA, Reijo Pera RA. Transplantation directs oocyte maturation from embryonic stem cells and provides a therapeutic strategy for female infertility. Hum Mol Genet 2009;18:4376–89.

Wernig M, Meissner A, Foreman R, Brambrink T, Ku M, Hochedlinger K, et al. In vitro reprogramming of fibroblasts into a pluripotent ES-cell-like state. Nature 2007;448:318–24.

123

Bilateral tubal ligation and future fertility

A 28-year-old woman, gravida 1, para 1, is interested in tubal sterilization. She requests information about the pros and cons of different approaches and tells you that she would prefer a method that is reversible. You inform her that the sterilization method most amenable to surgical reversal is

(A) hysteroscopic tubal occlusion
(B) laparoscopic bilateral electrocautery
* (C) laparoscopic application of a band or clip
(D) laparoscopic distal salpingectomy

Most women who choose sterilization do not regret their decision. Preoperative counseling should emphasize that the procedure is not intended to be reversible. Prospective analysis of U.S. Collaborative Review of Sterilization study data found that the cumulative probability of regret over 14 years of follow-up was 12.7%. By age group, the probability of regret was 20.3% for women aged 30 years or younger at the time of sterilization compared with 5.9% for women older than 30 years.

Counseling should be comprehensive with the intent to minimize regret among individual women. Patients who are younger than 30 years, who are of low parity, or who are in unstable relationships are at the highest risk of regretting the decision for permanent sterilization. Attention should be given to alternative methods such as vasectomy or long-acting reversible contraception.

Two factors are associated with successful reversal of sterilization: 1) the amount of tissue damage produced by the original procedure and 2) the length of proximal and distal tubal segments available for anastomosis. The spring-loaded clip is only 5 mm wide, produces the least amount of tubal damage, and leaves the longest tubal segments for reanastomosis. Similarly, the small inert synthetic ring produces minimal damage.

The method of sterilization that is most amenable to surgical reversal is laparoscopic application of a band or clip. Laparoscopic electrocautery produces considerable tissue damage along the length of both tubes, and distal salpingectomy will not conserve distal tubal segments for reanastomosis.

Hysteroscopic tubal occlusion can be performed in the office setting and has an excellent efficacy and safety profile. Patients who undergo this procedure must rely on another method of contraception for at least 3 months after the procedure. They also must undergo hysterosalpingography to confirm successful bilateral occlusion. This method of sterilization is designed to permanently occlude the tubal lumen and leaves no chance for reversal of sterilization.

Benefits and risks of sterilization. Practice Bulletin No. 133. American College of Obstetricians and Gynecologists. Obstet Gynecol 2013;121:392–404.

124

Pelvic inflammatory disease and reproductive dysfunction

A 27-year-old woman has a 2-year history of primary infertility. Her workup shows normal semen analysis in her partner and cyclic menses along with left tubal patency and right proximal tubal occlusion on hysterosalpingography (HSG). She recalls having several cervical or vaginal infections in her early 20s but does not believe she was ever diagnosed with pelvic inflammatory disease (PID). The next test to determine if she has had PID would be

    (A) diagnostic laparoscopy
    (B) culture for cervical *Chlamydia trachomatis* and *Neisseria gonorrhoeae*
*   (C) test for *C trachomatis* antibody levels
    (D) repeat HSG

Tubal disease is one of the most common causes of female infertility. It accounts for approximately 25–35% of cases of infertility in U.S. women, with more than one half of cases resulting from salpingitis. Pelvic inflammatory disease can be either an acute or chronic infection and is caused by the spread of pathologic bacteria from the lower to the upper genital tract with possible dissemination into the peritoneal cavity. The infection typically spreads to both fallopian tubes. According to the Centers for Disease Control and Prevention, acute PID affects more than 1 million U.S. women per year, and more than 100,000 women per year will become infertile as a result of the infection.

In addition to infertility, long-term sequelae of PID include ectopic pregnancy and chronic pelvic pain. The incidence of post-PID ectopic pregnancy is approximately 4–9%, six times higher than the general population. The likelihood of ectopic pregnancy rises with each subsequent episode, and it is estimated that 50% of ectopic pregnancies are a consequence of prior PID. Similarly, the incidence of infertility increases with the severity and number of prior tubal infections. A single episode of PID will cause infertility in approximately 8–13% of infected women. A second episode will increase the incidence to 20–36%, and a third episode causes infertility in up to 75% of infected women.

Salpingitis is usually a polymicrobial process. The most common organisms found in salpingitis are *C trachomatis*, *N gonorrhoeae*, *Staphylococcus*, *Streptococcus*, *Mycoplasma*, and *Ureaplasma urealyticum*.

In many women, PID is asymptomatic. Tubal disease or peritoneal adhesions often are found in women with infertility but no known prior history of PID. In one study of patients with tubal factor infertility, only 31% of patients could recall a prior episode of PID, but 79% of patients had serologic evidence of prior *C trachomatis* infection. This often is referred to as "silent" PID.

The most reliable test to determine if the described patient has had PID in the past is to check her *C trachomatis* antibody levels. The left tube is patent, but she has a history of secondary infertility for the past year. Her clinical decision making will be influenced by evidence of prior PID, eg, if *C trachomatis* antibodies are present, she may proceed to IVF.

Laparoscopy with chromotubation and HSG are the primary means of detecting tubal disease. Chlamydial antibody testing is an indirect means of screening for tubal disease. The test is noninvasive and relatively inexpensive. A repeat HSG is unlikely to be helpful. Unilateral proximal tubal occlusion is commonly caused by spasm and rarely indicates pathology. A detailed history may not help in the case of silent PID. Cervical cultures would detect current but not past infection. Laparoscopic surgery could be helpful but is more invasive than antibody testing.

Akande VA, Hunt LP, Cahill DJ, Caul EO, Ford WC, Jenkins JM. Tubal damage in infertile women: prediction using chlamydia serology. Hum Reprod 2003;18:1841–7.

Committee opinion: role of tubal surgery in the era of assisted reproductive technology. Practice Committee of the American Society for Reproductive Medicine. Fertil Steril 2012;97:539–45.

125
Obesity and pregnancy

A 31-year-old woman, gravida 2, para 1, aborta 1, with oligomenorrhea wishes to become pregnant. Her body mass index (BMI) is 36 (calculated as weight in kilograms divided by height in meters squared). Her thyroid-stimulating hormone level is 2.2 mU/mL and her prolactin level is 14 ng/mL. The best recommendation for her is the use of

  (A) clomiphene citrate
  (B) pharmacotherapy for weight loss
* (C) preconception nutrition assessment and consultation
  (D) bariatric surgery consultation
  (E) ovulation predictor kit

In 2008, 28.6% of U.S. women were overweight (BMI of 25–29.9) and 35.5% were obese (BMI greater than 30). Obesity is the most common chronic disease in the United States. The simplest measurement of weight is BMI. Obesity is associated with diabetes mellitus, hypertension, hypertriglyceridemia, hypercholesterolemia, and musculoskeletal disease.

Reproductive health is adversely affected by obesity. Anovulation and oligo-ovulation are common in overweight and obese women. Menstrual cycle irregularities are attributed to alterations in gonadotropin-releasing hormone secretion leading to a dysfunctional hypothalamic–pituitary–ovarian axis. Weight loss often will normalize ovulatory function and increase pregnancy rates in obese patients. Obese patients who become pregnant have an increased risk of miscarriage and birth defects. Causes for these adverse pregnancy outcomes may be attributable to undiagnosed metabolic disturbances. Pregnancy complications, including gestational diabetes, diabetes mellitus, preeclampsia, cesarean delivery, and hypertension, are common in overweight and obese patients. Fetuses of obese patients are at increased risk of congenital anomalies, stillbirth, macrosomia, difficult deliveries, and childhood obesity. Maternal and fetal risks can be a strong motivator for obese patients to lose weight before conception.

This patient is obese with oligomenorrhea and has come in for a preconception consultation. The best recommendation is for her to meet with a nutritionist to initiate a weight-loss program incorporating dietary and behavioral modification plus exercise. The most successful diets educate patients on healthy food choices and portion control. Nutritionists teach patients how to read food labels and substitute healthier food choices for those high in fat and glycemic content. Ovulatory function may return with weight loss, which will allow the patient to have control over her menstrual cycle and to time intercourse more precisely.

Clomiphene citrate is not the best initial recommendation for this patient. Clomiphene is an ovulation induction agent for the anovulatory patient with normal thyroid-stimulating hormone and prolactin levels. This patient is obese and, therefore, dietary and lifestyle changes are her first-line treatment strategy. Many obese patients who use clomiphene for ovulation induction do not respond to treatment. In addition, a multiple pregnancy is a greater risk for an obese patient because her comorbidities will be exacerbated. Obesity is a multifactorial problem with behavioral and neurohormonal complexities that need to be treated through a combination of lifestyle changes and sometimes by adding medications.

Bariatric surgery is an option for the described patient if lifestyle modifications prove unsuccessful. The two most common procedures are the Roux-en-Y gastric bypass and gastric banding. A number of studies have shown that 15% of patients who have undergone bariatric surgery maintain their reduced weight. Bariatric surgery is best reserved for patients with a BMI greater than 40 or greater than 35 with comorbidities. Surgery is the most effective therapy for morbid obesity, decreasing and even eliminating comorbidities while improving quality of life. Menstrual cycle regularity and ovulation often return in the oligo-ovulatory and anovulatory patient.

Fecundability peaks in the 2 days before ovulation. In women with regular menstrual cycles, increasing the frequency of intercourse after the cessation of menses increases the chances of pregnancy. In this patient, who has oligomenorrhea (menses occurring at intervals greater than 35 days), an ovulation predictor kit is unlikely to be accurate because her cycles are irregular and timing the luteinizing hormone surge will be cumbersome. In addition, she has anovulatory cycles even though she is menstruating. The best recommendation for this patient is lifestyle modification. Once she has become regular and ovulatory, timing intercourse or using an ovulation predictor kit can be instituted.

Holes-Lewis KA, Malcolm R, O'Neil PM. Pharmacotherapy of obesity: clinical treatments and considerations. Am J Med Sci 2013;345:284–8.

Kumbak B, Oral E, Bukulmez O. Female obesity and assisted reproductive technologies. Semin Reprod Med 2012;30:507–16.

Lash MM, Armstrong A. Impact of obesity on women's health. Fertil Steril 2009;91:1712–6.

Moran LJ, Norman RJ. The effect of bariatric surgery on female reproductive function. J Clin Endocrinol Metab 2012;97:4352–4.

Obesity and reproduction: an educational bulletin. Practice Committee of American Society for Reproductive Medicine. Fertil Steril 2008; 90:S21–9.

126
Donor oocyte use

A 26-year-old woman, gravida 2, para 2, comes to see you after undergoing laparoscopic tubal ligation. She is interested in serving as an oocyte donor. She has noticed that each clinic has a limit on the number of cycles for each donor. She also has learned that there is a limit to the number of times that a sperm donor may be used. You advise her that the most valid reason to limit the repetitive use of donor gametes from males and females is to avoid the risk of

- (A) vertical transmission of human immunodeficiency virus (HIV)
- (B) donor-to-recipient transmission of the hepatitis B virus
* (C) inadvertent consanguinity
- (D) donor-to-recipient pelvic infection

Data from the Centers for Disease Control and Prevention and the Society for Assisted Reproductive Technologies show a live-birth rate of 4.2% for patients older than 42 years who use in vitro fertilization (IVF) with their own oocytes. Oocyte donation has become an important option with delayed childbirth having become a significant trend over the last two decades.

Indications for use of a donor egg are diminished ovarian reserve, surgical or spontaneous menopause, genetic abnormalities in the female (eg, balanced translocation or autosomal disorder carrier status), history of gonadotoxic therapies, or multiple IVF failures using the patient's own oocytes. In 2011, the Society for Assisted Reproductive Technologies database recorded a 54.9% live-birth rate for patients who used donor oocytes in fresh embryo transfers. For the patient who has diminished ovarian reserve and who has not insisted on the use of her own oocytes, oocyte donation represents the best chance to achieve live birth. The literature showing gamete donation to be safe and effective for the recipients of donor eggs is well documented. The risks of repetitive oocyte donation are less well known.

Certain risks to society are known to emanate from gamete donation (either sperm or egg donation). Inadvertent consanguinity could occur if a donor has donated to two or more families in a relatively confined population and the offspring are unaware of their specific genetic heritage. For this reason, the American Society for Reproductive Medicine has published guidelines that advise an arbitrary limit of no more than 25 pregnancies per sperm or oocyte donor per population of 800,000 to limit inadvertent consanguinity. It seems reasonable to assume that these limitations are more directly applicable to sperm donors than egg donors. The most valid reason to limit the repetitive use of donor gametes from males and females is to avoid the risk of inadvertent consanguinity.

Specific risks to the egg donor include surgical risks from repeated oocyte retrieval procedures and exposure risks from repeated cycles of controlled ovarian hyperstimulation, notably the possible increased risk of ovarian cancer. At present, there is no documentation on any adverse long-term sequelae resulting from follicle aspiration. The patient who undergoes repetitive retrieval procedures is exposed to possible anesthetic complications and procedure-related pelvic infections, both of which are very small in number. Another risk is ovarian hyperstimulation syndrome. Donors are tested routinely for HIV and hepatitis B. As a result, transmission of either virus to the recipient is extremely rare.

Concern that ovarian stimulation might increase the risk of ovarian cancer was first reported in 1994. Additional data analysis has either shown no association between the use of ovulation-inducing agents and ovarian cancer or that the risk was only elevated when treatment exceeded 12 cycles. For this reason, the American Society for Reproductive Medicine has published guidelines to set a limit of six stimulated cycles in order to protect the safety of oocyte donors.

de Boer A, Oosterwijk JC, Rigters-Aris CA. Determination of a maximum number of artificial inseminations by donor children per sperm donor. Fertil Steril 1995;63:419–21.

Repetitive oocyte donation. Practice Committee of American Society for Reproductive Medicine. Fertil Steril 2008;90(suppl):S194–5.

Society for Assisted Reproductive Technology Clinic Outcome Reporting System. IVF success rate reports: clinic summary report. Birmingham (AL): SART; 2014. Available at: https://www.sartcorsonline.com/rptCSR_PublicMultYear.aspx?ClinicPKID=0. Retrieved March 14, 2014.

127
Breast cancer

A 38-year-old white woman, gravida 3, para 2, comes to your office for her annual well-woman examination. She informs you that her mother was treated for breast cancer when she was age 44 years and was found not to be a *BRCA1* or *BRCA2* mutation carrier. You decide to evaluate her for a possible increased risk of developing breast cancer using the online Breast Cancer Risk Assessment Tool. In addition to the age at which she gave birth to her firstborn child, the information you need from her medical history is the patient's

* (A) age at menarche
 (B) number of years of oral contraceptive (OC) use
 (C) smoking history
 (D) personal history of benign ovarian cysts
 (E) menstrual cycle pattern

In recent years, several tools have been developed to help patients and health care providers assess risk of future cancer. These tools provide risk estimates compared with the average risk of a specific age group over a certain period (often over the next 5–10 years). Most models combine many risk factors because the risk associated with a single factor is usually too small to provide significant guidance about screening plans.

The most widely used model to help predict risk of breast cancer is the Breast Cancer Risk Assessment Tool, often called the Gail model. This model combines several risk factors for breast cancer and allows health care providers to stratify patients into risk categories. The tool enables better screening strategies for the optimal management of individual patients. Although the Gail model performs well in populations of women, it has been shown to be less useful for the rare woman who has a very significant family history of breast cancer. In patients who have a known *BRCA* mutation in their family or who have a personal history of breast cancer, the Gail model is not appropriate to triage their screening or care. Such patients are at a significantly high risk of breast cancer development and should be referred directly to a genetic counselor or an oncology team that specializes in risk assessment, screening, and prevention strategies.

Many factors can influence the risk of breast cancer, such as menstrual and reproductive history, age, ethnicity or race, diet, and alcohol use (Box 127-1). If the model indicates that the 5-year risk of breast cancer is 1.7% or greater, patients are encouraged to consider risk-reduction strategies, such as lifestyle modification, medical therapy with tamoxifen citrate, or surgery with mastectomy or bilateral salpingo-oophorectomy. If a woman has a life expectancy of less than 10 years, there is probably minimal benefit to screening to determine eligibility for risk-reduction strategies.

Prior use of OCs has been studied to assess the association with breast cancer development. Many large case–control studies of low-dose OCs have consistently shown no relation between breast cancer and current or past use of OCs. Information about OC use is not used to estimate risk in the Gail model.

> **BOX 127-1**
>
> **Criteria Used to Calculate the 5-Year Risk of Breast Cancer According to the Modified Gail Model**
>
> - Current age
> - Age at menarche
> - Age at first live birth
> - Number of first-degree relatives with breast cancer
> - Number of previous breast biopsies
> - Whether any breast biopsy has shown atypical hyperplasia
> - Race
>
> National Cancer Institute. Breast cancer risk assessment tool. Bethesda (MD): NCI; 2011. Available at: http://www.cancer.gov/bcrisktool/. Retrieved March 20, 2014.

Current or prior smokers tend to have a higher risk of many types of cancer; however, the relation between smoking and breast cancer often is complicated by interaction with alcohol use and hormonal factors. To date, studies that have investigated the association between smoking and breast cancer have shown mixed results, with some studies indicating a mild increased risk and other studies showing no additional risk. Smoking history is not part of the Gail model.

Although a history of prior benign breast disease and biopsy is factored into the Gail model, prior benign ovarian cysts are not part of the model. Benign ovarian cysts are not associated with an increased risk of breast or ovarian cancer. Malignant ovarian tumors, especially in young women, may be associated with an increased risk of *BRCA1* or *BRCA2* gene mutations and would be a cause for concern for current or future breast cancer development.

A woman's menstrual cycle pattern is not a factor when completing the Gail model assessment. An irregular menstrual pattern may be associated with polycystic ovary syndrome (PCOS). However, studies of the possible association between PCOS and breast cancer have been inconclusive to date. Women with PCOS often are found to have chronically elevated androgens and estrogens, which may be associated with endometrial cancer risk. To date, no study has shown an increased lifetime risk of breast cancer in women with PCOS.

National Cancer Institute. Breast cancer risk assessment tool. Bethesda (MD): NCI; 2011. Available at: http://www.cancer.gov/bcrisktool/. Retrieved March 20, 2014.

National Comprehensive Cancer Network. Breast cancer risk reduction. NCCN Clinical Practice Guidelines in Oncology Version I.213 (after login). Fort Washington (PA): NCCN; 2013. Available at: http://www.nccn.org/professionals/physician_gls/pdf/breast_risk.pdf. Retrieved March 20, 2014.

Speroff L, Fritz MA. Amenorrhea. In: Speroff L, Fritz MA, editors. Clinical gynecologic endocrinology and infertility. 8th ed. Philadelphia (PA): Lippincott Williams and Wilkins; 2011. p. 483–4.

128–130

Evaluation of gonads in androgen insensitivity syndrome

For each clinical scenario (128–130), choose the likely diagnosis (A–F).

(A) 46,XX karyotype and serum testosterone in the female range
(B) 46,XX karyotype and serum testosterone in the male range
(C) 46,XY karyotype and serum testosterone in the female range
(D) 46,XY karyotype and serum testosterone in the male range
(E) 47,XXX karyotype and serum testosterone in the female range
(F) 47,XXY karyotype and serum testosterone in the male range

128. D.   Complete androgen insensitivity syndrome (CAIS)

129. C.   Pure gonadal dysgenesis (Swyer syndrome)

130. E.   Trisomy X syndrome

Androgen insensitivity syndrome (AIS) involves a mutation of the androgen receptor located on chromosome *Xq11-12*, resulting in complete loss of function (CAIS) or partial loss of function (PAIS), also referred to as Reifenstein syndrome. Approximately 2–5 women per 100,000 will exhibit CAIS. More than 500 mutations have been described, most of which are single-base substitutions. The androgen receptor is a transcription factor regulating gene transcription in response to androgens, including 5α-dihydrotestosterone and testosterone.

The presentation of AIS depends on the degree of loss of function of the androgen receptor. Typically, CAIS is seen in teenagers with appropriate breast development, reduced or absent pubic and axillary (underarm) hair, and primary amenorrhea. A less common presentation is the finding of a newborn female after a prenatal chromosome analysis showed a normal male karyotype. In a female infant or child, an inguinal hernia may suggest CAIS, with an incidence of approximately 1.1%. Age-appropriate evaluation should be considered in this circumstance. Partial AIS has a varied presentation depending on the degree of receptor function and number (Table 128–130-1).

Complete loss of androgen receptor activity causes lack of pigmented sexual hair in the pubic and axillary regions, lack of male external genitalia and prostate development primarily under the control of 5α-dihydrotestosterone, and lack of internal male reproductive organs, including epididymis, seminal vesicles, and vas deferens (wolffian ducts), primarily under the control of testosterone. Because the testes normally produce antimüllerian hormone, also called müllerian-inhibiting substance, the müllerian ducts that result in the fallopian tubes, uterus, cervix, and upper two thirds of the vagina are absent.

TABLE 128–130-1. Categorization of Androgen Insensitivity Syndrome

Grade	Type	Description
1	PAIS	Normal male phenotype, possibly gynecomastia or mild impairment of virilization. Possible infertility.
2	PAIS	Male phenotype, but small penis, penoscrotal hypospadias.
3	PAIS	Predominantly male phenotype with micropenis, perineal hypospadia, cryptorchidism and possibly bifid scrotum.
4	PAIS	Ambiguity of the external genitalia: very large clitoris, urogenital sinus with perineal opening and labioscrotal folds.
5	PAIS	Predominantly female phenotype: large clitoris, separate openings of the urethra and vagina.
6	PAIS	Female phenotype, androgen-induced pubic and axillar hair growth at the time of puberty. 50% inguinal hernia.
7	CAIS	Normal female phenotype. Lack of androgen-induced pubic and axillar hair growth at the time of puberty. 50% inguinal hernia.

Abbreviations: CAIS, complete loss of function; PAIS, partial loss of function.

Republished with permission of the Endocrine Society from Endocrine Reviews. Quigley CA, De Bellis A, Marschke KB, el-Awady MK, Wilson EM, French FS. Androgen receptor defects: historical, clinical, and molecular perspectives (Erratum in Endocr Rev 1995;16:546). Endocr Rev 1995;16:271–321. Copyright 1995.

Estrogen production in patients with CAIS is approximately twice that of normal males and originates from extraglandular formation and secretion from the testes. Estradiol is derived from direct secretion from the testes and from extraglandular conversion from testosterone via aromatization. Estrone is derived mainly from extraglandular conversion of androstenedione via aromatization and from metabolism of estradiol. An increase in testosterone production from the testes is responsible for the increased estrogen levels and is due to higher luteinizing hormone production by the pituitary. Increased pituitary luteinizing hormone production is secondary to lack of appropriate negative inhibition by testosterone on the hypothalamus.

With CAIS, physical examination generally shows Tanner stage 5 breast development, Tanner stage 1 pubic hair development, and a shortened, blind-ending vagina, typically 2.5–3.0 cm in length. Rectal examination or pelvic ultrasonography shows no evidence of a uterus. The gonads are typically in the pelvis with CAIS, but with grades 4–6 PAIS, they may be located in the inguinal canal or in the labia majora. Patients with CAIS have later closure of the epiphyses compared with normal women, most likely due to lower total estrogen levels. Thus, final height is greater than normal females but less than normal males. In general, patients with CAIS are not at risk of any other congenital anomalies. Evaluation of other organ systems is not necessary. Psychosexual function in patients with CAIS is the same as in normal females.

The diagnosis of CAIS is based on the physical examination findings along with supporting laboratory analysis. The karyotype is that of a normal male with a serum testosterone level in the normal or high-normal male range. In contrast, females born with pure gonadal dysgenesis have a male karyotype but female levels of testosterone because the gonad typically does not secrete hormones. Unrelated trisomies of the sex chromosomes include 47,XXX (Trisomy X syndrome) and 47,XXY (Klinefelter syndrome).

The differential diagnoses include imperforate hymen, transverse vaginal septum, and müllerian agenesis, also known as vaginal agenesis or Mayer–Rokitansky–Küster–Hauser syndrome. Imperforate hymen and transverse vaginal septum typically occur with episodic pelvic pain, and a uterus is identifiable on rectal examination. Patients with müllerian agenesis will differ from those with CAIS by the presence of normal pubic and axillary hair, a normal XX karyotype, and a serum testosterone level in the female range.

Treatment for CAIS is directed at hormone replacement after removal of the gonads and development of a functional vagina. An increased risk of gonadoblastoma and dysgerminoma in the gonads is evident, with risk greater in PAIS compared with CAIS (approximately 15% versus 1%, respectively). The risk increases with age and is approximately 33% at age 50 years, although this has not been analyzed separately for each condition. Given the low risk for most individuals, it is reasonable to await completion of breast development in women with CAIS before prophylactic gonadectomy. Earlier removal should be considered in women with PAIS, with consideration given to the degree of androgenization. After gonadectomy, age-appropriate hormone therapy should be instituted.

Creation of adequate vaginal caliber for intercourse should be based primarily on anticipated timing of sexual activity. Use of vaginal dilators should be the first choice. More than 75% of women will achieve adequate vaginal dimensions for successful intercourse by using vaginal dilators. Vaginal depth should be obtained first with more narrow dilators followed by dilators with increasing breadth. Surgical therapy may be necessary in women with failed attempts at dilation. A number of surgical approaches are available, including the development of a neovagina with a skin graft (McIndoe procedure).

The attitude toward disclosure of the diagnosis of AIS has changed with time. Until the late 1990s, many physicians suggested that there was no need to explain the pathophysiology. Patients were simply told that they had a birth anomaly that resulted in the lack of a uterus and the need for "ovary" removal due to increased risk of cancer. Attitudes have changed because of more emphasis on patient justice, nonmaleficence, beneficence, and autonomy. Another contributing factor has been the availability of health information on the Internet, which can lead to some patients discovering the diagnosis without the aid of counseling. At present, the recommendation is to discuss the diagnosis with the assistance of counselors trained in disorders of sexual development.

Hiort O. Clinical and molecular aspects of androgen insensitivity. Endocr Dev 2013;24:33–40.

MacDonald PC, Madden JD, Brenner PF, Wilson JD, Siiteri PK. Origin of estrogen in normal men and in women with testicular feminization. J Clin Endocrinol Metab 1979;49:905–16.

Oakes MB, Eyvazzadeh AD, Quint E, Smith YR. Complete androgen insensitivity syndrome—a review. J Pediatr Adolesc Gynecol 2008;21:305–10.

131–133

Hereditary cancer syndromes

For each clinical scenario (131–133), choose the most likely genetic disorder (A–E).

(A) *BRCA1* mutation
(B) *BRCA2* mutation
(C) Lynch II syndrome, ie, hereditary nonpolyposis colorectal cancer
(D) Peutz–Jeghers syndrome
(E) *RAD51C* mutation

131. B. A 53-year-old woman was recently diagnosed with breast cancer. She has a paternal grandmother with breast and ovarian cancer and a father with prostate cancer. She is most likely to have a mutation associated with a region in *exon 11* of the mutated gene.

132. D. Autosomal dominant disorder associated with mutations of the *STK11* gene that causes tumors of the gastrointestinal tract, breast, ovary, cervix, and testis as well as mucocutaneous pigmentation.

133. C. Germline mismatch mutation of the *MLH6* gene that confers a higher risk of endometrial cancer and a slightly lower risk of colorectal cancer compared with *MLH1* and *MSH2* mutations.

Approximately 10% of cases of breast cancer and 15% of cases of ovarian cancer are found to be associated with a hereditary gene mutation. Although the most common mutations found with breast and ovarian cancer are mutations of the *BRCA1* and *BRCA2* genes, other less common mutations, such as hereditary nonpolyposis colorectal cancer and Peutz–Jeghers syndrome, can confer a similar risk profile, although each has unique characteristics.

BRCA1 and *BRCA2* are tumor suppressor genes, which encode two types of proteins that have specific roles:

- Caretaker proteins maintain the integrity of the genome and gene stability.
- Gatekeeper proteins regulate tumor growth directly via mitosis and apoptosis.

In encoding these two protein types, the genes limit cell replication. Mutations in tumor suppressor genes can result in defects in the proteins' caretaker and gatekeeper roles.

The *BRCA1* gene is located on chromosome 17q21, and to date more than 800 mutations of the gene have been identified. The *BRCA2* gene is located on chromosome 13q12-13. Both types of gene mutations follow an autosomal dominant pattern. Cumulative risk of breast cancer by age 70 years is different for *BRCA1* mutations than for *BRCA2* mutations, approximately 57% (95% confidence interval [CI], 47–66%) and 49% (95% CI, 40–57%), respectively. A patient's personal risk must be individualized based on medical and reproductive history.

The normal *BRCA1* gene suppresses estrogen-dependent transcription pathways. Paradoxically, mutations of this gene are associated with tumors that are less likely to express estrogen and progesterone receptors. The *BRCA2* gene mutations are associated with an increased risk of prostate cancer in male family members. In addition, the *BRCA2* gene has a region in *exon 11* that appears to confer an increased risk of breast cancer (Clinical scenario 131).

Peutz–Jeghers syndrome is a rare autosomal dominant syndrome caused by germline mutations in a serine/threonine kinase (*STK11*) on chromosome 19p13.3 (Clinical scenario 132). This syndrome is commonly associated with hamartomatous polyps in the gastrointestinal tract and mucocutaneous melanin pigmentation of the lips, buccal mucosa, fingers, and toes (Fig. 131–133-1; see color plate). Carriers are at increased risk of gastrointestinal cancers (stomach, colorectal, small intestine, and pancreas) as well as cancers of the lung, breast, uterus, and ovary.

Lynch II syndrome is characterized by mutations of the *MLH1*, *MSH2*, *MSH6*, and *PMS2* genes, which are involved in the function of DNA mismatch repair enzymes. The syndrome has an autosomal dominant pattern and is associated with an increased risk of colon cancer, especially in carriers younger than 50 years. There is also an elevated risk of cancers of the endometrium, ovary, gastrointestinal tract, upper urinary tract, and bile duct (Clinical scenario 133). The risk of specific types of cancer varies based on which of the four DNA mismatch repair genes is mutated, ie, there is an increased risk of endometrial cancer in families where the *MLH6* gene is abnormal versus families with *MLH1* and *MSH2* mutations. The lifetime risk of colon cancer is 30–70% and the risk of ovarian cancer is 4–11%.

More recently, other genetic mutations have been discovered that are associated with an increased risk of ovarian and breast cancer. For example, the *RAD51C* gene is involved in DNA double-strand break repair. Mutations in this gene have been associated with increased rates of breast and ovarian cancer, with similar rates of occurrence as those seen in carriers of *BRCA1* and *BRCA2* mutations.

Chen S, Parmigiani G. Meta-analysis of *BRCA1* and *BRCA2* penetrance. J Clin Oncol 2007;25:1329–33.

Daniels MS. Genetic testing by cancer site: uterus. Cancer J 2012; 18:338–42.

National Comprehensive Cancer Network. NCCN Guidelines. Fort Washington (PA): NCCN; 2014. Available at: http://www.nccn.org professionals/physician_gls/f_guidelines.asp. Retrieved March 20, 2014.

134–138

Lifestyle choices and pregnancy outcome

Match the described patient (134–138) with the most likely outcome (A–E).

(A) Nutritional deficiencies
(B) Accelerated ovarian follicular depletion
(C) Resumption of regular menstrual cycles
(D) Congenital rickets
(E) Facial dysmorphia

134. A. A woman has undergone bariatric surgery and has had a 35.4-kg (78-lb) weight loss and frequent loose stools after eating.

135. C. An obese anovulatory woman has lost 5–10% of her body weight.

136. B. A 38-year-old woman has smoked one pack of cigarettes a day for 15 years.

137. D. A woman has severe maternal vitamin D deficiency during pregnancy.

138. E. A 37-year-old woman drank alcohol heavily in the first 10 weeks of gestation.

The preconception period provides a time to address and modify lifestyles that can increase the likelihood of a healthy mother and baby. During the patient's initial consultation, there should be an open dialogue concerning nutrition, exercise, and use of tobacco, alcohol, and vitamins.

Obesity continues to be the most common chronic disease in the United States; it is defined as a body mass index (BMI) of 30 or more (calculated as weight in kilograms divided by height in meters squared). In reproductive-aged patients, obesity often is associated with ovulatory dysfunction. Obese pregnant patients are at increased risk of pregnancy complications, including gestational diabetes mellitus, preeclampsia, hypertension, and cesarean delivery. Fetal risks of obese pregnant women include increased risks of pregnancy loss, prematurity, macrosomia, and congenital anomalies. Weight loss and regular exercise are the first recommendations for obese patients to reduce the incidence of these comorbidities. For patients who continue to have a BMI over 40 or a BMI over 35 with comorbidities after taking these steps, bariatric surgery is an option that decreases maternal and fetal pregnancy complications. Bariatric surgery patients are at increased risk of nutritional deficiencies, and an evaluation is indicated in the patient when there is a risk of a deficiency. Patient 134 has loose stools and a history of bariatric surgery, so an evaluation for iron, vitamin D, calcium, vitamin B_{12}, and folate deficiencies should be initiated.

More than 30% of obese patients have irregular menstrual cycles caused by ovulatory dysfunction. This dysfunction leads to infertility and has been linked to pregnancy loss and birth defects. Ovulation induction with clomiphene citrate or gonadotropins has been used to help anovulatory infertility patients. Before using these agents, loss of 5–10% of body weight in an obese patient can trigger resumption of regular menstrual cycles. The mechanism for cycle resumption is decreased aromatization of androgens to estrogen in adipose tissue (Patient 135).

Up to 30% of reproductive-aged women smoke cigarettes. Smoking is linked to infertility, decreased fecundity, pregnancy loss, and increased time to conception. Tobacco smoke has toxins that compromise oocyte quality. These toxins accelerate follicular depletion and decrease reproductive function. Menopause starts earlier in smokers. Patients who smoke should be counseled on the benefits of smoking cessation. Preconception care presents an opportunity for such counseling to be successful, given that it is a time when patients may be motivated to stop smoking (Patient 136).

Vitamin D is produced in the skin from sunlight exposure and obtained from dietary supplements, juice, fish oils, and fortified milk. Vitamin D deficiency is more prevalent in women who are vegetarians, women with limited exposure to sunlight, and women with darker skin. Prenatal vitamins usually contain 400 international units of vitamin D. For pregnant patients at risk of vitamin D deficiency, serum levels of 25-OH-D-vitamin D should be drawn. If levels are less than 32 ng/mL, most experts suggest daily supplementation with 1,000–2,000 international units of vitamin D. Severe deficiency of vitamin D in pregnancy has been associated with congenital rickets and fractures in the newborn (Patient 137).

Fetal alcohol syndrome results from the teratogenic effects of alcohol (Patient 138). This syndrome is associated with central nervous system abnormalities (impaired intellectual development) and facial dysmorphia. The U.S. Surgeon General and the American College of Obstetricians and Gynecologists advise pregnant women not to drink alcohol. Level of alcohol consumption can be ascertained during preconception with the use of the T-ACE questionnaire (Table 134–138-1). A patient who drinks heavily should be counseled during preconception and pregnancy to curtail alcohol consumption.

At-risk drinking and alcohol dependence: obstetric and gynecologic implications. Committee Opinion No. 496. American College of Obstetricians and Gynecologists. Obstet Gynecol 2011;118:383–8.

Bariatric surgery and pregnancy. ACOG Practice Bulletin No. 105. American College of Obstetricians and Gynecologists. Obstet Gynecol 2009;113:1405–13.

Obesity and reproduction: an educational bulletin. Practice Committee of American Society for Reproductive Medicine. Fertil Steril 2008; 90:S21–9.

Smoking and infertility: a committee opinion. Practice Committee of the American Society for Reproductive Medicine. Fertil Steril 2012; 98:1400–6.

Vitamin D: screening and supplementation during pregnancy. Committee Opinion No. 495. American College of Obstetricians and Gynecologists. Obstet Gynecol 2011;118:197–8.

TABLE 134–138-1. The T-ACE Problem Drinking Screen

Code	Factor	Cut-Points	Score	Question
"T"	Tolerance	≥2 or ≥3 drinks	2	How many drinks does it take to make you feel high?
"A"	Annoy	Yes/No	1	Has anybody ever annoyed you by complaining about your drinking?
"C"	Cut Down	Yes/No	1	Have you ever felt you ought to cut down on your drinking?
"E"	"Eye-Opener"	Yes/No	1	Have you ever needed a drink first thing in the morning to get going?

Total T-ACE Score Cut-Points: 2 or 3.

Chiodo LM, Sokol RJ, Delaney-Black V, Janisse J, Hannigan JH. Validity of the T-ACE in pregnancy in predicting child outcome and risk drinking. Alcohol 2010;44:595–603. Copyright 2010 by Elsevier.

139–142
Hormonal changes in pregnancy

Match the description (139–142) with the appropriate hormone (A–D).

(A) Antimüllerian hormone
(B) Human placental lactogen
(C) Corticotropin-releasing hormone (CRH)
(D) Thyroxine (T_4)

139. B. Increases significantly in pregnancy

140. B. Has growth hormone-like activity

141. C. Corticosteroids increase placental production of this hormone

142. D. Total hormone level is increased but free hormone level is normal.

Hormonal changes that take place during pregnancy are extensive. In pregnancy, a new endocrine organ, the placenta, plays a role in these hormonal changes by synthesizing large precursor hormones, steroid hormones, and various protein and peptide hormones.

Estrogen and progesterone levels increase significantly in pregnancy. Maternal progesterone levels rise from approximately 25 ng/mL in the luteal phase to 150 ng/mL at term. Fetal well-being does not affect progesterone levels because production of progesterone by the placenta is dependent on maternal supply of low-density lipoprotein cholesterol; the placenta lacks the ability to synthesize cholesterol from acetate. Progesterone is thought to inhibit smooth muscle contractility and cause uterine quiescence. Estrogens also increase markedly in pregnancy. The corpus luteum is the initial source of estrogen in the first few weeks of pregnancy, but later the trophoblast of the placenta produces the bulk of estrogen seen in pregnancy. Estradiol and estrone are synthesized by the placenta via conversion of dehydroepiandrosterone sulfate (DHEAS), which reaches the placenta from the maternal and fetal bloodstreams. However, estriol production is dependent upon conversion of DHEAS that is derived from the fetal adrenal gland to 16α-hydroxy DHEAS. This conversion occurs from the fetal liver. Because of the fetal contribution to estrogen production, estrogens, especially estriol, have been used historically to monitor fetal well-being because their production is dependent on fetal production of prehormones.

Human placental lactogen is a product of the placenta and has growth hormone-like activity. Its level increases with advancement of the pregnancy. Most actions of this hormone take place on the maternal side, causing lipolysis, which results in increased free fatty acids in maternal and fetal circulation, as well as development of insulin resistance and inhibition of gluconeogenesis favoring the transportation of glucose to the fetus. These effects may predispose the mother to gestational diabetes mellitus.

Maternal plasma cortisol increases in pregnancy secondary to an increase in cortisol-binding globulin. The placenta produces CRH, which causes an increase in maternal cortisol levels. Typically, corticosteroids cause a negative feedback effect on CRH production; however, in the placenta, corticosteroids have a positive feedback effect on the placental cell CRH output. It is believed that CRH promotes myometrial contractility and may play a role in parturition.

Thyroid gland size increases slightly in pregnancy. Despite marked increase in maternal serum total triiodothyronine and total T_4, the pregnant woman remains euthyroid because of a parallel increase in thyroxine-binding globulin. Therefore, the levels of free triiodothyronine and T_4 do not increase during pregnancy. It is believed that thyroxine-binding globulin is increased in response to estrogen exposure. Antimüllerian hormone levels decrease during pregnancy. This decrease may be related to ovarian suppression that normally takes place in pregnant women and is independent of the woman's age.

Beshay V, Carr BR. Fertilization, implantation, and endocrinology of pregnancy. In: Kovacs WJ, Ojeda SR, eds. Textbook of endocrine physiology. 6th ed. New York: Oxford University Press; 2012. p. 264–91.

Beshay VE, Carr BR, Rainey WE. The human fetal adrenal gland, corticotropin-releasing hormone, and parturition. Semin Reprod Med 2007;25:14–20.

Braunstein GD. Endocrine changes in pregnancy. In: Kronenberg HM, Melmed S, Polonsky KS, Larsen PR, editors. Williams textbook of endocrinology, 11th ed. Philadelphia (PA): Saunders Elsevier; 2008. p. 741–52.

Carr BR. The maternal–fetal–placental unit. In: Becker KL, editor. Principles and practice of endocrinology and metabolism. 3rd ed. Philadelphia (PA): Lippincott Williams & Wilkins; 2001. p. 1059–72.

Koninger A, Kauth A, Schmidt B, Schmidt M, Yerlikaya G, Kasimir-Bauer S, et al. Anti-Müllerian-hormone levels during pregnancy and postpartum. Reprod Biol Endocrinol 2013;11:60.

143–145
Statistical analysis

For each experiment (143–145), choose the most appropriate statistical analysis (A–E).

(A) Analysis of variance (ANOVA) with post-hoc Tukey honestly significant difference testing
(B) Student t-test
(C) Kaplan–Meier curve
(D) Fisher exact test
(E) Mann–Whitney U test

143. D. A study to compare whether the uterus increased in mass (yes or no) in six mice fed a soy diet versus six unrelated mice fed a regular diet

144. A. A laboratory study of breast cancer cell line proliferation rates after treatment with estradiol, estradiol plus norethindrone, or vehicle

145. C. Longitudinal pregnancy rates in women with endometriosis after diagnostic versus operative laparoscopy

Basic knowledge of statistical analyses includes understanding data types, distribution, and tests for comparisons of two groups and of multiple groups. Data may be qualitative or quantitative. Qualitative data are not expressed in numbers and may be either ordinal or nominal. Ordinal data can be placed in some type of order, such as size (eg, small, medium, and large). Nominal data cannot be ordered (eg, gender, ethnicity, and occupation). Ordinal and nominal data usually are described in proportions with frequency tables. Quantitative data require a numerical scale.

The most common distribution for quantitative data is normal or gaussian distribution. This is recognized as a bell-shaped curve when data are plotted in a histogram. In this case, the observations cluster around a mean value and decline when they are further away from the mean. The standard deviation (SD) describes the distribution of data around the mean. In a normal distribution, 68% of the data will be contained within 1 SD, 95% of data within 2 SD, and 99.7% of data within 3 SD. For example, in a test, the mean score was 50 points with a SD of 10 points. If the scores were normally distributed, 68% of students scored between 40 points and 60 points, 95% of students scored between 30 points and 70 points, and 99.7% scored between 20 points and 80 points. Specific tests to determine normal distribution include the Kolmogorov–Smirnov and Shapiro–Wilk tests. Normally distributed quantitative data are described with the mean and standard deviation, whereas nonnormal quantitative data are described with median and range.

Statistical testing can be defined as nonparametric and parametric. Nonparametric testing (distribution-free testing) is used in ordinal data, nominal data, and quantitative data that are not normally distributed. Parametric tests are used for normally distributed quantitative data. For quantitative data, nonparametric tests are less powerful than parametric tests, making it more difficult to find statistical significance. Thus, another option is to mathematically transform nonnormal data into normally distributed data so that parametric tests may be used. Various transformations include logarithms, square root, reciprocal, and arc–sine.

Statistical testing is used for many comparisons, including comparing a group with a hypothetical value, comparing two unmatched groups, comparing two matched groups, comparing three or more unmatched groups, comparing three or more matched groups, determining associations between two variables (correlation), and predicting the value of a variable from another measured variable (regression). Common statistical tests for each situation and each data type are shown in Table 143–145-1.

An example of comparing two groups of nominal data would be a "yes" or "no" observation of whether uterine mass increased in unmatched mice with a soy diet versus regular diet. The data can easily be placed into a 2 × 2 table for a Fisher exact test (Experiment 143). If the same study was done with exact measurements of the uterus, it would comprise continuous data. If the data were normally distributed, the means would be compared with an

TABLE 143–145-1. Common Statistical Analyses

	Type of Data			
Goal	Categorical: Nominal, Ordinal	Score, Rank, Nonnormal Continuous	Normally Distributed Continuous	Survival Time
Description	Proportions, frequency tables	Median, interquartile range	Mean, SD	Kaplan–Meier curve
Compare one group with a hypothetical value	Chi-square goodness of fit	Wilcoxon signed rank test	One sample t-test	
Compare two independent groups	Fisher exact test, chi-square test of independence	Mann–Whitney test	Unpaired t-test	Mantel–Haenszel or log-rank test
Compare two paired groups	McNemar test	Wilcoxon signed rank test	Paired t-test	Conditional proportional hazards regression
Compare three or more independent groups	Chi-square test of independence	Kruskal–Wallis test	One-way ANOVA	Cox proportional hazards regression
Examine association between two variables	Chi-square test of independence	Spearman correlation	Pearson correlation	

Abbreviations: ANOVA, analysis of variance; SD, standard deviation.

unpaired t-test. If the data were not normally distributed or transformed to normal data, a Mann–Whitney test would be appropriate.

Comparing the means of three or more groups is commonly performed with the parametric ANOVA test (Experiment 144). This is preferable to multiple comparisons of two means (t-test), which would increase the chance of concluding a significant difference when no significant difference exists (type 1 error). If there was no significant difference noted with the existing data, a true difference still may exist and could be detected with an increased sample size (type 2 error). A one-way ANOVA evaluates one factor whereas a two-way ANOVA looks for interactions between two factors. In the example of breast cancer cells being treated with three different hormones with determination of proliferation, a one-way ANOVA would be used. If the experiment was performed with a sample of breast cancer cells grown for 1 week and a sample grown for 1 year, then a two-way ANOVA may be used to look at the interaction of the age of the cells and the treatments. The ANOVA yields an overall statistical value to determine if there is a difference in the means. Subsequent post-hoc testing, such as the Tukey honestly significant difference test, is used to determine which means are different.

An outcome with time in one or more groups is commonly displayed with a Kaplan–Meier curve (Experiment 145). This often is referred to as a survival curve because of its common use in conditions in which the outcome is death. However, it can be used with any defined outcome, such as the percent of women that conceive a pregnancy after two types of laparoscopic surgery. Most commonly, two or more groups with an intervention are compared statistically on the Kaplan–Meier curve.

Bland JM, Altman DG. Survival probabilities (the Kaplan–Meier method). BMJ 1998;317:1572.

Gordi T, Khamis H. Simple solution to a common statistical problem: interpreting multiple tests. Clin Ther 2004;26:780–6.

Greenhalgh T. How to read a paper. Statistics for the non-statistician. I: Different types of data need different statistical tests [published erratum appears in BMJ 1997;315:675]. BMJ 1997;315:364–6.

146–148

Mechanism of action of hormonal contraceptives

For each levonorgestrel (LNG)-containing contraceptive method (146–148), select the most likely circulating serum level of LNG in pg/mL (A–E) in a 22-year-old woman with a body mass index of 21 (calculated as weight in kilograms divided by height in meters squared).

(A) 100–200
(B) 500–600
(C) 2,000–3,000
(D) 5,000–6,000
(E) 10,000–15,000

146. C. 30-microgram ethinyl estradiol with 150-microgram LNG oral contraceptive (OC) pill 24 hours after ingestion

147. E. Single-dose progestin-only emergency contraceptive 2 hours after ingestion

148. A. A 5-year LNG intrauterine device (IUD)

Circulating serum levels of LNG vary among different methods of LNG-containing contraceptives. Levels are lowest in women who use the first-generation LNG IUD compared with other LNG-containing contraceptive products. The first-generation LNG IUD was approved by the U.S. Food and Drug Administration in December 2000. It consists of a T-shaped polyethylene vertical stem with a reservoir containing 52 mg of LNG. Two monofilament threads are attached to a loop at the end of the vertical stem. The device contains barium sulfate, making it radiopaque. Initially, LNG is released at a concentration of 20 micrograms/d. The amount decreases to 11 micrograms/d after the device's 5-year period of approved use. The average circulating serum level of LNG in women who use these IUDs is 100–200 pg/mL. Despite the low circulating serum levels of LNG, the intrauterine location allows for the progestin to exert an antiproliferative effect on the endometrium, leading to decidualization and endometrial atrophy. The progestin further thickens the nearby cervical mucus, which decreases sperm penetration into the uterine cavity. These two actions contribute to the contraceptive efficacy of the LNG IUD. Although it is possible to have systemic adverse effects from the intrauterine release of LNG, circulating serum levels are at a minimum compared with other contraceptive methods. Serum LNG levels further decrease with the patient's increasing body mass index. Levonorgestrel is a lipophilic molecule that partitions into the fat tissue. Obese individuals have been shown to have lower circulating serum LNG levels. Despite these lower levels, the incidence of vaginal bleeding does not correlate with the circulating serum LNG level.

In comparison with the LNG IUD, the average circulating serum levels of LNG in other contraceptives are much higher. In normal-weight women, the average circulating LNG level is 2,500 pg/mL 24 hours after ingestion of a combination OC containing 30 micrograms of ethinyl estradiol and 150 micrograms of levonorgestrel. This level is 12.5–25 times higher than the circulating level of LNG caused by the LNG IUD. In a comparison group of obese women in the same pharmacokinetic study, the average peak LNG level 2 hours after ingestion of the combination OC was slightly lower than normal-weight women (5,600 versus 7,000 pg/mL, respectively, $P=.19$), but their trough level 24 hours after OC ingestion was similar to that of the normal-weight women (2,600 pg/mL versus 2,500 pg/mL, respectively, $P=.56$). Recently, a 3-year LNG IUD was developed that contains 13.5 mg of LNG, a lower dose than the first-generation 5-year LNG IUD. This amount does not appreciably change the circulating level of LNG.

The progestin-only emergency contraceptive was introduced in the United States in July 2009; it is a Food and Drug Administration-approved method of emergency contraception available over the counter to women at risk for an unintended pregnancy. The average peak serum level of LNG 2 hours after ingestion of the progestin-only emergency contraceptive is 14,600 pg/mL, approximately 100-fold greater than the average circulating serum level of LNG associated with the LNG IUD. The progestin-only emergency contraceptive is a single dose of 1.5 mg of LNG. When taken orally within 72 hours of unprotected sexual intercourse or contraceptive failure, this medication is effective in preventing pregnancy.

In a large multicenter trial for pregnancy prevention, a single dose of 1.5 mg of LNG was associated with a 1.5% failure rate.

Kives S, Hahn PM, White E, Stanczyk FZ, Reid RL. Bioavailability of the Yuzpe and levonorgestrel regimens of emergency contraception: vaginal vs. oral administration. Contraception 2005;71:197–201.

Mansour D. The benefits and risks of using a levonorgestrel-releasing intrauterine system for contraception. Contraception 2012;85:224–34.

Seeber B, Ziehr SC, Gschließer A, Moser C, Mattle V, Seger C, et al. Quantitative levonorgestrel plasma level measurements in patients with regular and prolonged use of the levonorgestrel-releasing intrauterine system [published erratum appears in Contraception 2013;88:194]. Contraception 2012;86:345–9.

von Hertzen H, Piaggio G, Ding J, Chen J, Song S, Bartfai G, et al. Low dose mifepristone and two regimens of levonorgestrel for emergency contraception: a WHO multicentre randomised trial. Lancet 2002;360:1803–10.

Westhoff CL, Torgal AH, Mayeda ER, Pike MC, Stanczyk FZ. Pharmacokinetics of a combined oral contraceptive in obese and normal-weight women. Contraception 2010;81:474–80.

149–151

Differential diagnosis for hirsutism conditions

For each clinical scenario (149–151), choose the likely diagnosis for a woman with hirsutism (A–E).

(A) Polycystic ovary syndrome (PCOS)
(B) Cushing syndrome
(C) Sertoli–Leydig cell tumor
(D) Adult-onset congenital adrenal hyperplasia (CAH)
(E) Adrenal adenoma

149. B. High 24-hour urinary free cortisol

150. C. Total testosterone level of 300 ng/dL

151. D. Follicular phase morning blood tests showing 17-hydroxyprogesterone level of 488 ng/dL

Hirsutism is the abnormal growth of pigmented hair either in excessive amounts in normal locations or in locations not typically seen in most women. Significant ethnic differences exist in body hair growth, ie, hirsutism is scarce among East Asians but common in Southern Europeans. Differences in skin 5α-reductase activity partially explain these differences.

The Ferriman–Gallwey score is the most frequently used quantitative method to diagnose hirsutism (Appendix D). The amount of pigmented hair is scored 0–4 (0 is no terminal hair and 4 is equivalent to a hairy male) in nine body regions, including the upper lip, chin, chest, arms, upper abdomen, lower abdomen, upper back, lower back, and thighs. A score of greater than 8 is diagnostic of hirsutism. A score of 2 or more on the chin or lower abdomen appears to be most predictive of hirsutism.

Up to 80–90% of women with hirsutism will have an androgen-excess disorder such as PCOS, adult-onset CAH, Cushing disease, or adrenal or ovarian androgen-producing neoplasia. Occasionally, hirsutism will be seen with primary hypothyroidism and hyperprolactinemia. In most patients with hirsutism, the underlying diagnosis will be PCOS. Thus, hirsutism may be the presenting symptom for significant health issues. Associated health problems may include insulin resistance with increased risk of type 2 diabetes mellitus as well as dyslipidemia and increased vascular inflammation with increased risk of arteriosclerosis. Less commonly, women may have idiopathic hirsutism characterized by normal androgen levels, normal ovulatory cycles, and normal-appearing ovaries on ultrasonography. Idiopathic hyperandrogenism refers to women with clinical and biochemical androgen excess but normal ovulatory cycles and normal-appearing ovaries on ultrasonography.

A complete medical history and physical examination are the keys to evaluation of hirsutism. Functional disorders, such as PCOS and adult-onset CAH, characteristically have a peripubertal onset with slow, gradual worsening of symptoms. Cushing disease is rare and often has other symptoms of corticosteroid excess, including relatively rapid weight gain with truncal obesity, moon facies, buffalo hump, violaceous striae, and peripheral myopathy with weakness. Androgenization due to tumors would only coincidentally appear at puberty and commonly has a more rapid course. Ovarian androgen-producing tumors are vastly more common than adrenal tumors, with Sertoli–Leydig cell tumors being the most common (Fig. 149–151-1; see color plate). Physical examination should focus on acanthosis nigricans (consistent with insulin resistance) and signs of virilization (clitoromegaly, balding). Virilization is concerning for a neoplasm because it is almost never seen with PCOS and is rarely observed with adult-onset CAH.

Laboratory testing should include dehydroepiandrosterone sulfate (DHEAS), testosterone, morning follicular phase 17-hydroxyprogesterone, thyroid-stimulating hormone, and prolactin levels. Some experts recommend taking a free testosterone level instead of a total testosterone level because the free testosterone level has a greater sensitivity for a diagnosis of hyperandrogenism. Approximately 10% of patients with PCOS will have isolated increases in DHEAS with a normal testosterone level; DHEAS is not typically elevated in adult-onset CAH and is not useful as a screening test. A screening 17-hydroxyprogesterone level greater than 200 ng/dL requires an adrenocorticotropic hormone stimulation test for further evaluation of adult-onset CAH. A screening 17-hydroxyprogesterone level of 800 ng/dL does not require further testing. Initial screening tests for Cushing syndrome include 24-hour urinary collection for free cortisol, overnight 1-mg dexamethasone suppression test, and late-night salivary cortisol test. Laboratory findings suggestive of an ovarian androgen-producing tumor include a total testosterone level greater than 200 ng/dL and a reversal in the androstenedione-to-testosterone ratio. In normal women and women with functional androgen excess (not caused by a tumor), total androstenedione will be greater than total testosterone. This relation often is reversed in cases of ovarian tumors.

Treatment for hirsutism depends on the underlying pathology. With idiopathic hirsutism and functional disorders such as PCOS and adult-onset CAH, the goals are to decrease free androgen levels and block androgen action. Decreasing free androgen levels consists of decreasing androgen production by the ovary and decreasing the free fraction by increasing sex hormone-binding globulin. Combination hormonal contraceptives remain the most cost-effective treatment for this purpose, although to date, no studies have compared the efficacy of combination oral contraceptives (OCs). An OC with a third-generation progestin (desogestrel and norgestimate) or drospirenone will result in the greatest increase in sex hormone-binding globulin. Decreased androgen action involves blockage of the androgen receptor with or without decreasing 5α-reductase activity converting testosterone to dihydrotestosterone. Spironolactone blocks the androgen receptor and decreases 5α-reductase activity. Spironolactone treatment can result in more frequent uterine bleeding. Uncommon adverse effects include hyperkalemia and hypotension. Use of spironolactone is contraindicated in women with renal insufficiency. Androgen receptor antagonists include flutamide and bicalutamide. These drugs have received less use because of potential hepatotoxicity and the need to monitor liver function. Examples of 5α-reductase inhibitors are finasteride (inhibits 5α-reductase type 1) and dutasteride (inhibits 5α-reductase type 1 and 2). Finasteride has been shown to be effective in the treatment of hirsutism. By contrast, dutasteride has not been studied adequately for the treatment of this condition.

None of the mentioned drugs, including OCs, are approved by the U.S. Food and Drug Administration for the treatment of hirsutism, although some OCs are approved for the treatment of acne. Antiandrogens (spironolactone, flutamide, finasteride) sometimes are used with combination OCs to increase efficacy if OC use alone is not effective. All androgen receptor blockers and 5α-reductase inhibitors are teratogenic and must be used with adequate contraceptive methods to prevent pregnancy.

Eflornithine hydrochloride 13.9% cream is the only FDA-approved product for the treatment of unwanted facial hair. The chemical inhibits the enzyme ornithine decarboxylase, which is important in cell growth. Thus, eflornithine does not remove hair but decreases hair growth. Other topical therapies for hirsutism include shaving, plucking, waxing, bleaching, depilatories, electrolysis, laser, and intense pulsed light.

Escobar-Morreale HF, Carmina E, Dewailly D, Gambineri A, Kelestimur F, Moghetti P, et al. Epidemiology, diagnosis and management of hirsutism: a consensus statement by the Androgen Excess and Polycystic Ovary Syndrome Society [published erratum appears in Hum Reprod Update 2013;19:207]. Hum Reprod Update 2012;18:146–70.

Paparodis R, Dunaif A. The Hirsute woman: challenges in evaluation and management. Endocr Pract 2011;17:807–18.

Yildiz BO, Bolour S, Woods K, Moore A, Azziz R. Visually scoring hirsutism. Hum Reprod Update 2010;16:51–64.

152–154

Testing for thyroid disease

Match the patient (152–154) with the thyroid function test results (A–D) that would help diagnose her condition.

(A) Elevated thyroid-stimulating hormone (TSH) level and normal free thyroxine (T_4) level
(B) Low TSH level and elevated T_4 level with thyroid antibodies
(C) High TSH level and low T_4 level
(D) Low TSH, normal T_4, and normal triiodothyronine (T_3) level

152. C. A 33-year-old woman, gravida 1, para 1, with 40-day menstrual cycles has dry skin and constantly feels tired. She is trying to become pregnant but is having difficulty interpreting her ovulation predictor kit.

153. A. A 32-year-old woman, gravida 2, para 0, has had recurrent pregnancy loss. She is in good health and has regular menses. Her recurrent pregnancy loss evaluation result is negative except for subclinical hypothyroidism.

154. B. A 28-year-old nulligravid woman has a 6-month history of amenorrhea. She is feeling frustrated because she wants to become pregnant and is not cycling. Her symptoms are heat intolerance and palpitations. On physical examination, she has lid lag and patellar tendon hyperreflexia. Her thyroid gland is diffusely enlarged.

The thyroid gland secretes T_4 at 20 times the rate of T_3. Free T_4 is converted to T_3 mainly in the liver by removal of an iodine atom. Triiodothyronine is the hormone responsible for most of the thyroid's action on peripheral organs. Free T_4 is controlled by the pituitary hormone TSH. Changes in T_4 will modulate changes in TSH in order to keep the appropriate balance of circulating thyroid hormone. The thyroid axis, in addition to the thyroid gland and pituitary, includes TSH, which is produced by the hypothalamus. These three glands work together to maintain normal thyroid function and peripheral action. Free T_4 and T_3 circulate bound and unbound to binding proteins. It is the unbound hormones (free T_4 and free T_3) that exert effects on target tissues. Thyroid dysfunction is diagnosed by measuring TSH, free T_4, and free T_3.

The thyroid antibody tests for antithyroid peroxidase and antithyroglobulin can help diagnose the etiology of thyroid dysfunction. In a patient with hypothyroidism, antibodies help make the diagnosis of autoimmune thyroiditis, also known as Hashimoto thyroiditis. When hyperthyroidism is suspected, TSH receptor autoantibodies point to autoimmune thyroid disease.

The signs and symptoms of hypothyroidism (low thyroid hormone) include dry skin, cold sensitivity, fatigue, muscle cramps, constipation, and menstrual irregularities. Patient 152 has elevated TSH and low T_4 levels, which confirm the diagnosis (Table 152–154-1). In the United States, where iodine is sufficient, the most common cause of hypothyroidism is chronic autoimmune thyroiditis, which is 5–10 times more common in women than in men. Antithyroid antibodies infiltrate the thyroid because of abnormal regulation of the immune response. This impairs free T_4 and free T_3 production. In response to

TABLE 152–154-1. Clinical Characteristics of Hypothyroidism and Hyperthyroidism

Hypothyroidism	Hyperthyroidism
Firm irregular goiter	Enlarged homogenous thyroid gland
Slow reflexes, decreased energy	Hyperreflexia
Bradycardia	Tachycardia, palpitations
Depression	Nervousness, irritability, anxiety, difficulty sleeping
Constipation	Frequent bowel movements
Muscle cramps, voice changes	Hand tremors, muscle weakness (upper arms/ thighs)
Weight gain	Weight loss despite a good appetite
Dry skin	Sweating, brittle hair, warm moist skin
Irregular menstrual periods	Irregular menstrual periods
Cold intolerance	Heat intolerance

low circulating thyroid hormone levels, the pituitary will try to compensate by increasing TSH levels to restore thyroid hormone levels. When the pituitary is no longer able to compensate by increasing TSH, thyroid hormone levels will decrease and the diagnosis of hypothyroidism is confirmed. Once hypothyroidism is treated, irregular menstrual cycles often will become regular, allowing for more accurate timed intercourse.

Subclinical hypothyroidism is early hypothyroidism before signs and symptoms present. It is a compensated state detected by an elevated TSH level and normal free T_4 level as observed in Patient 153. Subclinical hypothyroidism progresses to overt hypothyroidism at a rate of 2.6–4.3% per year depending on whether antithyroid antibodies are present. Many experts recommend measuring thyroid antibodies in the patient with subclinical hypothyroidism to help predict and follow progression to hypothyroidism. Treatment is recommended to avoid goiter development and decrease the risk of pregnancy loss.

Graves disease is the most common cause of hyperthyroidism. It is 5–10 times more common in women than in men. It is an autoimmune disorder caused by thyroid antibody stimulation of the TSH receptor, increasing release of thyroid hormone with concomitant growth of the thyroid gland. Symptoms include heat intolerance, diarrhea, weight loss, sweating, palpitations, and nervousness. As seen in Patient 154, physical examination includes lid lag, goiter, hyperreflexia, moist skin, tachycardia, and tremor. Laboratory evaluation for Patient 154 is classic for hyperthyroidism: low TSH level and elevated free T_4 with thyroid antibodies. Treating her hyperthyroidism will most likely normalize her menstrual cycles. Patients with hyperthyroidism can have normal menstrual cycles, oligo-ovulation, or amenorrhea.

Bahn (Chair) RS, Burch HB, Cooper DS, Garber JR, Greenlee MC, Klein I, et al. Hyperthyroidism and other causes of thyrotoxicosis: management guidelines of the American Thyroid Association and American Association of Clinical Endocrinologists. American Thyroid Association and the American Association of Clinical Endocrinologists [published errata appear in Thyroid 2011;21:1169. Thyroid 2012;22:1195]. Thyroid 2011;21:593–646.

Fritz MA, Speroff L. Reproduction and the thyroid. In: Fritz MA, Speroff L, editors. Clinical gynecologic and infertility. 8th ed. Philadelphia (PA): Lippincott Williams & Wilkins; 2011. p. 885–905.

Garber JR, Cobin RH, Gharib H, Hennessey JV, Klein I, Mechanick JI, et al. Clinical practice guidelines for hypothyroidism in adults: cosponsored by the American Association of Clinical Endocrinologists and the American Thyroid Association. American Association Of Clinical Endocrinologists And American Thyroid Association Taskforce On Hypothyroidism In Adults [published errata appear in Thyroid 2013;23:251. Thyroid 2013;23:129]. Thyroid 2012;22:1200–35.

155–158

Obesity and pregnancy

For each set of medical conditions in a morbidly obese patient of reproductive age (155–158), select the therapy (A–D) that is most likely to accomplish long-term improvement.

(A) Bariatric surgery
(B) Dopamine agonist
(C) Clomiphene citrate
(D) Myomectomy

155. A. Hirsutism, anovulation, glucose intolerance, and morbid obesity

156. C. Infertility and infrequent ovulation; normal follicle-stimulating hormone (FSH) level, thyroid-stimulating hormone level, and prolactin level

157. D. Anemia, abnormal uterine bleeding, and several 1-cm submucosal leiomyomas

158. B. Recurrent pregnancy loss, galactorrhea, and a shortened luteal phase

As of 2004, approximately 66% of all U.S. adults were either overweight or obese, making overweight and obesity problems of epidemic proportions. The National Health and Nutrition Examination Survey published in 2009 found that more than one half of all pregnant women in the United States are overweight or obese. Obesity is highest among non-Hispanic black women (50%) and Mexican American women (45%) compared with non-Hispanic white women (33%).

Numerous consequences result from obesity. It is widely recognized that obesity has a central role in the pathogenesis of cardiovascular disease, hypertension, dyslipidemia, obstructive sleep apnea, and cancer. In females, obesity can cause anovulation, menstrual irregularities, infertility, sexual dysfunction, and fetal loss. In pregnancy, overweight and obese women are at increased risk of preeclampsia, gestational diabetes, hypertension, cesarean delivery, and postpartum weight retention. Delivery is complicated by increased rates of infectious morbidity, anesthetic complications, increased operative times, greater blood loss, and higher rates of thromboembolism. Children born to obese women are at increased risk of prematurity, stillbirth, congenital anomalies, macrosomia, birth injury, and childhood obesity.

Nonsurgical approaches to weight loss have traditionally consisted of behavioral modification along with changes in diet and exercise. These strategies have been time-consuming and success rates have been low. Bariatric surgery has emerged as an option for rapid weight loss in reproductive-aged women. The annual rate of bariatric surgery in the United States increased 10-fold from 1999 to 2005. Patients with a body mass index of 40 or more (calculated as weight in kilograms divided by height in meters squared) or those with a body mass index of 35 or more with comorbidities are the optimal candidates to benefit from the procedure.

Bariatric surgery produces rapid weight loss and can lead to an improvement in glucose tolerance and hirsutism as well as normalization of menstrual cycles in obese women who have abnormalities as a consequence of polycystic ovary syndrome. Therefore, bariatric surgery would be the best choice for Patient 155. The hormonal profile associated with morbid obesity appears to reverse after surgical weight loss. Higher fertility rates have been consistently reported, but bariatric surgery should not yet be considered as a primary treatment for infertility. Bariatric surgery is now being used to treat adolescents with obesity.

As obesity becomes more prevalent in reproductive-aged women, practitioners will encounter many patients who have had or are considering bariatric surgery. Preconception counseling and assessment are encouraged so the patient can be informed of specific maternal and fetal risks in pregnancy. Initial steps should include encouragement to follow a weight-reduction program. Bariatric surgery is now recognized as the most effective weight-loss intervention for morbid obesity. It has improved fertility, sexuality, quality of life, and pregnancy outcomes. However, bariatric surgery does carry specific risks. The procedure can lead to malabsorption and nutritional deficiencies. A broad evaluation for micronutrient deficiencies at the beginning of pregnancy is recommended. After bariatric surgery with a malabsorptive component, patients and practitioners should be aware that oral contraceptives may have an increased failure rate. If contraception is desired, a nonhormonal method should be used.

Patient 156 has infertility with infrequent ovulation but normal FSH, thyroid-stimulating hormone, and prolactin levels. The best therapy for this patient would be clomiphene citrate, a selective estrogen receptor modulator that increases production of FSH by inhibiting negative feedback on the hypothalamus.

Patient 157 has anemia, abnormal uterine bleeding, and several 1-cm submucosal leiomyomas. Abnormal uterine bleeding that results in anemia would be an indication for surgery. Thus, Patient 157 would benefit from hysteroscopic myomectomy.

Patient 158 has experienced recurrent pregnancy loss, galactorrhea, and a shortened luteal phase. These are symptoms of hyperprolactinemia, which, once confirmed, would best be treated by a dopamine agonist, such as cabergoline or bromocriptine.

Bariatric surgery and pregnancy. ACOG Practice Bulletin No. 105. American College of Obstetricians and Gynecologists. Obstet Gynecol 2009;113:1405–13.

Merhi ZO. Impact of bariatric surgery on female reproduction. Fertil Steril 2009;92:1501–8.

Obesity in pregnancy. Committee Opinion No. 549. American College of Obstetricians and Gynecologists. Obstet Gynecol 2013;121:213–7.

Pories WJ. Bariatric surgery: risks and rewards. J Clin Endocrinol Metab 2008;93(suppl):S89–96.

Teitelman M, Grotegut CA, Williams NN, Lewis JD. The impact of bariatric surgery on menstrual patterns. Obes Surg 2006;16:1457–63.

159–161

Laparoscopic surgery complications

For each patient's set of symptoms (159–161), select the most likely complication (A–E) after laparoscopic surgery.

(A) Ureteral injury
(B) Bowel injury
(C) Gas embolism
(D) Vascular injury
(E) Bladder injury

159. B.   Abdominal distention, diarrhea, ileus, and leukocytosis

160. C.   Tachycardia, hypoxia, hypotension, and decreased end tidal carbon dioxide (CO_2)

161. A.   Fever, flank pain, leukocytosis, and elevated C-reactive protein

Laparoscopic procedures are being performed with greater frequency by gynecologists. Potentially preventable surgical errors have received a great deal of attention after the publication in 2010 by the Joint Commission of universal protocols for prevention of wrong-site, wrong-patient, and wrong-person surgery. A systems approach has been adopted that includes checklists, multiple safety stops, improved communication, and involvement of the patient in order to avoid errors.

Complications are an inevitable part of laparoscopic surgery. The incidence of complications has been estimated to be 3–6 per 1,000 cases, with up to one half of all major complications shown to occur at the time of surgical entry. The surgical literature does not provide an obvious consensus when the safety of direct trocar entry and Veress needle entry are compared. Open laparoscopy was first described in 1971 and continues to be a safe procedure with regard to the avoidance of major vascular injury or death. Lifting the umbilicus with towel clips placed close to and inverting the umbilicus allows significant elevation from the great vessels and also has been shown to be effective for safe trocar placement. The reported rates of intestinal injury with open laparoscopy are similar to those with the use of blind trocar entry.

Approximately 20–25% of laparoscopic complications are not recognized intraoperatively. Early detection and rapid response are the keys to reducing morbidity and preventing mortality from laparoscopic injury. Laparoscopic injuries to different organ systems may have certain characteristic features that will aid in diagnosis.

The close proximity of the abdominal wall to the retroperitoneal vessels (aorta, vena cava, and internal and external iliac) is a risk factor, especially in thin patients. The surgeon should have proper training and experience with patient position, angle of insertion, abdominal wall elevation, and degree of pneumoperitoneum. Blood on aspiration of the Veress needle, intraperitoneal blood, retroperitoneal hematoma, or unexplained sudden change in vital signs are all indicative of major vessel injury.

Bowel injuries account for almost one half of all major complications of laparoscopy. Early recognition and repair aids in the prevention of sequela, but less than 50% of bowel injuries are recognized at the time of surgery. Postoperative recognition of bowel injury is associated with higher morbidity and significant mortality. Delayed diagnosis can lead to peritonitis, septicemia, and multi-organ failure. Typical symptoms include pain at the trocar site nearest to the site of injury, fever, diarrhea, distention, and leukocytosis (Patient 159). Presenting signs may be subtle and varied. As a result, the surgeon should err on the side of safety and admit any patient with a suspected bowel injury for observation and further assessment.

Gas embolism is a rare but devastating complication arising from the use of CO_2 insufflation. Decreased venous return and elevated pulmonary vascular pressures are known conditions that occur during laparoscopy and elevate CO_2 levels during surgery. Gas embolism can occur when large volumes of CO_2 gain entry into the arterial or venous system. This life-threatening complication requires immediate release of pneumoperitoneum and resuscitation. Presenting symptoms include increased pulmonary artery pressure, tachycardia, hypotension, mill-wheel murmur, decreased end tidal CO_2, and rapidly falling oxygen saturation (Patient 160).

Bladder injuries are much more common than ureteral injuries and are easier to recognize and more likely to be repaired intraoperatively. Ureteral injury can be difficult to identify intraoperatively; more than 70% of ureteral injuries are diagnosed postoperatively. This injury is more likely to occur with laparoscopic hysterectomy or removal of an entrapped ovary. Presenting signs and symptoms include hematuria, flank pain, leukocytosis, oliguria, and an elevated C-reactive protein level (Patient 161).

Baggish M. Major laparoscopic complications: a review in two parts. J Gynecol Surg 2012;28:315–32.

Bhoyrul S, Vierra MA, Nezhat CR, Krummel TM, Way LW. Trocar injuries in laparoscopic surgery. J Am Coll Surg 2001;192:677–83.

Goldberg JM, Chen CCG, Falcone T. Complication of laparoscopic surgery. In: Falcone T, Goldberg JM, editors. Basic, advanced and robotic laparoscopic surgery. Philadelphia (PA): Saunders Elsevier, 2010. p. 221–40.

Magrina JF. Complications of laparoscopic surgery. Clin Obstet Gynecol 2002;45:469–80.

Makai G, Isaacson K. Complications of gynecologic laparoscopy. Clin Obstet Gynecol 2009;52:401–11.

Roy GM, Bazzurini L, Solima E, Luciano AA. Safe technique for laparoscopic entry into the abdominal cavity. J Am Assoc Gynecol Laparosc 2001;8:519–28.

Appendix A
Normal Values for Laboratory Tests*

Analyte	Conventional Units
Alanine aminotransferase, serum	8–35 units/L
Alkaline phosphatase, serum	15–120 units/L
Menopause	
Amniotic fluid index	3–30 mL
Amylase	20–300 units/L
Greater than 60 years old	21–160 units/L
Aspartate aminotransferase, serum	15–30 units/L
Bicarbonate	
Arterial blood	21–27 mEq/L
Venous plasma	23–29 mEq/L
Bilirubin	
Total	0.3–1 mg/dL
Conjugated (direct)	0.1–0.4 mg/dL
Newborn, total	1–10 mg/dL
Blood gases (arterial) and pulmonary function	
Base deficit	Less than 3 mEq/L
Base excess, arterial blood, calculated	−2 mEq/L to +3 mEq/L
Forced expiratory volume (FEV_1)	3.5–5 L
	Greater than 80% of predicted value
Forced vital capacity	3.5–5 L
Oxygen saturation (So_2)	95% or higher
Pao_2	80 mm Hg or more
Pco_2	35–45 mm Hg
Po_2	80–95 mm Hg
Peak expiratory flow rate	Approximately 450 L/min
pH	7.35–7.45
Pvo_2	30–40 mm Hg
Blood urea nitrogen	
Adult	7–18 mg/dL
Greater than 60 years old	8–20 mg/dL
CA 125	Less than 34 units/mL
Calcium	
Ionized	4.6–5.3 mg/dL
Serum	8.6–10 mg/dL
Chloride	98–106 mEq/L
Cholesterol	
Total	
Desirable	140–199 mg/dL
Borderline high	200–239 mg/dL
High	240 mg/dL or more
High-density lipoprotein	40–85 mg/dL
Low-density lipoprotein	
Desirable	Less than 130 mg/dL
Borderline high	140–159 mg/dL
High	Greater than 160 mg/dL
Total cholesterol-to-high-density lipoprotein ratio	
Desirable	Less than 3
Borderline high	3–5
High	Greater than 5
Triglycerides	
20 years and older	Less than 150 mg/dL
Younger than 20 years old	35–135 mg/dL

*Values listed are specific for adults or women, if relevant, unless otherwise differentiated.

(continued)

Normal Values for Laboratory Tests* (*continued*)

Analyte	Conventional Units
Cortisol, plasma	
8 AM	5–23 micrograms/dL
4 PM	3–15 micrograms/dL
10 PM	Less than 50% of 8 AM value
Creatinine, serum	0.6–1.2 mg/dL
Dehydroepiandrosterone sulfate	60–340 micrograms/dL
Erythrocyte	
Count	3,800,000–5,100,000/mm^3
Distribution width	10 plus or minus 1.5%
Sedimentation rate	
Wintrobe method	0–15 mm/hour
Westergren method	0–20 mm/hour
Estradiol-17β	
Follicular phase	30–100 pg/mL
Ovulatory phase	200–400 pg/mL
Luteal phase	50–140 pg/mL
Child	0.8–56 pg/mL
Ferritin, serum	18–160 micrograms/L
Fibrinogen	150–400 mg/dL
Follicle-stimulating hormone	
Premenopause	2.8–17.2 mIU/mL
Midcycle peak	15–35 mIU/mL
Postmenopause	24–170 mIU/mL
Child	0.1–7 mIU/mL
Glucose	
Fasting	70–105 mg/dL
2-hour postprandial	Less than 120 mg/dL
Random blood	65–110 mg/dL
Hematocrit	36–48%
Hemoglobin	12–16 g/dL
Fetal	Less than 1% of total
Hemoglobin A$_{1c}$ (nondiabetic)	5.5–8.5%
Human chorionic gonadotropin	0–5 mIU/mL
Pregnant	Greater than 5 mIU/mL
17α-Hydroxyprogesterone	
Adult	50–300 ng/dL
Child	32–63 ng/dL
25-Hydroxyvitamin D	10–55 ng/mL
International Normalized Ratio	Greater than 1
Prothrombin time	10–13 seconds
Iron, serum	65–165 micrograms/dL
Binding capacity total	240–450 micrograms/dL
Lactate dehydrogenase, serum	313–618 units/L
Leukocytes	
Total	5,000–10,000/cubic micrometers
Differential counts	
Basophils	0–1%
Eosinophils	1–3%
Lymphocytes	25–33%
Monocytes	3–7%
Myelocytes	0%
Band neutrophils	3–5%
Segmented neutrophils	54–62%

*Values listed are specific for adults or women, if relevant, unless otherwise differentiated.

(*continued*)

Normal Values for Laboratory Tests* (*continued*)

Analyte	Conventional Units
Lipase	
60 years or younger	10–140 units/L
Older than 60 years	18–180 units/L
Luteinizing hormone	
Follicular phase	3.6–29.4 mIU/mL
Midcycle peak	58–204 mIU/mL
Postmenopause	35–129 mIU/mL
Child	0.5–10.3 mIU/mL
Magnesium	
Adult	1.6–2.6 mg/dL
Child	1.7–2.1 mg/dL
Newborn	1.5–2.2 mg/dL
Mean corpuscular	
mCH Hemoglobin	27–33 pg
mCHC Hemoglobin concentration	33–37 g/dL
mCV Volume	80–100 cubic micrometers
Partial thromboplastin time, activated	21–35 seconds
Phosphate, inorganic phosphorus	2.5–4.5 mg/dL
Platelet count	140,000–400,000/mm^3
Potassium	3.5–5.3 mEq/L
Progesterone	
Follicular phase	Less than 3 ng/mL
Luteal phase	2.5–28 ng/mL
On oral contraceptives	0.1–0.3 ng/mL
Secretory phase	5–30 ng/mL
Older than 60 years	0–0.2 ng/mL
1st trimester	9–47 ng/mL
2nd trimester	16.8–146 ng/mL
3rd trimester	55–255 ng/mL
Prolactin	0–17 ng/mL
Pregnant	34–386 ng/mL by 3rd trimester
Prothrombin time	10–13 seconds
Reticulocyte count	Absolute: 25,000–85,000 cubic micrometers
	0.5–2.5% of erythrocytes
Semen analysis, spermatozoa	
Antisperm antibody	% of sperm binding by immunobead technique; greater than 20% = decreased fertility
Count	Greater than or equal to 20 million/mL
Motility	Greater than or equal to 50%
Morphology	Greater than or equal to 15% normal forms
Sodium	135–145 mEq/L
Testosterone, female	
Total	6–86 ng/dL
Pregnant	3–4 × normal
Postmenopause	One half of normal
Free	
20–29 years old	0.9–3.2 pg/mL
30–39 years old	0.8–3 pg/mL
40–49 years old	0.6–2.5 pg/mL
50–59 years old	0.3–2.7 pg/mL
Older than 60 years	0.2–2.2 pg/mL
Thyroid-stimulating hormone	0.2–3 microunits/mL
Thyroxine	
Serum free	0.9–2.3 ng/dL
Total	1.5–4.5 micrograms/dL

*Values listed are specific for adults or women, if relevant, unless otherwise differentiated.

(*continued*)

Normal Values for Laboratory Tests* (*continued*)

Analyte	Conventional Units
Triiodothyronine uptake	25–35%
Urea nitrogen, blood	
Adult	7–18 mg/dL
Older than 60 years	8–20 mg/dL
Uric acid, serum	2.6–6 mg/dL
Urinalysis	
Epithelial cells	0–3/HPF
Erythrocytes	0–3/HPF
Leukocytes	0–4/HPF
Protein (albumin)	
Qualitative	None detected
Quantitative	10–100 mg/24 hours
Pregnancy	Less than 300 mg/24 hours
Urine specific gravity	
Normal hydration and volume	1.005–1.03
Concentrated	1.025–1.03
Diluted	1.001–1.01

*Values listed are specific for adults or women, if relevant, unless otherwise differentiated.

Appendix B

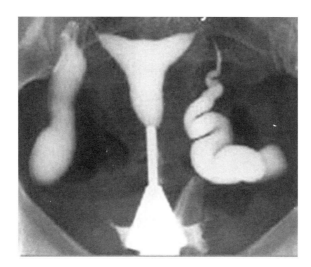

Appendix C

```
┌─────────────────────────────────────────┐
│      Abnormal Uterine Bleeding (AUB)    │
│   • Heavy menstrual bleeding (AUB/HMB)  │
│   • Intermenstrual bleeding (AUB/IMB)   │
└─────────────────────────────────────────┘
```

PALM: Structural Causes
Polyp (AUB-P)
Adenomyosis (AUB-A)
Leiomyoma (AUB-L)
  Submucosal myoma (AUB-L_{SM})
  Other myoma (AUB-L_O)
Malignancy and hyperplasia (AUB-M)

COEIN: Nonstructural Causes
Coagulopathy (AUB-C)
Ovulatory dysfunction (AUB-O)
Endometrial (AUB-E)
Iatrogenic (AUB-I)
Not yet classified (AUB-N)

Appendix C. Basic PALM–COEIN classification system for the causes of abnormal uterine bleeding in nonpregnant reproductive-aged women. This system, approved by the International Federation of Gynecology and Obstetrics, uses the term "abnormal uterine bleeding" paired with terms that describe associated bleeding patterns ("heavy menstrual bleeding" or "intermenstrual bleeding"), a qualifying letter (or letters) to indicate its etiology (or etiologies), or both. (Data from Munro MG, Critchley HO, Broder MS, Fraser IS. FIGO classification system [PALM-COEIN] for causes of abnormal uterine bleeding in nongravid women of reproductive age. FIGO Working Group on Menstrual Disorders. Int J Gynaecol Obstet 2011;113:3–13.)

Appendix D

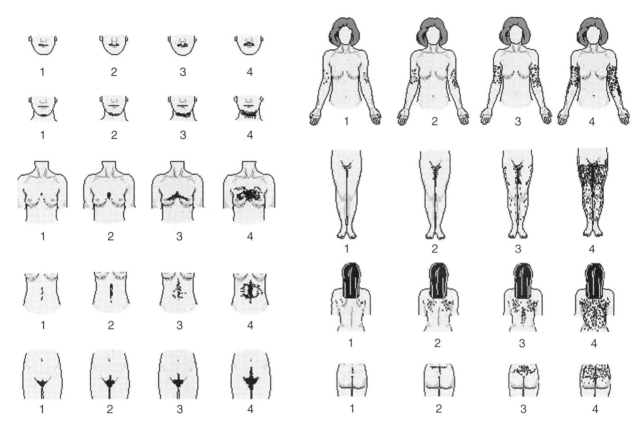

Appendix D. The Ferriman–Gallwey score is used to quantify the degree of hirsutism by the use of representative images as shown. The lowest score is 0, indicating no excessive hair growth, and the highest score is 32. (Hatch R, Rosenfield RL, Kim MH, Tredway D. Hirsutism: implications, etiology, and management. Am J Obstet Gynecol 1981; 140:815–30. Copyright Elsevier 1981.)

Index

A

Ablation, uterine nerve, 42
Abnormal uterine bleeding (AUB), 18, 109. *See also* Heavy menstrual bleeding
   in adolescents, 92
   causes of, 18
   leiomyomas and, 56
   PALM–COEIN classification system for, 18, 92
   polycystic ovary syndrome-related, 109
   serum reproductive hormone levels in, 109
   treatment for, 18, 56
Abortion, 17, 26
Absolute risk
   described, 57
   relative risk versus, 57
Acanthosis nigricans, in women with insulin resistance, 91
Acetaminophen, for dysmenorrhea, 42
Acne, 84
ACTH. *See* Adrenocorticotropic hormone
Acupuncture, for infertility-related issues, 85
Adhesion(s), labial, 30
Adolescent(s)
   abnormal uterine bleeding in, 92
   acne in, 84
   low bone mineral density in, 101
Adoption, for patients with müllerian dysgenesis, 27
Adrenal cortex, adrenal steroid biosynthesis in, 62
Adrenal hyperplasia. *See* Congenital adrenal hyperplasia
Adrenal steroid biosynthesis, in adrenal cortex, 62
Adrenarche
   normal, 62
   premature, 62
Adrenocorticotropic hormone (ACTH), in Cushing syndrome diagnosis, 110
Adrenocorticotropic hormone deficiency, Sheehan syndrome and, 70
Adrenocorticotropic hormone stimulation test, in androgen-secreting ovarian tumor evaluation, 98
Air embolism, hysteroscopic resection and, 24
AIS. *See* Androgen insensitivity syndrome
Alcohol use
   breast cancer associated with, 127
   pregnancy outcomes related to, 134–138
Alendronate sodium, for female athlete triad, 108
Alpha error, described, 57
Ambiguity, sexual, 115
Ambiguous genitalia, 115
Amenorrhea, 97
   functional hypothalamic, 108, 117
   hypothalamic, 117
   primary, 27, 86, 103
   secondary, 10, 110, 117
American College of Cardiology, on metabolic syndrome, 104
American College of Obstetricians and Gynecologists. *See* The American College of Obstetricians and Gynecologists
American Diabetes Association, on metabolic syndrome, 104
American Gastroenterological Association, on osteoporosis in celiac disease, 101
American Society for Reproductive Medicine, 32
   on bioidentical hormones, 113
   on endometriosis, 37
   "Guidelines for Gamete and Embryo Donation" of, 39
   on in vitro fertilization, 1
   on limiting number of embryos transferred at time of in vitro fertilization, 116
   on oocyte donor safety, 126
   on polycystic ovary syndrome, 59
   on varicocele repair in treatment of male factor infertility, 77
American Society for Reproductive Medicine Ethics Committee guidelines, on donor sperm for same-sex couples, 39
   "Guidelines for Gamete and Embryo Donation," 39
American Society of Andrology, on male factor infertility management, 77
American Thyroid Association, on hypothyroidism screening, 114
American Urological Association, on varicocele repair in treatment of male factor infertility, 77
Analysis of variance with post-hoc Tukey honestly significant difference testing, 143–145
Androgen(s)
   main action of, 84
   maternal virilization due to, 68
   synthetic, 68
Androgen disorders, 61, 103
Androgen excess, prevalence of, 81
Androgen insensitivity syndrome (AIS), 48
   categorization of, 128–130
   complete, 128–130
   described, 128–130
   features of, 103
   gonad evaluation in, 128–130
   partial, 128–130
   presentation of, 128–130
Androgen production, sexual hair growth related to, 62
Androgen-secreting ovarian tumors, 98
Anesthesia, for hysteroscopy, 24
Aneuploidy screening, 71
Anorexia nervosa, nutritional management of, 54
Anti-β_2-glycoprotein I antibodies, in antiphospholipid syndrome, 3
Antiandrogens, for hirsutism, 112
Antibiotics, for acne, 84
Antibody(ies), anticardiolipin, 3
Anticardiolipin antibodies, 3
Anticoagulant(s), lupus, 3
Antidepressant(s)
   for anorexia nervosa, 54
   for infertility-related issues, 85
Antimüllerian hormone level, in testing for ovarian reserve, 95
Antiphospholipid syndrome (APS), 3, 69
APS. *See* Antiphospholipid syndrome
Aromatase inhibitors
   for androgen disorders, 61
   for leiomyoma in infertility, 36
   for ovulation induction, 90
ART. *See* Assisted reproductive technology
Artificial insemination, for same-sex couples, 39
Assisted reproductive technology (ART), ovarian hyperstimulation syndrome and, 8
AUB. *See* Abnormal uterine bleeding
Aura, migraine with, 7
Autoimmune thyroiditis, 104, 114, 152–154
Autonomy, patient, 1
Autosomal recessive inherited disorder, congenital adrenal hyperplasia as, 60
Autotransplantation, tissue cryopreservation with, 75
AZF. *See* Azoospermia factor
Azoospermia, 5, 119
   nonobstructive, 119
   obstructive, 119
Azoospermia factor (AZF), 119

B

β-hCG, 40, 64
Bariatric surgery, 9, 125
   annual rate of, 155–158
   described, 155–158
   indications for, 78
Basal body temperature monitoring, in conception, 121
Beck Depression Inventory, on infertility, 85
Beckwith–Wiedemann syndrome, intracytoplasmic sperm injection and, 11
Benzoyl peroxide, for acne, 84
Biestrogen, for menopausal symptoms, 113
Biguanide, for ovulation induction in patient with polycystic ovary syndrome, 59
Bilateral salpingectomy, 12
Bilateral tubal ligation, fertility after, 123
Binge-eating disorder, 52
Bioidentical hormones, 113
Bisphosphonates
   in bone loss treatment, 94
   for female athlete triad, 108
Bladder injuries, laparoscopic surgery and, 159–161
Bleeding. *See also* Hemorrhage
   abnormal uterine, 18, 56, 92, 109
   after uterine fibroid embolization, 63
   estrogen breakthrough, 73
   heavy menstrual, 18, 80, 106
   hysteroscopic resection and, 24
   postmenopausal uterine, 34
BMD. *See* Bone mineral density
Bone, physiology of, 94
Bone loss, depot medroxyprogesterone acetate and, 58
Bone mass
   low, 94
   screening for, 94
Bone mineral density (BMD)
   in children and adolescents with newly diagnosed celiac disease, 101
   low, 58

NOTE: Numbers refer to questions, not pages.

Bowel injury, laparoscopic surgery and, 159–161
BRCA gene mutations, 22, 43
BRCA1 gene mutations, 13, 131–133
BRCA2 gene mutations, 131–133
Breakthrough bleeding, estrogen, 73
Breast cancer, 127
　alcohol use and, 127
　risk factors for, 127
　screening for, 29
　in smokers, 127
Breast Cancer Prevention Trial, of National Surgical Adjuvant Breast and Bowel Project, 2
Breast Cancer Risk Assessment Tool, in breast cancer risk prediction, 127
Breast development
　premature, 20
　Tanner staging of, 20
Bromocriptine, in hyperprolactinemia management, 97
Bulimia nervosa, 52

C

Cabergoline, in hyperprolactinemia management, 97
CAH. *See* Congenital adrenal hyperplasia
CAIS. *See* Complete androgen insensitivity syndrome
Calcitonin, in bone loss treatment, 94
Calcium supplementation, in bone loss treatment, 94
Calendar timing, in conception, 121
Cancer(s). *See also specific types*
　breast, 29, 127
　cervical, 29
　endometrial, 22, 34
　fertility preservation in patient undergoing treatment for, 32
　ovarian, 13, 22
　screening for, 29
Cardiovascular disease, Framingham Point Score for, 104
CDC. *See* Centers for Disease Control and Prevention
Celiac disease, 101
Centers for Disease Control and Prevention (CDC)
　on in vitro fertilization with own oocytes, 126
　on pelvic inflammatory disease, 124
Central precocious puberty, causes of, 82
Cerebral palsy, in triplet pregnancies, 116
Cervical cancer, screening for, 29
Cervical laceration, hysteroscopic resection and, 24
Cervical mucus, peak production of, 121
Children
　labial adhesions in, 30
　low bone mineral density in, 101
Chlamydia trachomatis antibody levels, in pelvic inflammatory disease detection, 124
Chronic malodorous vaginal discharge, after uterine fibroid embolization, 63
Chronic pelvic pain, 96
Cigarette smoking, bone mineral density loss related to, 58
Cleavage-stage embryo biopsy, 71
Clinical hypothyroidism, 114
Clomiphene citrate
　for anorexia nervosa, 54
　in male factor infertility management, 77

Clomiphene citrate *(continued)*
　for obese patients, 125, 155–158
　in ovarian reserve testing, 95
　for ovulation induction, 59, 90
　for prolactin-induced sexual dysfunction, 38
　for unexplained infertility, 16
Clonidine, for menopausal symptoms, 50
Coagulopathy, abnormal uterine bleeding in adolescents related to, 92
Cognitive–behavioral therapy, in bulimia nervosa management, 52
Colles fractures, osteoporosis and, 94
Combination contraceptives
　for hirsutism, 149–151
　nonoral, 19
　for patient with systemic lupus erythematosus, 66
Combination estrogen–progestin therapy, for *BRCA1*-positive patient with ovarian cancer, 13
Combination hormonal contraceptives
　for hirsutism, 149–151
　in hyperprolactinemia management, 97
Combination oral contraceptives (OCs)
　for hirsutism, 112
　for patient with venous thromboembolism, 19
"Common migraine," 7
Complete androgen insensitivity syndrome (CAIS), 128–130
　differential diagnosis of, 128–130
　treatment of, 128–130
Computed tomography (CT), in androgen-secreting ovarian tumor evaluation, 98
Conception, monitoring during period of, 121
Condom(s), failure of, 65
Congenital adrenal hyperplasia (CAH), 45, 60
　causes of, 115
　described, 60
　late-onset, 81, 104
　management of, 115
　newborn screening for, 115
　nonclassic, 81
　prevalence of, 115
　prevention of, 60
Congenital rickets, 134–138
Contraception. *See also* Contraceptives; Oral contraceptives; *specific types*
　condoms, 65
　for diabetes mellitus, 44, 67
　emergency, 65
　morning-after, 65
　for older adult smoker, 55
　for patient with *BRCA* gene mutation, 22
　for patient with systemic lupus erythematosus, 66
　for patient with venous thromboembolism, 19
Contraceptive(s). *See also* Contraception; Oral contraceptives
　combination, 13, 19, 66, 112, 149–151
　hormonal, 97, 146–148, 149–151
　in hyperprolactinemia management, 97
　levonorgestrel-containing, 146–148
　long-acting reversible, 17
　obesity effects on, 17
　progestin-only, 19, 57, 84, 146–148
Copper intrauterine device, for patient with systemic lupus erythematosus, 66
Corticosteroids, for acne, 84
Cortisol, Sheehan syndrome and, 70

Cortisol deficiency, 115
Counseling, infertility-related, 85
Cryopreservation techniques, 32
　embryo cryopreservation, 75
　oocyte cryopreservation, 75
CT. *See* Computed tomography
Cushing disease versus Cushing syndrome, 110
Cushing syndrome, 104, 110
Cyst(s), dermoid, 100
Cystic fibrosis, 6, 88

D

D&E. *See* Dilation and evacuation
Dehydroepiandrosterone (DHEA), premature adrenarche and, 62
Dehydroepiandrosterone sulfate (DHEAS)
　elevated, 93
　in nonclassic adrenal hyperplasia evaluation, 81
　in pregnancy, 139–142
　premature adrenarche and, 62
Depot medroxyprogesterone acetate (DMPA)
　bone loss and, 58
　levonorgestrel, 19
　for patient with systemic lupus erythematosus, 66
　unscheduled uterine bleeding related to, 49
Depression, infertility-related, 85
Dermoid cysts, 100
Dexamethasone, in congenital adrenal hyperplasia prevention, 60
DHEA. *See* Dehydroepiandrosterone
DHEAS. *See* Dehydroepiandrosterone sulfate
Diabetes mellitus
　contraception for, 44, 67
　diagnosis of, 91
　impaired glucose tolerance and, 91
　type 2
　　impaired glucose tolerance and, 51, 91
　　metformin hydrochloride for, 78
Dienogest, in pregnancy termination, 26
Dilation and evacuation (D&E), pregnancy termination by, 26
Disordered proliferative endometrium, polycystic ovary syndrome and, 73
Distal radial fractures, osteoporosis and, 94
Distention media, for diagnostic hysteroscopy, 24
Distribution-free testing, 143–145
DMPA. *See* Depot medroxyprogesterone acetate
DNA fragmentation testing, 71
Donor egg in vitro fertilization (IVF), 29
Donor oocytes
　müllerian dysgenesis and, 27
　uses for, 126
Donor sperm
　insemination, 119
　for same-sex couples, 39
　use, 119
Dopamine agonists
　for macroadenoma, 46
　for obese patients, 155–158
　for prolactin-induced sexual dysfunction, 38
Dual energy X-ray absorptiometry (DXA) scan
　of hip, 94
　of spine, 94
Dwarfism, Laron, 84
DXA scan. *See* Dual energy X-ray absorptiometry scan
Dysmenorrhea, 42
Dysmorphia, facial, 134–138

NOTE: Numbers refer to questions, not pages.

E

Early pregnancy, human chorionic gonadotropin in, 64
Early pregnancy failure, complications of, 99
Eating disorders, treatment of, 52, 54. *See also specific types*
Echocardiography, before donor egg in vitro fertilization, 29
Ectopic pregnancy, 64, 99, 124
ED. *See* Erectile dysfunction
Eflornithine hydrochloride, for hirsutism, 149–151
Egg donors, surgical risks to, 126
Ejaculation, retrograde, 111
Ejaculatory dysfunction, 111
Electrocautery, laparoscopic, 123
Embolism
 hysteroscopic resection and, 24
 laparoscopic surgery and, 159–161
Embryo biopsy, cleavage-stage, 71
Embryo cryopreservation, in fertility preservation, 75
Embryonal carcinoma cells, in reproduction, 122
Embryonic stem cells (ESCs), in reproduction, 122
Emergency contraception
 types of, 65
 U.S. Food and Drug Administration on, 65
Emergency contraceptives, mechanism of action of, 146–148
Endocrine Society Diagnosis of Cushing Syndrome Task Force, 110
Endometrial abnormalities, in asymptomatic women, 34
Endometrial biopsy results, interpretation of, 118
Endometrial cancer, 22, 34
Endometrial polyps, 4
Endometrioma cyst wall resection, in endometriosis patient, 14
Endometriosis, 37
 abnormal uterine bleeding in adolescents related to, 92
 described, 14
 infertility related to, 14
 management of, 37
 prevalence of, 37, 96
 treatment of, 96
Endometrium
 cancer of, 34
 disordered proliferative, 73
 polycystic ovary syndrome effects on, 73
Erectile dysfunction (ED), male factor infertility related to, 38
ESCs. *See* Embryonic stem cells
Estradiol, 25
17β-Estradiol, for menopausal symptoms, 113
Estrogen(s)
 in bone loss treatment, 94
 breakthrough bleeding, polycystic ovary syndrome and, 73
 levels, in pregnancy, 139–142
 low-dose, 72
 replacement therapy, 72
 in venous thromboembolism management, 19
Estrogen-containing oral contraceptives (OCs), venous thromboembolism related to, 57
Ethamsylate, for abnormal uterine bleeding, 18
Ethical issues, in vitro fertilization-related, 1
Ethinyl estradiol, for menopausal symptoms, 113
Etonogestrel, in pregnancy termination, 26

Etonogestrel implant, for patient with venous thromboembolism, 19
European Academy of Andrology, on male factor infertility management, 77
European Society for Human Reproduction and Embryology, on polycystic ovary syndrome, 59

F

Facial dysmorphia, 134–138
FDA. *See* U.S. Food and Drug Administration
Female athlete triad, treatment of, 108
Female infertility, tubal disease and, 124
Female sterilization, with hysteroscopically placed permanent contraceptive micro-insert, 35
Ferriman–Gallwey score, in hirsutism diagnosis, 149–151
Fertility, after tubal ligation, 47, 123
Fertility preservation
 in patient undergoing cancer treatment, 32
 techniques, 75
Fertility therapy, multifetal pregnancies related to, 116
Fertilization, window of, 121
Fetal alcohol syndrome, 134–138
Fibrosis
 cystic, 6, 88
 luminal tubal, 79
Finasteride
 for acne, 84
 for hirsutism, 112
Finger numbness, oral contraceptives for migraine based on, 7
FISH analysis. *See* Fluorescence in situ hybridization analysis
Fisher exact test, 143–145
Fluorescence in situ hybridization (FISH) analysis, in preimplantation genetic screening, 71
Fluoxetine, for menopausal symptoms, 50
Flutamide, for hirsutism, 112
Follicle-stimulating hormone (FSH) level, 95, 119
Follicular phase, in menstrual cycle length, 41
Folliculogenesis, 41
Food and Drug Administration. *See* U.S. Food and Drug Administration
Fracture(s), osteoporosis and, 94
Fracture Risk Assessment Tool, of World Health Organization, 94
Fragile X mutation, 83
Fragile X premutation, 83
Fragile X syndrome, 6, 83
Framingham Point Score, for cardiovascular disease, 104
Frank method, 23
Fructose, in sperm, 5
FSH. *See* Follicle-stimulating hormone
Functional hypothalamic amenorrhea, 108, 117
Functional vagina, creation for patients with Müllerian anomalies, 23

G

Gabapentin, for menopausal symptoms, 50
Gail model, in breast cancer risk prediction, 127
Galactorrhea, 10, 81, 97
Gamete donation, risks associated with, 126
Gas embolism, laparoscopic surgery and, 159–161
Gastric banding, for obese patients, 125

Gastric bypass, 9
Gaucher disease, 6
Genetic screening, preimplantation, 71
Genitalia, ambiguous, 115
German measles, 29
Gestational carrier and in vitro fertilization, for müllerian dysgenesis, 27
Ghrelin, in reproductive function, 15
Glucose tolerance, impaired, 51
Glucose tolerance test, in insulin resistance evaluation in patients with polycystic ovary syndrome, 91
GnRH agonists. *See* Gonadotropin-releasing hormone agonists
Gonad(s), in androgen insensitivity syndrome, 128–130
Gonadal dysgenesis
 46,XY, 86, 103
 primary amenorrhea and, 86
Gonadectomy, for primary amenorrhea due to gonadal dysgenesis, 86
Gonadotropin(s)
 menopausal, 90
 for ovulation induction in patient with polycystic ovary syndrome, 59
Gonadotropin abnormalities, in Sheehan syndrome, 70
Gonadotropin-releasing hormone (GnRH) agonists
 after laparoscopic surgery in endometriosis patient, 14
 for androgen disorders, 61
 in fertility preservation, 75
 for leiomyoma in infertility, 36
 for ovulation induction, 90
 for precocious puberty, 82
 analog, for heavy menstrual bleeding, 56
 antagonists, for ovulation induction, 90
Granulosa cell tumors, 100
Graves disease
 diagnosis of, 74
 hyperthyroidism and, 74, 152–154

H

Hair growth, sexual, 62
Hashimoto thyroiditis, 114, 152–154
hCG. *See* Human chorionic gonadotropin
Heavy menstrual bleeding, 18, 80, 106. *See also* Abnormal uterine bleeding
 defined, 80, 92
 described, 106
 leiomyomas and, 56
 management of, 106
Hematopoietic stem cells (HSCs), 122
Hemorrhage, after uterine fibroid embolization, 63. *See also* Bleeding
Hereditary cancer syndromes, 131–133
High-order multiple gestations, defined, 116
Hip(s), dual-energy X-ray absorptiometry scan of, 94
Hip fractures, osteoporosis and, 94
Hirsutism, 112, 149–151
 atypical, 98
 disorders associated with, 149–151
Hormonal changes, in pregnancy, 139–142
Hormonal contraceptives, mechanism of action of, 146–148
Hormone(s), bioidentical, 113
Hormone therapy (HT)
 for acne, 84
 as emergency contraception, 65

NOTE: Numbers refer to questions, not pages.

Hormone therapy (HT) *(continued)*
    for menopausal symptoms, 113
    postmenopausal, 72
    postmenopausal uterine bleeding and, 34
    risks and benefits of, 31
    types of, 72
    Women's Health Initiative study on, 31
HSCs. *See* Hematopoietic stem cells
HSG. *See* Hysterosalpingography
HT. *See* Hormone therapy
Human chorionic gonadotropin (hCG)
    β-, 40
    in early pregnancy, 64
Human placental lactogen, in pregnancy, 139–142
Hydrosalpinges, in vitro fertilization–embryo transfer and, 12
11β-Hydroxylase deficiency, 60
17α-Hydroxylase deficiency, 60
21-Hydroxylase deficiency, congenital adrenal hyperplasia related to, 60, 115
17α-Hydroxyprogesterone, morning follicular serum, 81
3β-Hydroxysteroid dehydrogenase type 2, 60
Hyperandrogenism, in pregnancy, 100
Hypergonadism, hypergonadotropic, 117
Hypergonadotropic hypogonadism, 117
Hyperinsulinemic euglycemic clamp, in insulin resistance evaluation in patients with polycystic ovary syndrome, 91
Hyperlipidemia, 104
Hyperplasia. *See* Congenital adrenal hyperplasia
Hyperprolactinemia, 97
    erectile dysfunction and, 38
    galactorrhea due to, 10
    menstrual irregularity and galactorrhea related to, 81
    non-tumor-related, 97
Hyperreactio luteinalis
    defined, 68
    maternal virilization in, 68, 100
Hyperstimulation, ovarian, 100
Hyperthyroidism, 74
    causes of, 74
    clinical characteristics of, 152–154
    Graves disease and, 74, 152–154
    during pregnancy, 53
    prevalence of, 74
    transient, 53
Hypogonadism
    primary hypogonadotropic, 88
    secondary hypogonadotropic, 88
Hypogonadotropic hypogonadism
    primary, 88
    secondary, 88
Hypothalamic amenorrhea, 108, 117
Hypothyroidism, 114
    causes of, 114
    clinical, 114
    during pregnancy, 53
    screening for, 114
    signs and symptoms of, 152–154
    subclinical, 114, 152–154
Hysterosalpingography (HSG), 79
    complications of, 89
    described, 89
Hysteroscope(s), 24
Hysteroscopic proximal tubal catheterization, 79
Hysteroscopic resection, of endometrial polyp, 4

Hysteroscopic tubal occlusion, 123
Hysteroscopically placed permanent contraceptive microinsert, female sterilization with, 35
Hysteroscopy
    anesthesia for, 24
    complications of, 24
    diagnostic, 24
    for endometrial polyps, 4
    operative, 24

I
ICSI. *See* Intracytoplasmic sperm injection
Idiopathic premature adrenarche, 62
IGF-1. *See* Insulin-like growth factor 1
IGT. *See* Impaired glucose tolerance
Impaired glucose tolerance (IGT), diabetes mellitus and, 51, 91
Imprinting disorders, 11
In vitro fertilization (IVF), 1
    donor egg, 29
    goal of, 11
    intracytoplasmic sperm injection and, 11
    limiting number of embryos transferred at time of, 116
    ovarian hyperstimulation syndrome and, 11
    with own oocytes, 126
    patient autonomy in, 1
    for unexplained infertility, 16
In vitro fertilization and gestational carrier, for müllerian dysgenesis, 27
In vitro fertilization–embryo transfer, hydrosalpinges and, 12
In vitro maturation, 32
Induced pluripotent, in reproduction, 122
Infection(s)
    intrauterine device-related, 102
    after uterine fibroid embolization, 63
Infertility
    counseling related to, 85
    depression related to, 85
    endometriosis and, 14
    female, 124
    leiomyoma in, 36
    male factor, 5, 38, 77, 119
    prevalence of, 85
    psychologic effects of, 85
    tubal disease and, 124
    unexplained, 16
Insemination
    donor sperm, 119
    intrauterine, 16, 111
    for same-sex couples, 39
Insulin-like growth factor 1 (IGF-1), 10, 84
Insulin resistance
    described, 91
    obesity and, 78
    in polycystic ovary syndrome, 91
    syndrome, 104
Insulin-sensitizing agents
    for hirsutism, 112
    for ovulation induction in patient with polycystic ovary syndrome, 59
Intracytoplasmic sperm injection (ICSI)
    Beckwith–Wiedemann syndrome and, 11
    in vitro fertilization and, 11
Intrauterine devices (IUDs)
    complications of, 102
    copper, 66
    levonorgestrel, 19, 36, 66, 80, 96, 106, 146–148

Intrauterine devices (IUDs) *(continued)*
    mechanism of action of, 146–148
    for obese women, 17
Intrauterine insemination, 16, 111
Intrauterine microinsert
    follow-up, 87
    pregnancy and, 35
Irregular menstrual cycles, 120, 127
Isosexual precocious puberty, evaluation of, 82
Isotretinoin, for acne, 84
IUDs. *See* Intrauterine devices
IVF. *See* In vitro fertilization

K
Kallman syndrome, 90, 103
Kaplan–Meier curve, 143–145
Karotypic abnormalities, oligospermia related to, 11
Karyotype, in premature ovarian failure evaluation, 83
Karyotype abnormalities, premature ovarian failure related to, 83
Karyotype analysis, in recurrent pregnancy loss evaluation, 69
Klinefelter syndrome, 88
Kolmogorov–Smirnov test, 143–145

L
Labial adhesions, in children, 30
Labial agglutination, in children, 30
Laceration(s), cervical, 24
Laparoscopic applications of band or clip, in surgical reversal, 123
Laparoscopic electrocautery, 123
Laparoscopic oophorectomy, for ovarian androgen-secreting tumor, 98
Laparoscopic surgery, complications of, 159–161
Laparoscopy, in endometriosis, 14
Laron dwarfism, 84
Laser hair removal, 112
Late-onset congenital adrenal hyperplasia, 81, 104
Legal issues, in vitro fertilization-related, 1
Leiomyoma(s)
    abnormal uterine bleeding in adolescents related to, 92
    heavy menstrual bleeding and, 56
    in infertility, 36
Leptin
    negative energy balance effects on, 15
    in reproductive function, 15
Letrozole
    for leiomyoma in infertility, 36
    for ovulation induction in patient with polycystic ovary syndrome, 59
Leuprolide, indications for, 89
Levonorgestrel-containing contraceptive methods, 146–148
Levonorgestrel intrauterine devices
    in chronic pelvic pain management, 96
    for heavy menstrual bleeding, 80
    in heavy menstrual bleeding management, 106
    for leiomyoma in infertility, 36
    mechanism of action of, 146–148
    in systemic lupus erythematosus patient, 66
    in venous thromboembolism patient, 19
Levonorgestrel-only method, for emergency contraception, 65
Lifestyle choices, pregnancy outcomes related to, 134–138

NOTE: Numbers refer to questions, not pages.

Long-acting reversible contraceptives, 17
Low bone mineral density, depot medroxyprogesterone acetate and, 58
Low-dose estrogen therapy, in postmenopausal women, 72
Luminal tubal fibrosis, 79
Lupus anticoagulant, 3
Luteal phase deficiency, 118
Luteoma(s), pregnancy, 100
Lynch II syndrome, 131–133

M

Macroadenoma, 10, 46
Male factor infertility, 38, 119
   erectile dysfunction and, 38
   evaluation of, 5
   management of, 77
   varicoceles and, 77
Massachusetts Women's Health Study, on menopause, 120
Maternal plasma cortisol, in pregnancy, 139–142
Maternal virilization, in hyperreactio luteinalis, 68, 100
Maturation, in vitro, 32
Mature cystic teratomas, 100
Mayer–Rokitansky–Küster–Hauser syndrome, 23, 27
McCune–Albright syndrome, 61
   peripheral precocious puberty due to, 82
   treatment of, 61
McIndoe operation, 23, 128–130
Measles, German, 29
Mechanical hair removal techniques, for hirsutism, 112
Medroxyprogesterone acetate
   for androgen disorders, 61
   for *BRCA1*-positive patient with ovarian cancer, 13
   in pregnancy termination, 26
Mefenamic acid, for AUB, 56
Menarche, age at, 127
Menopausal gonadotropins, for ovulation induction, 90
Menopausal symptoms, early, 120
Menopausal transition, normal, 120
Menopause
   age at, 120
   alternative therapies for, 50
   Massachusetts Women's Health Study on, 120
Menstrual bleeding. *See* Heavy menstrual bleeding
Menstrual cycle(s)
   described, 41
   irregular, 28, 120
   normal, 25, 41
   obesity effects on, 134–138
   phases of, 41
   physiology of, 25
   variability in length of, 28
Menstrual history, in breast cancer risk prediction, 127
Menstrual irregularity, hyperprotactinemia and, 81
Menstruation, described, 25
Metabolic syndrome, 104
Metformin
   for hirsutism, 112
   for impaired glucose tolerance and type 2 diabetes mellitus, 51

Metformin hydrochloride
   for acne, 84
   for ovulation induction in patient with polycystic ovary syndrome, 59
   for type 2 diabetes mellitus, 78
Methotrexate
   contraindications to, 64
   for tubal pregnancy, 64
Microadenoma, 97
Microinsert, intrauterine, 35, 87
Microsurgical tubal resection, 79
Mifepristone
   for abnormal uterine bleeding, 49
   for leiomyoma in infertility, 36
Migraine
   oral contraceptives for, 7
   types of, 7
Mineralocorticoid deficiency, congenital adrenal hyperplasia and, 60
Minocycline, for acne, 84
Misoprostol
   for abnormal uterine bleeding, 56
   in pregnancy termination, 26
Modified Pomeroy tubal ligation, 79
Montpellier conference, on consideration of ovarian reserve before surgery in women experiencing infertility, 14
Morning-after contraception, 65
Morning follicular serum, 17α-hydroxyprogesterone, in nonclassic adrenal hyperplasia, 81
Mosaic Turner syndrome, 103
Mucus, cervical, 121
Müllerian agenesis, 27. *See also* Müllerian dysgenesis
Müllerian anomalies, 23, 76
Müllerian dysgenesis, 23, 27. *See also* Müllerian agenesis
   adoption for patients with, 27
   prevalence of, 27
Multifetal pregnancy, fertility therapy and, 116
Multifetal pregnancy reduction, 116
Multiple gestation, high-order, 116
Myoma(s), vagina passage of, 63
Myomectomy, for obese patients, 155–158

N

National Aeronautics and Space Administration, on robotic platform, 105
National Cholesterol Education Program, on metabolic syndrome, 104
National Health and Nutrition Examination Survey III
   on obesity, 155–158
   on TSH levels, 114
National Osteoporosis Foundation, 94
National Surgical Adjuvant Breast and Bowel Project, Breast Cancer Prevention Trial of, 2
Negative predictive value (NPV), 57
Neovagina, creation of, 23
Neurectomy, presacral, 42
Newborn screening, for congenital adrenal hyperplasia, 115
Niemann–Pick disease, 6
Nonclassic adrenal hyperplasia, 81
Nonmaleficence, 1
Nonobstructive azoospermia, 119
Nonparametric testing, 143–145

Nonsteroidal antiinflammatory drugs (NSAIDs)
   for abnormal uterine bleeding, 18
   for dysmenorrhea, 42
Norethindrone
   for abnormal uterine bleeding, 56
   for patient with venous thromboembolism, 19
   in pregnancy termination, 26
Normal adrenarche, 62
Normal menopausal transition, 120
NPV. *See* Negative predictive value
NSAIDs. *See* Nonsteroidal antiinflammatory drugs
Nutritional deficiencies
   after bariatric surgery, 9
   pregnancy outcomes related to, 134–138
Nutritional status, reproductive function related to, 15
Nutritional support, for anorexia nervosa, 54

O

Obesity
   consequences resulting from, 155–158
   contraceptive choices and, 17
   diseases associated with, 125
   insulin resistance effects of, 78
   maternal complications of, 78
   menstrual cycles effects of, 134–138
   oligomenorrhea and, 125
   polycystic ovary syndrome and, 51, 78
   pregnancy effects of, 51, 125, 134–138, 155–158
   prevalence of, 17, 125, 155–158
   reproductive health effects of, 125
   thrombosis related to, 17
Obstructive azoospermia, 119
OCs. *See* Oral contraceptives
Older adults, contraception for smoker, 55
Oligo-ovulation, abnormal uterine bleeding in adolescents related to, 92
Oligoasthenoteratozoospermia, 11
Oligomenorrhea
   obesity with, 125
   ovulation predictor kit and, 125
Oligospermia, 11, 77
Oocyte(s)
   donor, 27, 126
   in vitro fertilization with own, 126
Oocyte cryopreservation, 32, 75
Oophorectomy
   in endometriosis, 14
   for ovarian androgen-secreting tumor, 98
Operative hysteroscopy, 24
Oral contraceptives (OCs). *See also* Contraception; Contraceptive(s)
   for acne, 84
   for *BRCA1*-positive patient with ovarian cancer, 13
   in breast cancer risk prediction, 127
   in chronic pelvic pain management, 96
   combination, 19, 112
   in decreasing cancer risk of endometrium and ovaries, 22
   for dysmenorrhea, 42
   estrogen-containing, 57
   for female athlete triad, 108
   for functional hypothalamic amenorrhea, 108
   for hirsutism, 149–151
   for migraine, 7
   for patient with venous thromboembolism, 19
   progestin-only, 57
   venous thromboembolism related to, 57

NOTE: Numbers refer to questions, not pages.

Oral corticosteroids, for acne, 84
Osteopenia, primary ovarian insufficiency and, 83
Osteoporosis
 celiac disease and, 101
 fractures related to, 94
 functional hypothalamic amenorrhea and, 108
 primary ovarian insufficiency and, 83
Ovarian androgen-secreting tumor, 98
Ovarian cancer
 in *BRCA1*-positive patient, 13
 oral contraceptives in decreasing risk of, 22
Ovarian failure, premature, 83
Ovarian follicular depletion, pregnancy outcomes related to, 134–138
Ovarian hyperstimulation, 100
Ovarian hyperstimulation syndrome
 assisted reproductive technology and, 8
 in vitro fertilization and, 11
Ovarian insufficiency, 117
Ovarian progesterone, in implantation and early embryonic development, 118
Ovarian reserve, testing for, 95
Ovarian Sertoli–Leydig cell tumor, 107
Ovarian stimulation, risks associated with, 126
Ovarian tissue freezing, 32
Ovarian transposition, 32
Ovarian tumors
 androgen-secreting, 98
 during pregnancy, 68
Ovary(ies), polycystic, 8
Overnight dexamethasone suppression test
 in androgen-secreting ovarian tumor evaluation, 98
 in Cushing syndrome diagnosis, 110
Ovulation induction
 methods of, 90
 polycystic ovary syndrome and, 59
Ovulation predictor kit, oligomenorrhea and, 125

P
Pain, chronic pelvic, 96
PAIS. *See* Partial androgen insensitivity syndrome
PALM–COEIN classification system, for abnormal uterine bleeding, 18, 92
Parathyroid hormone, for female athlete triad, 108
Partial androgen insensitivity syndrome (PAIS), 128–130
Patient autonomy, 1
PCOS. *See* Polycystic ovary syndrome
Pelvic inflammatory disease (PID)
 Centers for Disease Control and Prevention on, 124
 long-term sequelae of, 124
 reproductive dysfunction related to, 124
Pelvic pain, chronic, 96
Perforation, hysteroscopic resection and, 24
Perimenopausal changes, 28
Perimenopause, 28
Peripheral precocious puberty, 82
Perrault syndrome, 48
Peutz–Jeghers syndrome, 131–133
Phantom β-hCG, results of, 40
PID. *See* Pelvic inflammatory disease
Pioglitazone, for ovulation induction in patient with polycystic ovary syndrome, 59
Pituitary gland, postpartum infarction of, 70

Placental aromatase deficiency, 68
Pluripotent, induced, 122
Pluripotent stem cells, in reproduction, 122
Polycystic ovary(ies), assisted reproductive technology and, 8
Polycystic ovary syndrome (PCOS), 104
 acne and, 84
 abnormal uterine bleeding and, 109
 clinical manifestations of, 78
 cyclic progesterone administration for, 25
 described, 117
 estrogen breakthrough bleeding in, 73
 evaluation of, 78, 98
 hirsutism and, 112
 insulin resistance testing in, 91
 irregular menstrual pattern with, 127
 obesity and, 51, 78
 ovulation induction in patient with, 59
 pathophysiology of, 109
 premature adrenarche and, 62
 prevalence of, 59, 73, 78, 109
Polyp(s)
 endometrial, 4
 tamoxifen citrate and, 2
Positive predictive value (PPV), 57
Postembolization syndrome, after uterine fibroid embolization, 63
Postmenopausal uterine bleeding, hormone therapy and, 34
Postmenopausal women
 low bone mass in, 94
 low-dose estrogen therapy in, 72
Postpartum infarction of pituitary, 70
Postpartum thyroiditis, 53
PPV. *See* Positive predictive value
Precocious puberty, 61, 82
Preconception evaluation
 before donor egg in vitro fertilization, 29
 for obese patients, 125
Pregnancy
 abnormal uterine bleeding in adolescents related to, 92
 alcohol consumption during, 134–138
 complications of, 99
 early, 64, 99
 ectopic, 64, 99, 124
 hormonal changes in, 139–142
 hyperandrogenism in, 100
 hyperthyroidism in, 53
 hypothyroidism in, 53
 intrauterine microinsert and, 35
 lifestyle choices effects on, 134–138
 multifetal, 116
 obesity effects on, 51, 125, 134–138, 155–158
 ovarian tumors during, 68
 termination of, 26
 tubal, 64
Pregnancy failure, early, 64
Pregnancy loss, recurrent, 21, 69
Pregnancy luteomas, 100
Pregnancy reduction, multifetal, 116
Preimplantation genetic screening, 71
Premature adrenarche, 62
Premature development of secondary sexual characteristics, evaluation of, 61
Premature ovarian failure, 83
Premature pubarche, 62
Premature thelarche, 20
Presacral neurectomy, for dysmenorrhea, 42
Preterm delivery, in triplet pregnancy, 116

Primary amenorrhea, 103
 gonadal dysgenesis and, 86
 müllerian dysgenesis and, 27
Primary dysmenorrhea, 42
Primary hypogonadotropic hypogonadism, 88
Primary ovarian insufficiency, 6, 83
Primary subclinical hypothyroidism, 114
Progesterone
 cyclic, 25
 in implantation and early embryonic development, 118
 ovarian, 118
Progesterone levels, in pregnancy, 139–142
Progestin, 19, 25
Progestin implant, for obese women, 17
Progestin-only contraceptives
 for acne, 84
 mechanism of action of, 146–148
 for patient with venous thromboembolism, 19
Progestin-only oral contraceptives, venous thromboembolism related to, 57
Prolactin, in nonclassic adrenal hyperplasia evaluation, 81
Prolactin-induced sexual dysfunction, dopamine agonist therapy for, 38
Prolactin levels, irregular menstrual cycles and, 120
Prolactinomas, 10
Protein levels, reproductive function related to, 15
Proximal tubal occlusion, 79
Proximal tubal repair, 79
Psychologic counseling, infertility-related, 85
Psychologic effects, of infertility, 85
Pubarche, premature, 62
Puberty, precocious, 61, 82
Pure gonadal dysgenesis, 128–130

Q
Quality of life, after bariatric surgery, 9

R
RAD51C gene mutation, 131–133
Radioiodine uptake, in hyperthyroidism evaluation, 74
Reciprocal translocations, in couples with recurrent pregnancy loss, 21
Recurrent pregnancy loss, 21, 69
Relative risk
 absolute risk versus, 57
 described, 57
Reproduction
 pelvic inflammatory disease effects on, 124
 stem cells in, 122
Reproductive function, nutritional status and protein levels related to, 15
Reproductive health, obesity effects on, 125
Reproductive history, in breast cancer risk prediction, 127
Reproductive Medicine Network, on polycystic ovary syndrome, 59
Retinoids, for acne, 84
Retrograde ejaculation, 111
Reversible contraceptives, long-acting, 17
Rickets, congenital, 134–138
Risk(s), absolute versus relative, 57
Robertsonian translocations, in couples with recurrent pregnancy loss, 21
Robotic-assisted surgery, complications of, 105
Rosiglitazone, for ovulation induction in patient with polycystic ovary syndrome, 59

NOTE: Numbers refer to questions, not pages.

Roux-en-Y gastric bypass, 9, 78, 125
Rubella, 29

S
Salpingectomy, bilateral, 12
Salpingitis, 124
Salpingitis isthmica nodosa, 79
Salt wasting, 115
Same-sex couples, donor sperm for, 39
SCOFF questionnaire, 54
Sebum, 84
Secondary amenorrhea, 6, 10, 110, 117
Secondary dysmenorrhea, 42
Secondary hypogonadotropic hypogonadism, 88
Secondary sexual characteristics, premature development of, 61
Sensitivity, described, 57
Sequential vaginal dilators, in creating functional vagina, 23
Sertoli–Leydig cell tumor
    ovarian, 107
    virilization due to, 107
Serum reproductive hormone levels, abnormal uterine bleeding-related, 109
Sex steroid-producing tumor, 61
Sexual ambiguity, 115
Sexual dysfunction, prolactin-induced, 38
Sexual hair growth, androgen production and, 62
Shapiro–Wilk test, 143–145
Sheehan syndrome, 70
Sildenafil citrate, for prolactin-induced sexual dysfunction, 38
SLE. See Systemic lupus erythematosus
Small-bowel obstruction, robotic-assisted surgery and, 105
Smoker(s)
    bone mineral density loss in, 58
    breast cancer in, 127
    contraception for older adult, 55
Smoking, pregnancy outcomes related to, 134–138
Smoking cessation, in bone mineral density loss prevention, 58
Society for Assisted Reproductive Technologies, on in vitro fertilization with own oocytes, 126
Society of Gynecologic Oncologists, on *BRCA* gene mutation screening, 43
Specificity, described, 57
Sperm
    donor use of, 119
    fructose in, 5
Spine, dual-energy X-ray absorptiometry scan of, 94
Spironolactone
    for acne, 84
    for hirsutism, 112, 149–151
Stages of Reproductive Aging Workshop, 28
Stanford Research Institute, on robotic platform, 105
Statistical analysis, 143–145
Statistical testing, 143–145
Stem cell(s)
    in reproduction, 122
    types of, 122
Sterilization
    female, 35
    reversal of, 123
    tubal, 67

Sterilization decision, U.S. Collaborative Review of Sterilization study on, 123
Subclinical hypothyroidism, 114, 152–154
Surgical reversal, sterilization-related, 123
Swyer syndrome, 48, 86, 103, 128–130
Syndrome X, 68, 104
Synthetic androgen
Synthetic ethinyl estradiol, for patient with venous thromboembolism, 19
Systemic lupus erythematosus (SLE), 66

T
T-ACE Problem Drinking Screen, 134–138
Tamoxifen citrate
    polyps due to, 2
    safety of, 2
    ultrasonographic findings in patient taking, 2
Tanner staging of breast development, 20
Teratoma(s), mature cystic, 100
Teriparatide, in bone loss treatment, 94
Testosterone
    for prolactin-induced sexual dysfunction, 38
    virilization and ovarian mass effects on, 107
Tetracycline, for acne, 84
The American College of Obstetricians and Gynecologists
    on alcohol consumption during pregnancy, 134–138
    on bioidentical hormones, 113
    on emergency contraception, 65
Thelarche
    described, 20
    premature, 20
Thiazolidinediones, for ovulation induction in patient with polycystic ovary syndrome, 59
Three-dimensional ultrasonography, in müllerian anomaly evaluation, 76
Thromboembolism, venous, 19
Thrombosis, in obese women, 17
Thyroid antibody tests, 152–154
Thyroid disease, 53, 114
    oligo-ovulation screening for, 54
    testing for, 152–154
Thyroid gland, in pregnancy, 139–142
Thyroid-stimulating hormone (TSH) levels
    in hypothyroidism screening, 114
    irregular menstrual cycles and, 120
    in Sheehan syndrome, 70
Thyroiditis
    autoimmune, 104, 114, 152–154
    Hashimoto, 114, 152–154
    postpartum, 53
Thyrotoxicosis, causes of, 74
Thyroxine (T_4), in pregnancy, 139–142
Tissue cryopreservation with later autotransplantation, in fertility preservation, 75
Tissue freezing, ovarian, 32
Total testosterone level, in nonclassic adrenal hyperplasia, 81
Tranexamic acid
    for abnormal uterine bleeding, 18, 56
    for heavy menstrual bleeding, 56
Transient hyperthyroidism, 53
Translocation(s), in couples with recurrent pregnancy loss, 21
Triestrogen, for menopausal symptoms, 113
Triiodothyronine, 152–154
Triplet pregnancy, diseases associated with, 116
Trisomy X syndrome, 128–130

Trophectoderm biopsy, in preimplantation genetic screening, 71
TSH levels. See Thyroid-stimulating hormone levels
Tubal disease
    detection methods, 124
    female infertility due to, 124
Tubal ligation
    by clip, 79
    fertility after, 47, 123
    in vitro fertilization after, 1
    modified Pomeroy, 79
Tubal obstruction, hysterosalpingography and, 89
Tubal occlusion, proximal, 79
Tubal pregnancy, intramuscular methotrexate use in, 64
Tubal repair, proximal, 79
Tubal sterilization, for patient with diabetes mellitus, 67
Tumor(s). *See also specific types*
    granulosa cell, 100
    ovarian, 68, 98
    Sertoli–Leydig cell, 107
    sex steroid-producing, 61
Turner syndrome, 33, 103
    described, 33
    mosaic, 103
    phenotypic findings in, 33
    premature ovarian failure related to, 83
    presentation of, 33
Type 2 diabetes mellitus. See Diabetes mellitus
Type I error, described, 57

U
Ulipristal acetate, as morning-after contraception, 65
Ultrasonographic follicular monitoring, in conception, 121
Ultrasonography
    in androgen-secreting ovarian tumor evaluation, 98
    in hypothyroidism screening, 114
    tamoxifen citrate findings on, 2
    three-dimensional, 76
Unexplained infertility, 16
Unscheduled uterine bleeding, depot medroxyprogesterone acetate and, 49
Ureteral injury, laparoscopic surgery and, 159–161
U.S. Collaborative Review of Sterilization study, on sterilization decision, 123
U.S. Department of Defense, on robotic platform, 105
U.S. Food and Drug Administration (FDA), 35
    on combination OCs for acne, 84
    on donor sperm for same-sex couples, 39
    on emergency contraception, 65
    on estrogen therapy in postmenopausal women, 72
    on hirsutism treatment, 149–151
    on robotic platform, 105
    on tranexamic acid for heavy menstrual bleeding, 56
U.S. Preventive Services Task Force
    on nutritional supplementation in bone loss management, 94
    on thyroid-stimulating hormone levels, 114
U.S. Surgeon General, on alcohol consumption during pregnancy, 134–138
Uterine artery embolization, 63

NOTE: Numbers refer to questions, not pages.

Uterine bleeding
  abnormal, 18, 109
  postmenopausal, 34
  unscheduled, 49
Uterine fibroid embolization, 63
Uterine nerve ablation, for dysmenorrhea, 42
Uterine perforation, hysteroscopic resection and, 24

V

Vagina, sequential vaginal dilators in creating functional, 23
Vaginal atrophy
  management of, 72
  vaginal estradiol tablets for, 72
Vaginal dilators, sequential, 23
Vaginal discharge, after uterine fibroid embolization, 63
Vaginal estradiol tablets, in vaginal atrophy management, 72

Varicocele(s), 77, 88
Varicocelectomy, 77
Vascular endothelial growth factor (VEGF), 8
Vascular endothelial growth factor receptor 2 (VEGF-2), 8
Vecchietti operation, 23
VEGF. *See* Vascular endothelial growth factor
VEGF-2. *See* Vascular endothelial growth factor receptor 2
Venlafaxine, for menopausal symptoms, 50
Venous thromboembolism
  contraception for patient with, 19
  oral contraceptives and, 57
Vertebral fractures, osteoporosis and, 94
Virilization, osteoporosis and, 94
Vitamin D, in bone loss treatment, 94
Vitamin D deficiency, pregnancy outcomes related to, 134–138
Vitrification, in fertility preservation, 75

W

Weight loss, nonsurgical approaches to, 155–158
WHI study. *See* Women's Health Initiative study
WHO. *See* World Health Organization
"Window of fertilization," 121
Women's Health Initiative (WHI) study
  on estrogen in bone loss management, 94
  on hormone therapy, 31
  on menopausal symptoms management, 113
  on postmenopausal hormone therapy, 72
World Health Organization (WHO)
  Fracture Risk Assessment Tool of, 94
  normal semen parameters of, 5

X

X chromosome, genetic abnormalities of, 6

Y

Y chromosome microdeletion, 88

NOTE: Numbers refer to questions, not pages.

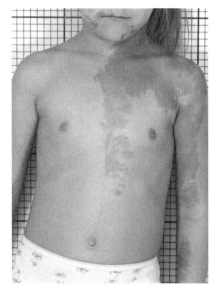

FIG. 61-1

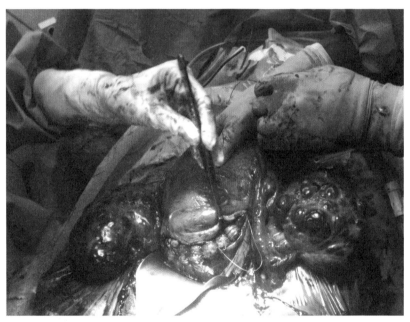

FIG. 68-1

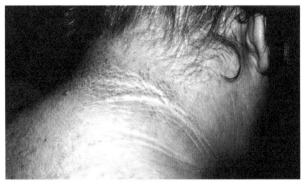

FIG. 91-1

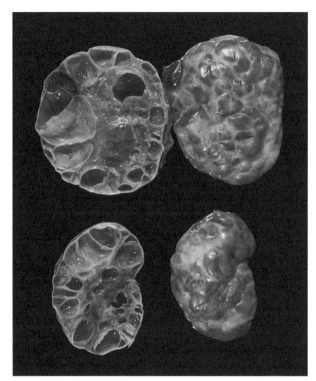

FIG. 100-1. Hyperreactio luteinalis showing ovaries enlarged by multiple thin-walled cysts. (Courtesy PathologyOutlines.com, Inc., AFIP Fascicle 3rd Series, Vol. 23.)

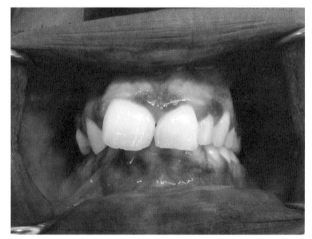

FIG. 131–133-1. Mucocutaneous pigmentation. (Image reprinted with permission from Talib Najjar, DMD, MDS, PhD, Rutgers School of Dental Medicine, published by Medscape Reference [http://emedicine.medscape.com/], 2014, available at: http://emedicine.medscape.com/article/1078143-overview.)

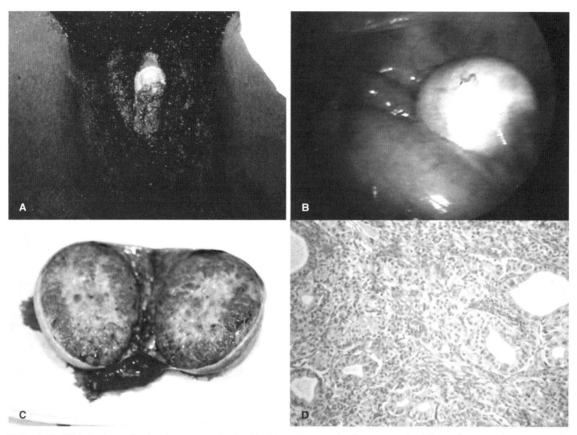

FIG. 149–151-1. A patient who presented with hirsutism and clitoromegaly **(A)** underwent laparoscopic oophorectomy **(B)** showing a solid tumor **(C)** with final pathology **(D)** consistent with a Sertoli–Leydig cell tumor.

Acknowledgments

Fig. 10-1 was originally published in Beshay VE, Beshay JE, Halvorson LM. Pituitary tumors: diagnosis, management, and implications for reproduction. Semin Reprod Med 2007;25:388–401.

Fig. 61-1 was originally published in Dumitrescu CE, Collins MT. McCune–Albright syndrome. Orphanet J Rare Dis 2008;3:12.

Fig. 68-1 provided courtesy of Jason S. Yeh, MD, Duke University Medical Center.

Fig. 70-1 was originally published in Sarafoff N, Baur DM, Gaa J, von Beckerath N. Images in cardiovascular medicine. Recurrent syncope due to torsades de pointes in a 41-year-old woman with an empty sella, anterior pituitary insufficiency, and a long-QT interval. Circulation 2009;120:e127–9.

Fig. 91-1 was supplied courtesy of Task Force member J. Ricardo Loret de Mola, MD.

Fig. 149–151-1 was supplied by Task Force member Thomas M. Price, MD.

Appendix B figure showing hydrosalpinx was reproduced courtesy of Gary S. Berger, MD. This figure previously appeared at http://www.tubal-reversal.net/blog/2011/dr-berger/diagnosing-and-fixing-blocked-tubes.html.